W9-DDZ-874

Surface Science

Surface Science

An Introduction

John B. Hudson
Rensselaer Polytechnic Institute
Troy, New York

Butterworth-Heinemann
Boston London Oxford Singapore Sydney Toronto Wellington

Recognizing the importance of preserving what has been written, it is the policy of Butterworth-Heinemann to have the books it publishes printed on acid-free paper, and we exert our best efforts to that end.

Library of Congress Cataloging-in-Publication Data
Hudson, John B.
Surface science: an introduction / John B. Hudson.
p. cm.
ISBN 0-7506-9159-X (case bound)
1. Surfaces (Physics) 2. Surface chemistry. 3. Surfaces (Technology) I. Title.
QC173.4.S94H84 1991
530.4′27—dc20 91-21912
CIP

British Library Cataloguing in Publication Data
Hudson, John B.
Surface science: an introduction.
I. Title
530.41
ISBN 0-7506-9159-X

Butterworth-Heinemann
80 Montvale Avenue
Stoneham, MA 02180

10 9 8 7 6 5 4 3 2 1

Printed in the United States of America

To my parents,
Aurie and Ken Hudson

Contents

Preface

The term *surface science* came into common parlance less than thirty years ago, in the early 1960s. Although there had been many previous studies of surfaces and surface phenomena, some of which led to the Nobel prize winning work of Langmuir for his studies of adsorption and of Davison and Germer on the demonstration of low energy electron diffraction, these studies were generally classified under various other scientific subdisciplines, such as physical chemistry or electron physics.

Since its inception, the field of surface science has undergone an explosive expansion. This expansion has been driven by the combination of the ready availability of ultrahigh vacuum environments, the development of techniques for the preparation of macroscopic single crystal surfaces, and the application of an increasingly complex array of surface analytical techniques, which have made possible characterization of the structure and reactivity of a wide range of surfaces.

This rapid development poses both opportunities and challenges as far as the writing of a textbook is concerned. The surface science literature is vast, as is the range of materials and phenomena covered. Thus, there is no lack of material to be used as examples. The challenges arise in selecting from this available material that which is most central to a conceptual understanding of the field and that which will not be outdated before the book appears in print. What I have tried to do is to choose subject matter and examples that illustrate the basics of surface structure and surface reactivity, to include state-of-the-art techniques where relevant, but also to include as many cases as possible of the classical work that first illuminated a given phenomenon.

The book is aimed at advanced undergraduate and graduate students in engineering and the physical sciences who want a general overview of the field of surface science, and at beginning researchers in the field who require a background in its development. As such, the coverage in many areas tends to be broad rather than deep. References to original work, or to more complete summaries, are included with each chapter. The background assumed is an understanding at the senior undergraduate level of thermodynamics, solid state physics, and physical chemistry.

The book is divided into four major parts. It begins with descriptions of the structure, thermodynamics, and mobility of clean surfaces, to establish the basis for the study of processes occurring at surfaces that are discussed later. The second part deals with the interaction of gas molecules with solid surfaces. This part is introduced by two chapters on the behavior of gases in equilibrium and the formation of molecular beams and continues with a description of increasingly complex interactions, ranging from elastic scattering to surface chemical reactions. The third part discusses the energetic particle interactions that are the basis of the majority of the techniques that have been developed since about 1960 to reveal the structure and chemistry of surfaces. A final part presents the background material involved in crystal nucleation and growth.

The book has grown out of notes that I have developed over a period of more than twenty years for a course that I have taught, at various times, to both advanced undergraduate and graduate students in physics, chemistry, and various engineering disciplines. Because of the diversity of backgrounds and changes in my own areas of interest, the book contains more material than may be desirable for a one-semester course. Accordingly, the emphasis of the course may be adjusted by selective deletion of certain chapters. For example, the material in Chapters 3–5 will be of more interest to chemists and chemical or materials engineers, while the material in Chapters 14–16 may be of more interest to physicists or electrical engineers.

The presentation is fairly heavily slanted toward the experimental approach to surface science. This emphasis is based on my own prejudice that an intuitive understanding of the phenomena involved is more easily gained from physical examples than from theory. The inclusion of both general and detailed references at the end of each chapter is intended to provide a guide to those who wish to pursue a given topic in more depth or with more rigor. I have also included problems with each chapter, chosen from those that I have developed as homework or examination problems over the years. Again, the intent here is to provide concrete examples of the phenomena and structures described in the text.

I close with thanks to those whose work I have used in developing this book and apologies to those whose work I have been unable to include. I owe a special debt of gratitude to Professor Jack Blakely, of Cornell University, whose pioneering book *Introduction to the Properties of Crystal Surfaces* was invaluable in the early days of my teaching and which has served as a model for much of what I have written. I also thank Professor A. Peter Jardine of the State University of New York at Stony Brook, and Professor William W. Mullins of Purdue University for the suggestions arising from their reviews of the manuscript in progress. I am also greatly indebted to Professor John T. Yates, Jr., and to his student Rex D. Ramsier, for their thorough and thoughtful review of the manuscript.

Surface Science

PART I

Surface Structure, Thermodynamics, and Mobility

CHAPTER 1

Atomic Structure of Surfaces

This chapter presents a discussion of the structure of surfaces, both ideal and real, at the atomic level.

Definition of a Surface

The discussion of the structure of surfaces begins by addressing two questions: What is a surface, or interface, and how can we form a surface or interface?

The most inclusive way to define a surface is to say that a surface or interface exists in a system in any case where there is an abrupt change in the system properties with distance. Note that there are many degrees of abruptness. At one extreme is the case of a crystalline solid in contact with its own vapor at low temperature. In this case, the interface is effectively one atomic distance in width. At the other extreme is a system near its critical point, such as a liquid in contact with its own vapor at high temperature and pressure. In this case, the thickness of the interface approaches infinity as the critical point is approached. Note that a change in any property of the system may be used to define an interface. Typical properties showing an abrupt change at an interface are density, crystal structure, crystal orientation, chemical composition, and ferromagnetic or paramagnetic ordering.

Consider next how to form an interface. In particular, consider the formation of the interface between a crystalline solid and a vapor phase, or simply, a crystal surface. There are basically two ways to form such a surface, both in principle and in practice. One can carry out a process of comminution, by subdividing a large piece of material into smaller pieces, or a process of synthesis, by forming a crystalline aggregate from a vapor of individual atoms or molecules. Experimentally, comminution would involve taking a large crystal and cleaving it, crushing it, or grinding it to make two or more smaller pieces of crystal, with a concomitant increase in the total surface area. A common experimental example of synthesis is the growth of crystals from the vapor, either by homogeneous precipitation or by growth on an existing surface, as in thin film growth.

These two basic approaches to interface formation also carry over into the theoretical description of surfaces. In the comminution approach, a crystal is separated along a predetermined plane of atoms and then allowed to relax to its equilibrium configuration. In the synthesis approach, the crystal is treated as a molecule of increasing size, with the equilibrium configuration being calculated for each incremental increase in size. Surface theorists are actively pursuing both these approaches and the discussion in Chapter 2 shows that both treatments have merit in various situations.

Description of Surface Structure—The TLK Model

Consider next how to describe the structure of a crystal surface in contact with a vapor phase and how to determine whether real crystals have structures in accord with this description. The simplest possible description of the surface—the so-called ideal surface—consists of forming the surface by comminution of the bulk crystal along the desired crystal plane, without allowing any relaxation to take place. That is, the surface is represented by the ideal termination of the bulk lattice, in which the atoms in the resulting surface have the same position relative to other atoms, in the surface and in the bulk of the crystal, that they had before formation of the surface.

Ideal surfaces, and other surfaces to be discussed later in this chapter, can be described in terms of a model called the *terrace-ledge-kink*, or *TLK*, model. According to this model, all surfaces fall into one of three categories: singular, vicinal, and rough. Singular surfaces correspond to surface orientations lying along a low index plane of the bulk lattice, for example, the (100), (111), or (110) planes of cubic crystals. These surfaces are essentially *smooth* on an atomic scale. Surfaces are customarily represented, using the TLK model, as an array of cubes or spheres, each cube or sphere representing one atom in the lattice. The TLK model representation of the (100) plane of a simple cubic solid is shown in Figure 1.1.

Vicinal surfaces correspond to orientations finitely removed from the singular orientations. In the TLK model, these surfaces are considered to be made up of flat terraces of the closest singular orientation, separated by ledges of monatomic height, spaced so as to account for the surface orientation relative to the singular orientation. On vicinal

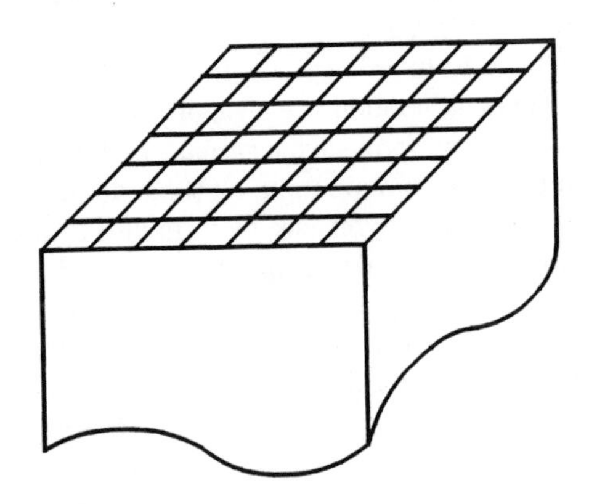

Figure 1.1 TLK model of the (100) surface of a simple cubic crystal.

surfaces that are misoriented in two directions relative to the singular orientation, the ledges will contain regularly spaced jogs or kinks, one atom position deep, in order to accommodate the misorientation in the second direction. Examples of TLK representations of vicinal surfaces in the simple cubic system are shown in Figure 1.2a for misorientation in one direction and in Figure 1.2b for misorientations in two directions.

For the sake of completeness, we must consider rough surfaces. In order to be defined as rough in terms of the TLK model, a surface would have to be disordered on the atomic scale, even at very low temperature, to the extent that a description in terms of terraces, ledges, and kinks was not possible. There is no evidence that surfaces of this type exist as equilibrium structures, and this classification will not be considered further.

In the application of the TLK model, keep in mind that the model does not allow for any relaxation of the structure, although small relaxations would not change the picture very much. In addition, the structures pictured by the model are essentially 0 K structures. No allowance is made for thermally induced defects. We will consider the complications associated with both of these effects in the sections on real surfaces and surface point defects.

Field Ion Microscopy

A number of experimental observations are relevant to the question of whether or not the structures of the surfaces of real crystals resemble those predicted by the TLK model. The results of two techniques are considered here: *field ion microscopy* (*FIM*) and *low-energy electron diffraction* (*LEED*).

Consider first field ion microscopy. The physics of the technique will not be discussed in detail at this point. Instead, this section will simply illustrate the way the technique is set up and show the observed results as they apply to the structural description of crystal surfaces. The experimental setup of an FIM is shown in Figure

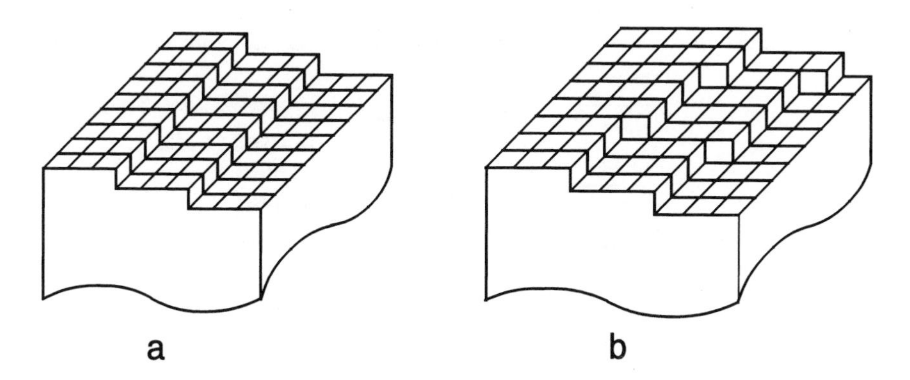

Figure 1.2 TLK models of surfaces vicinal to the (100) surface of a simple cubic crystal showing (a) misorientation in one direction and (b) misorientation in two directions.

1.3. The sample under study is present as a fine wire, etched to produce a tip on the order of 10 nm in radius. The sample assembly is held at a high positive potential, and an atmosphere of helium gas at 10^{-4} Pa pressure is present in the tube. Under these conditions, the electric field strength near the tip will be high enough to cause helium atoms very close to the tip to lose an electron and become positive helium ions. These ions are then immediately accelerated away from the tip along the electric field lines, which emerge essentially radially from the sample. When these ions strike the phosphor screen, held at ground potential, they cause phosphorescence, leading to a highly magnified image of the sample tip. The magnification, M, is given by

$$M = \frac{R_{screen}}{R_{tip}}, \tag{1.1}$$

where R_{screen} and R_{tip} are the mean radii of curvature of the screen and tip, respectively. Typical values for M are on the order of 10^6. Figure 1.4 shows a typical FIM image of a metal tip. This image shows the regularly spaced arrays of atoms in terraces, with the terraces separated by ledges, as predicted by the model. Note that the terraces are narrow, and the ledges closely spaced, due to the extremely sharp radius of curvature of the tip. The similarity between the FIM image and the TLK model may be seen even more clearly by comparing the image to a ball model of a tip of the same crystal orientation, such as that shown in Figure 1.5.

Low-Energy Electron Diffraction

The most commonly used method of determining the structure of crystal surfaces is by observation of the diffraction of electrons, either of low or high energy, from the crystal surface. These techniques based on electron diffraction are similar in concept to the x-ray diffraction techniques used to determine bulk crystal structures. A major difference, however, is the much smaller penetrating power of electrons relative to x-rays. For example, the 30 keV x-rays typically used in x-ray diffraction penetrate into the crystal lattice on the order of 10^5 nm. The 100 eV electrons used in low energy electron diffraction (LEED) penetrate on the order of 1 nm before being scattered. This is the reason that LEED is such a sensitive technique for surface structure measurement—

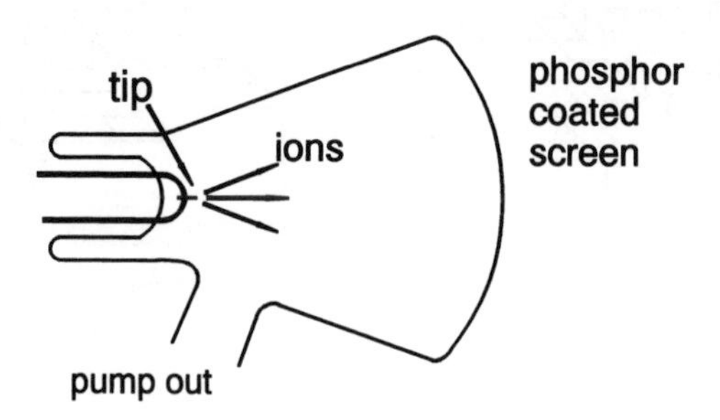

Figure 1.3 Experimental setup for field ion microscopy.

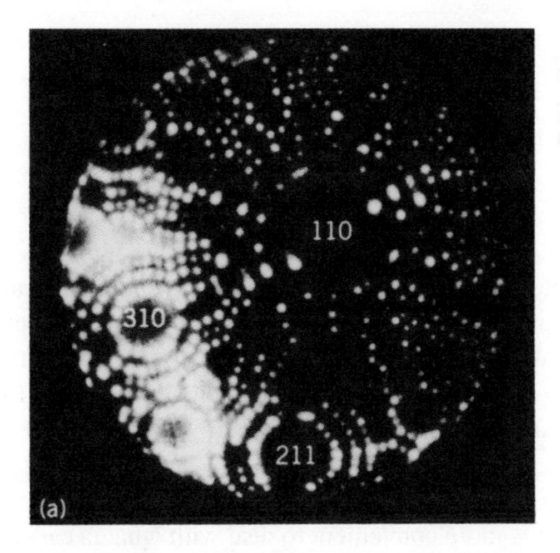

Figure 1.4 Field ion micrograph of a tungsten tip. The (110) plane of the crystal is perpendicular to the axis of the tip. (Courtesy of Gert Ehrlich)

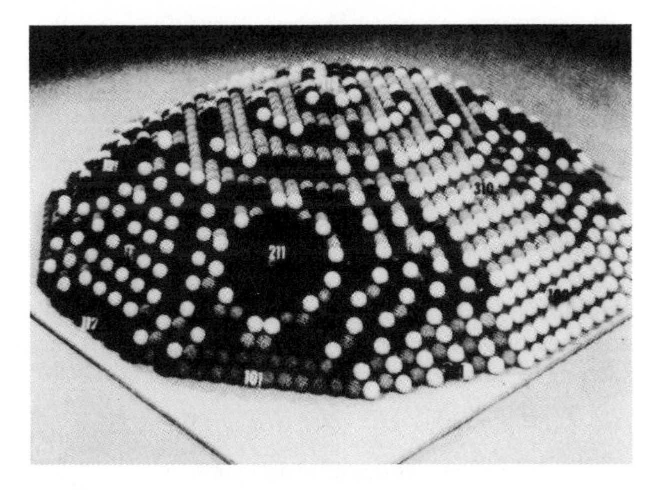

Figure 1.5 Ball model of a field emitter tip surface. The (110) surface of a BCC crystal is shown perpendicular to the axis of the tip. (Courtesy of Gert Ehrlich)

practically all of the elastic collisions that lead to diffraction peaks take place in the one or two atom layers closest to the surface.

The diffraction of electrons at a surface can be treated by considering the scattering of the electron, which is presumed to act like a wave, from a two-dimensional array of scattering sites. This is an oversimplified picture, but it will serve to illustrate the principle of surface structure determination by LEED. The electron wave incident on the surface can be represented by a vector $\mathbf{k}_0$, whose amplitude is given by the deBroglie equation:

$$\mathbf{k}_0 = \frac{2\pi}{\lambda} = 2\pi\left(\frac{2\text{meV}}{h^2}\right)^{\frac{1}{2}} \approx 2\pi\left(\frac{V}{150}\right)^{\frac{1}{2}}. \qquad 1.2$$

If V is taken in volts, then the units of $\mathbf{k}_0$ will be (Angstroms)$^{-1}$.

The array of atoms with which the electron wave interacts can be represented by a two-dimensional lattice having a unit cell defined by lattice vectors **a, b** (similar to the three-dimensional case with vectors, **a, b, c**). For example, for the (100) surface orientation of a face centered cubic (FCC) crystal, shown in Figure 1.6a, the surface unit cell is given by the small rectangle shown. Figure 1.6b shows the (111) surface of an FCC crystal.

In order to characterize the interaction between the electron wave and the two-dimensional lattice, it is more convenient to deal with what is called a reciprocal lattice than with the real-space lattice already defined. To do this, we again draw an analogy to the three-dimensional case. In three dimensions, a reciprocal lattice is constructed by translations of reciprocal lattice vectors, defined relative to the real-space lattice vectors by

$$\mathbf{a}^* = \frac{2\pi\mathbf{b}\times\mathbf{c}}{\mathbf{a}(\mathbf{b}\times\mathbf{c})},$$
$$\mathbf{b}^* = \frac{2\pi\mathbf{c}\times\mathbf{a}}{\mathbf{a}(\mathbf{b}\times\mathbf{c})}, \qquad 1.3$$
$$\mathbf{c}^* = \frac{2\pi\mathbf{a}\times\mathbf{b}}{\mathbf{a}(\mathbf{b}\times\mathbf{c})}.$$

For this case, the criterion for constructive interference in the scattered waves (that is, the criterion for a diffraction maximum) is simply that the change in the electron wave vector in the scattering event must be equal to a reciprocal lattice vector. That is

$$(\mathbf{k}'-\mathbf{k}_0) = \mathbf{g} = h\mathbf{a}^* + k\mathbf{b}^* + l\mathbf{c}^* \; (h, k, l = 0, 1, 2 \ldots). \qquad 1.4$$

Note that because the diffraction features arise as a consequence of the elastic scattering of the electrons, $k' = k_0$. Only the direction of the wave vector is changed in the scattering process; the amplitude of the wave vector is unchanged.

The criterion for a diffraction maximum can be shown graphically using a construction known as the Ewald sphere. One first constructs the reciprocal lattice by repeated translations along the reciprocal lattice vectors. A sphere is then drawn in this lattice, having a radius $k_0 = 2\pi/\lambda$, with the vector $\mathbf{k}_0$ extending from the center of the sphere to terminate on a reciprocal lattice point. The condition $(\mathbf{k}' - \mathbf{k}_0) = \mathbf{g}$ will be met for any case in which a reciprocal lattice point lies on the surface of the sphere. An example of this construction is shown in Figure 1.7.

The extension of this argument to two dimensions essentially involves letting the lattice vector c approach infinity. This will lead to $\mathbf{c}^* \rightarrow 0$. In this case, if the lattice

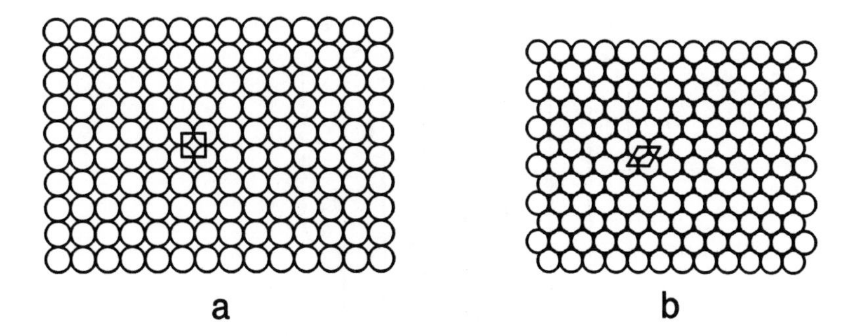

Figure 1.6 Ball models of surface showing the surface unit mesh on (a) FCC (100) and (b) FCC (111).

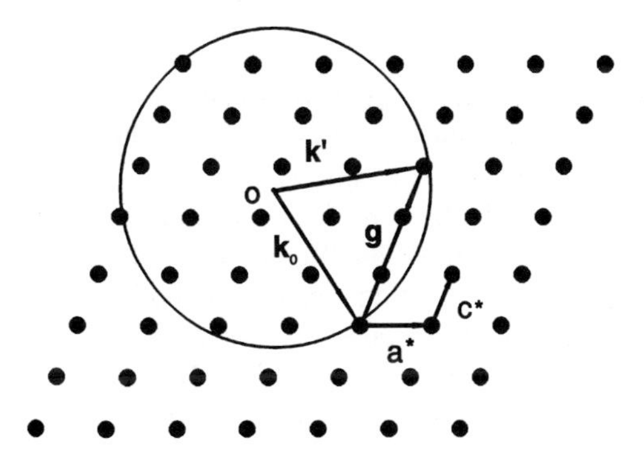

Figure 1.7 Ewald sphere construction for diffraction from a three-dimensional structure.

were strictly two-dimensional, for example, a single layer of atoms, the reciprocal lattice would look like a series of rods in a two-dimensional array, rather than a three-dimensional array of points. The unit translations along **c*** would produce a solid line for the case where $\mathbf{c}^* \rightarrow 0$. The Ewald sphere construction for this case is shown in Figure 1.8. One would observe a diffracted beam for every point at which a rod intersects the surface of the sphere. In actual practice, since the surface is not strictly two dimensional, some scattering of electrons occurs from atom layers below the first. The diffraction pattern thus shows a combination of two-dimensional and three-dimensional effects: the diffracted beam represented by a given rod is always present, for any value of the sphere radius, but its intensity varies periodically with sphere radius (i.e., with electron energy). Qualitatively, as the electron energy is increased, the wavelength decreases and the radius of the Ewald sphere increases. One result of this is that the rods are cut at different positions along their length, and consequently the diffracted intensity will be a function of electron energy. A second consequence of changing

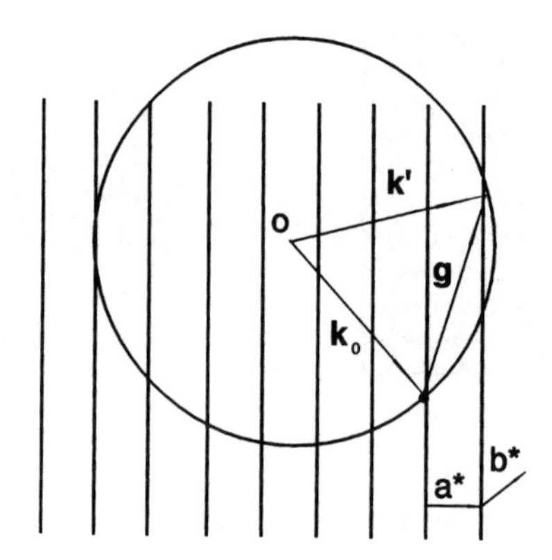

Figure 1.8 Ewald sphere construction for diffraction from a two-dimensional structure.

sphere size is that as the electron energy increases, the direction of $\mathbf{k}'$ moves toward the specular beam direction. In addition, as the size of the sphere increases, more rods will be cut and more diffracted beams will appear.

In the application of LEED, consider how the measurement is carried out and what the diffraction information obtained looks like. The foregoing discussion indicated that there is structural information contained in both the diffracted peak positions and in the peak intensities as a function of electron energy. Both of these kinds of data may be obtained from commercially available LEED systems of the so-called display type. The basic configuration of such a system is shown schematically in Figure 1.9. The system consists of an electron gun used to produce the primary electron beam, which generally is about 1 mm in diameter; the sample being examined, mounted perpendicular to the electron beam; a series of high transparency hemispherical grids, with the sample at the origin of the spheres; and a phosphor screen, held at a high positive potential.

The innermost screen is held at sample potential (usually ground) to provide a field-free flight path for the diffracted electrons. The middle grid is biased a few volts positive relative to the filament of the electron gun, to reject all but the elastically scattered electrons. The third grid is generally held at ground and serves to make the cutoff of the inelastic electrons sharper by shielding the retarding (middle) grid from the phosphor screen potential. The screen potential is sufficiently high (several kilovolts) that electrons that strike it cause a visible fluorescence. Each diffracted beam thus gives rise to a bright spot on the phosphor screen. The resulting spot pattern is observed through a window placed opposite the phosphor screen. The whole apparatus is mounted in an ultrahigh vacuum system, both to avoid scattering of the electrons by background gas and to avoid contamination of the sample during the measurement. The apparatus is also mounted to avoid stray electric or magnetic fields.

From the geometry of the system, the known primary electron energy, and the observed spot positions, one can determine $\mathbf{g}$, and thus the reciprocal lattice vectors

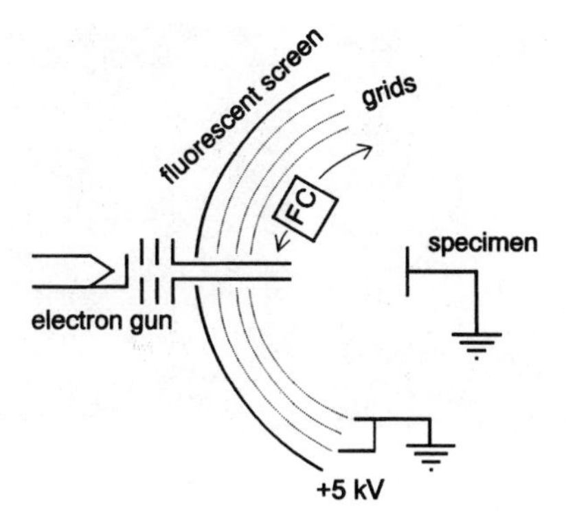

Figure 1.9 Display type LEED apparatus. FC is a Faraday cup used to collect the electron current in a diffracted beam.

associated with each of the observed spots. The diffracted beam intensities can be measured either photometrically, to determine the spot intensity on the phosphor screen, or by mounting a *Faraday cup* in the system, as shown in Figure 1.9, to collect the electron current in the diffracted beam. In this latter case the cup must be mounted so that it can be moved to follow changes in the diffracted beam position with changing primary electron energy.

Surface Structure Analysis with LEED

LEED studies can provide information on both the symmetry of the surface structure and on the absolute positions of the surface and near-surface atoms relative to one another. Consider first the question of surface symmetry. Typical LEED spot patterns are shown in Figures 1.10 and 1.11. Note that the symmetry of the patterns reflects the symmetry of the atomic arrangement at the surface. Moreover, since each spot spacing, relative to the origin, is associated with a surface reciprocal lattice vector $\mathbf{g} = h\mathbf{a}^* + k\mathbf{b}^*$, the spot spacing reveals the size of the surface mesh as well. (Note that since the spacing in the reciprocal lattice is inversely proportional to the spacing in the real-space lattice, close-together spots are associated with a large surface unit cell and vice versa). The patterns also show the changes in spot spacing associated with reconstruction of the clean surface, a topic that will be considered in detail later in this chapter, in the section on "real surfaces." The spot positions *per se*, however, do not give us any information about the atom positions in the vicinity of the surface, only the symmetry of the atomic arrangement. For example, a relaxation in which the first atomic layer was displaced normal to the surface relative to layers underneath would not affect spot positions.

In order to make a unique determination of the atom positions, it is necessary to resort to analysis of the curves of spot intensity *vs.* primary electron energy (the so-called I-V curves). This analysis is both complicated and tedious because of the high

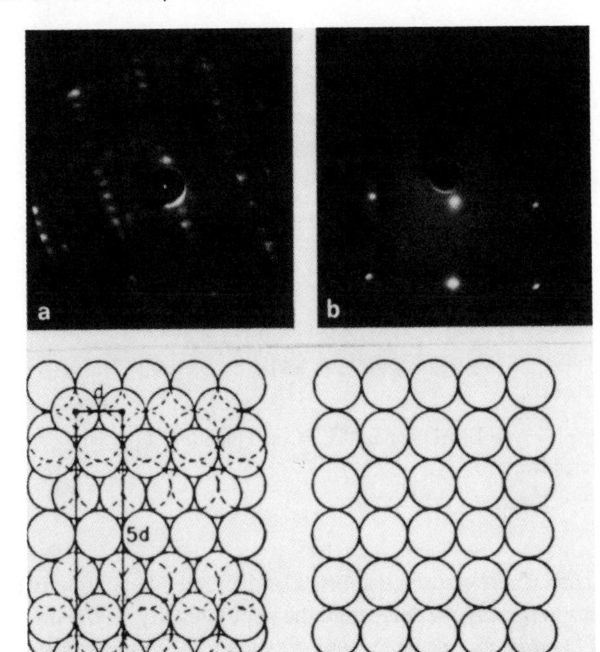

Figure 1.10 LEED patterns from clean Pt (100) surfaces showing (a) 5 × 1 reconstruction and (b) 1× 1 (unreconstructed) surface. The ball models below the patterns indicate the surface atomic arrangement. (Source: G.A. Somorjai, in R. Vanselow, ed., *Chemistry and Physics of Solid Surfaces V*. Berlin: Springer-Verlag, 1982, p. 3. Reprinted with permission.)

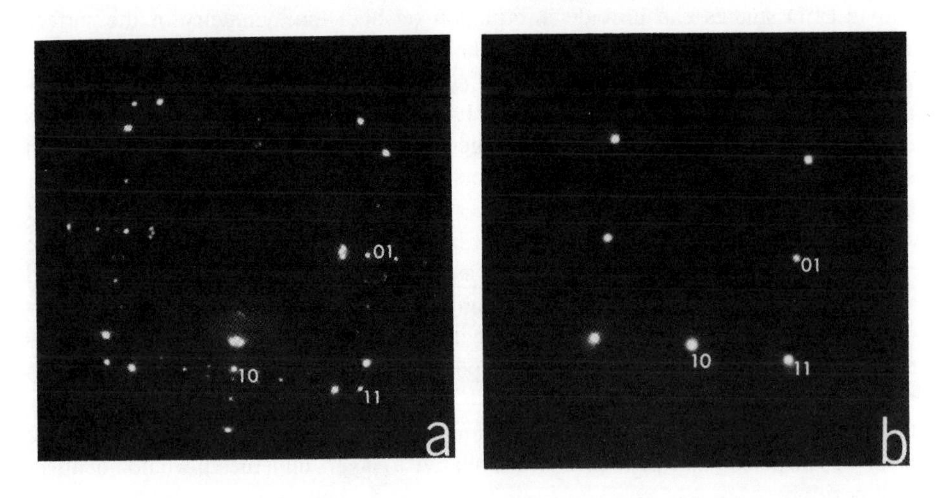

Figure 1.11 LEED patterns from Au (100) surfaces showing (a) 5 × 20 reconstruction on clean surface and (b) 1 × 1 (unreconstructed) surface (after oxygen ion bombardment). (Source: J.F. Wendelken and D.M. Zehner, *Surface Sci.* 1978; 71:178. Reprinted with permission.)

probability of multiple scattering of the very low-energy electrons used. In spite of this, many surface structures have been worked out, and research on methods for the analysis of I-V curves continues.

The results of many LEED studies bear on the question of whether or not the ideal surfaces postulated by the TLK model exist in fact. With few exceptions, it appears that most metal surfaces are essentially as predicted by the TLK model. There are some cases in which the topmost layers have been shown to be spaced differently (normal to the surface) than the bulk, and there are a few cases of significant *reconstruction* of the surface, but these appear to be exceptions to the rule. We will discuss some of these exceptions in the section on real surfaces. Semiconductor surfaces, on the other hand, show many examples of reconstruction, in some cases leading to quite complicated surface structures with large surface unit cells. These, too, will be discussed in the section on scanning tunneling microscopy.

Surface Crystallography

The preceding section presented the ways in which LEED can determine the crystallographic structure of surfaces. In this section the nomenclature used to describe surface structures will be developed, and the field of surface crystallography will be considered in general. The subject of surface crystallography has been developed extensively along the same lines as bulk crystallography. Concepts such as the various symmetry operations that can be performed on a given structure, the exhaustive list of possible symmetry combinations, and surface Bravais lattices have been developed at length. This subject will not be discussed in detail here, as it is not necessary to an understanding of the material to be considered in this book.

An understanding of the nomenclature used to describe surface structures is important, however, as this nomenclature will be used repeatedly in the description of surface processes. Surface structures are always described in terms of their relationship to the bulk structure beneath the surface. The substrate lattice parallel to the surface is taken as the reference net. The surface net is indexed with respect to the substrate. That is, for any translation between lattice points on the substrate net we have

$$\mathbf{T} = n\mathbf{a} + m\mathbf{b}\ (n, m, = 0, 1, 2\ldots). \tag{1.5}$$

For translations between lattice points on the surface net we have

$$\mathbf{T}_s = n'\mathbf{a}_s + m'\mathbf{b}_s\ (n', m' = 0, 1, 2, \ldots). \tag{1.6}$$

The relationship between $\mathbf{T}$ and $\mathbf{T}_s$ is in many cases quite simple, as one often finds that

$$\mathbf{a}_s = p\mathbf{a},\ \mathbf{b}_s = q\mathbf{b}\ (p, q = 0, 1, 2, \ldots). \tag{1.7}$$

That is, the surface unit cell vectors are parallel to those of the substrate. In the simple case of a clean ideal surface, in which by definition no relaxation has taken place, p and q are both unity. In the general case, the notation used to describe the surface is

$$M(hkl)\ p \times q - A. \tag{1.8}$$

Here M is the chemical species making up the substrate, (hkl) are the Miller indices of the substrate orientation, p and q the integers relating surface to bulk periodicity, and A is the chemical species, if any, that is adsorbed on the surface. For example, a surface that was formed by cutting a nickel crystal parallel to the (100) plane of the crystal and adsorbing oxygen atoms on the surface to form a structure that had a periodicity twice that of the substrate in both a and b directions would be called

$$\text{Ni}\ (100)\ 2 \times 2 - O. \tag{1.9}$$

In the more complicated case in which the surface and substrate vectors are not parallel to one another, we may write

$$\mathbf{a}_s = p_1\mathbf{a} + q_1\mathbf{b} \quad \mathbf{b}_s = p_2\mathbf{a} + q_2\mathbf{b}. \tag{1.10}$$

In most cases where this behavior is observed, it is found that the angle between $\mathbf{a}_s$ and $\mathbf{b}_s$ is the same as the angle between $\mathbf{a}$ and $\mathbf{b}$. In this case, the surface structure can be described by the form

$$M(hkl)\ \frac{\mathbf{a}_s}{\mathbf{a}} \times \frac{\mathbf{b}_s}{\mathbf{b}}\ R\alpha - A, \tag{1.11}$$

where α is the angle of rotation between the surface and substrate nets. For example, a commonly observed structure for a Ni (111) surface with adsorbed oxygen atoms is

$$\text{Ni}(111)\sqrt{3} \times \sqrt{3}\, R30° - O. \tag{1.12}$$

This notation may be applied to any surface, be it clean or covered with an adsorbed layer. Note that the unrelaxed ideal surface would always be indexed as 1×1.

Let us now return to the question of what LEED spot patterns look like, in general, and what structures are commonly observed. Recall that the spots observed on the LEED pattern will be those determined from the Ewald sphere construction for the reciprocal lattice of the surface net in question. In the strictly two-dimensional case, the reciprocal lattice vectors are defined by

$$\mathbf{a}^*_s = \frac{2\pi}{\mathbf{a}_s},\ \mathbf{b}^*_s = \frac{2\pi}{\mathbf{b}_s}. \tag{1.13}$$

Spots will thus be observed for $\mathbf{a}^*_s$, $\mathbf{b}^*_s$ combinations that satisfy the condition

$$\mathbf{k}' - \mathbf{k}_0 = \mathbf{g} = h\mathbf{a}^*_s + k\mathbf{b}^*_s. \tag{1.14}$$

The observed spots can be indexed in terms of the values of h and k. For example, the specularly reflected beam, for which $\mathbf{k}' = \mathbf{k}_0$, is indexed as the (0,0) spot, the beam associated with $\mathbf{g} = \mathbf{b}^*_s$ is the (0,1) spot, and so forth. An example of the LEED pattern expected for a 1×1 structure on a face centered cubic (100) surface orientation is shown in Figure 1.12 (assuming a primitive square for the unit mesh). Recall that the number of spots seen in a given pattern will depend on the size of the Ewald sphere, whose size increases with increasing primary electron energy. The higher the electron energy, the more spots will be seen and the closer a given spot will be to the specular beam.

Note finally the way in which the LEED spot pattern changes if an adsorbed overlayer is added to the surface. For example, if a 2 × 1 overlayer is adsorbed on a (100) 1 × 1 structure, a pattern such as that shown in Figure 1.13 will be observed. The larger the integers in the description of the overlayer, the more closely the overlayer spots are spaced. Many examples of overlayer patterns appear throughout this book. Note again that the spot pattern gives only the symmetry of the structure but not the atom positions. The pattern shown in Figure 1.13 would have the same spot spacing irrespective of whether the adsorbed atoms were adsorbed directly on top of the surface atoms or in the interstices between atoms. In order to determine which of the above configurations is in fact formed, one must use the LEED I-V curves or some independent analytical technique.

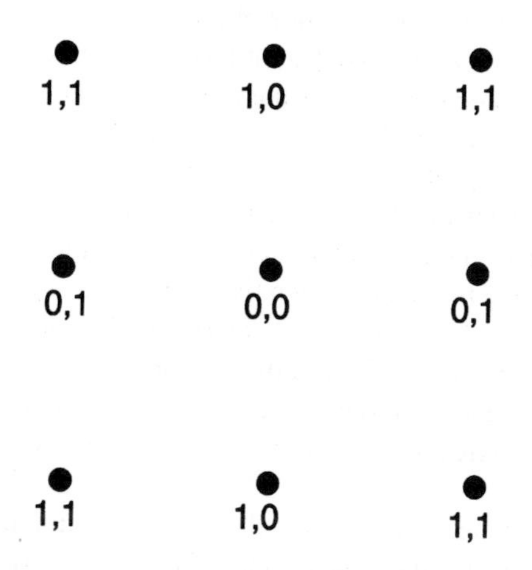

Figure 1.12 Schematic picture of the LEED spot pattern for a clean FCC (100) 1 × 1 surface.

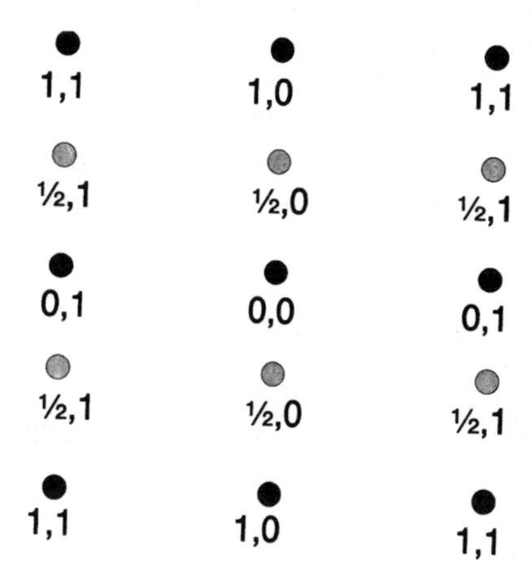

Figure 1.13 Schematic picture of the LEED spot pattern for an FCC (100) surface with an ordered adsorbed layer in a 2 × 1 structure.

Real Surfaces

With the background material presented thus far, we are now ready to look at the differences observed in practice between real surfaces and the ideal surfaces represented by the TKL model. We will consider two kinds of differences, namely, large-scale relaxations and surface point defects.

Consider first surface relaxations. The simplest kind of relaxation to consider is the motion of the outermost layers of atoms relative to the bulk of the crystal, that is, a progressive change in spacing along the c direction near the surface of the crystal, as shown schematically in Figure 1.14. Whether or not such relaxations occur was a matter of controversy for some time. Theoretical estimates of the magnitude of the expected relaxation give results in the range of plus or minus a few percent in c. Typical theoretical estimates for various copper surfaces are summarized in Table 1.1.

Most experimental attempts to measure this relaxation have involved analysis of LEED I-V curves, such as those shown in Figure 1.15 for the Cu (100) surface. Note that the curves show major peaks in intensity near the expected positions for diffraction maxima for the three-dimensional lattice, that is, electron energies such that the Ewald sphere intersects a three-dimensional reciprocal lattice point. If the outer layers of the crystal are relaxed relative to the bulk, then the positions and intensities of these maxima will be changed but in a complex way. Experimental evidence of the sort presented in Figure 1.15 indicates that relaxations of this sort are present on many metal surfaces. Results obtained in different studies, however, or using different methods for analyzing the I-V curves, often disagree as to both the magnitude and sometimes the sign of the relaxation.

A second type of relaxation involves the relative motion of various atoms in the surface with respect to one another, either in the plane of the surface or perpendicular to the surface, or motion of a plane of surface atoms laterally, relative to the bulk of the crystal. Relaxations of this type are usually referred to as surface reconstructions. Experimental evidence, primarily from LEED spot patterns, indicates that most clean metal surfaces are not reconstructed. The observed LEED patterns are usually the 1×1 type. Notable exceptions to this rule are the (100) faces of gold and platinum, the (100) faces of tungsten and molybdenum, and the (110) faces of platinum and iridium. Examples of these reconstructions are shown in the LEED patterns of Figures 1.10 and 1.11. In the case of gold and platinum, the reconstructed structure has been interpreted in

Table 1.1 Calculated surface relaxations for various copper surfaces

Surface	$\delta d_1/d$	$\delta d_2/d$	$\delta d_3/d$	$\delta d_4/d$
(100)	0.129	0.033	0.008	0.001
(110)	0.196	0.047	0.019	0.003
(111)	0.055	0.009	0.001	

δi is the displacement of the ith layer and d the interplanar spacing normal to the surface. These calculations were made using a Morse type of pairwise interaction potential. (Source: P. Wynblatt and N.A. Gjostein, *Surface Sci.* 1968;12:109. Reprinted with permission.)

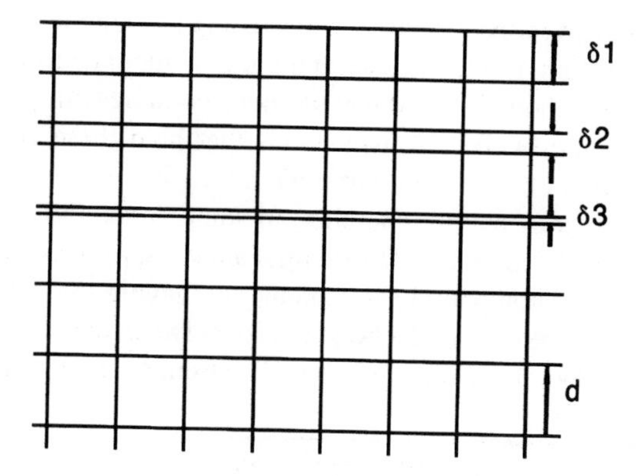

Figure 1.14 Surface relaxation normal to the plane of the surface, showing the change in a c-axis spacing with distance.

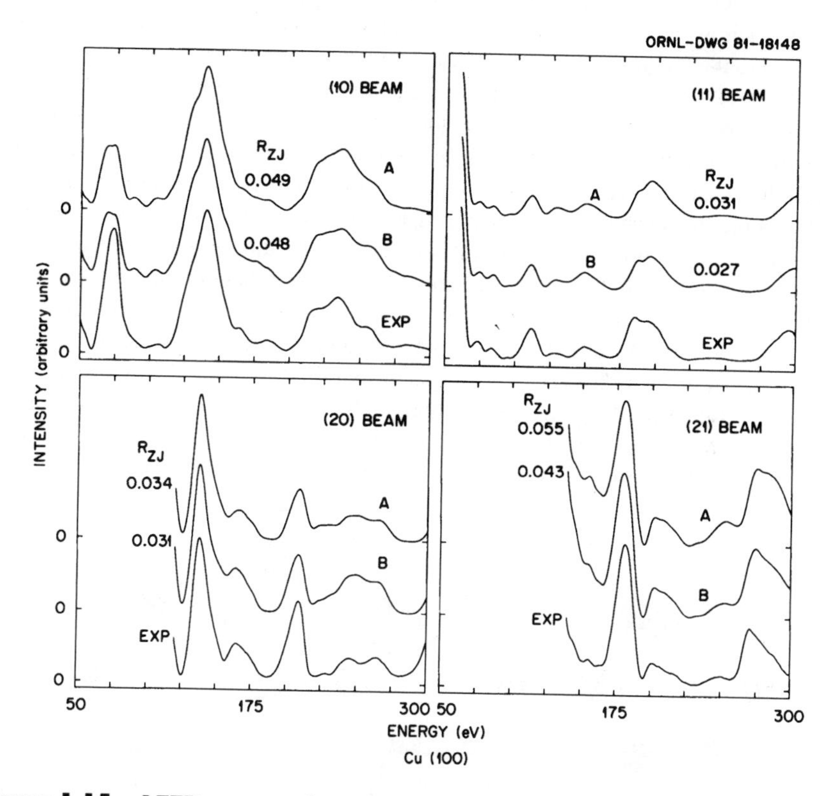

Figure 1.15 LEED current–voltage (I–V) curves for the Cu (100) surface. Each set of curves compares the experimental results with the predictions of a theory which assumes changed interlayer spacing for the top two atomic layers. Curves *A* calculated for $\Delta d_{12} = -0.90\%$ and $\Delta d_{23} = 0.00\%$; Curves *B* calculated for $\Delta d_{12} = -1.45\%$ and $\Delta d_{23} = 2.25\%$. (Source: H.L. Davis and J.R. Noonan, *J. Vac. Sci. Technol.* 1982; 20:842. Reprinted with permission.)

terms of a rearrangement of the top-layer atoms to a (111) symmetry. The tungsten and molybdenum structures are formed by lateral motion of alternate rows of surface atoms.

The case of semiconductor surfaces is more complicated, probably due to the highly directional nature of the covalent bonds that hold these crystals together. In some cases 1 × 1 structures are seen, but often the LEED patterns of clean elemental and compound semiconductors are complex. In some cases, the nature of the reconstruction has been elucidated by LEED structural analysis. A classic case of such a system is the (110) face of compound semiconductors having the zincblende structure, shown in Figure 1.16. In this case, the bond angles between the atoms in the outermost layers of atoms are markedly changed by the relaxation, as shown in the figure.

Scanning Tunneling Microscopy

Major advances in our understanding of both surface reconstructions and surface point defects have been made using a second technique that permits imaging the surface at the atomic level, namely, use of the scanning tunneling microscope (STM) and related instruments such as the atomic force microscope (AFM). Developed around 1980, this technique led to the Nobel prize in physics for its inventors, G. Binnig and H. Rohrer.

In this technique, a fine metal tip, similar in concept to the tip in a field ion microscope, is positioned within 1 nm of the surface to be studied, and a voltage is applied between sample and tip. Under these conditions, a small electronic current will flow from tip to sample (or vice versa) due to electron tunneling. Because the magnitude of this tunneling current is extremely sensitive to the distance between the tip and the surface, the magnitude of the current is a measure of the tip-to-sample distance, with a distance resolution of less than 0.01 nm. A schematic diagram of a typical STM is shown in Figure 1.17. In operation, either the tip is moved over the sample and the variation in tunneling current is used to make a topographic map of the surface or the feedback signal to the tip position control system required to maintain a constant tunneling current is measured to provide the topographical information. One of the early triumphs of this technique was the definitive characterization of a complex relaxation of the Si (111) surface, known as the Si (111) 7 × 7 reconstruction. This surface had

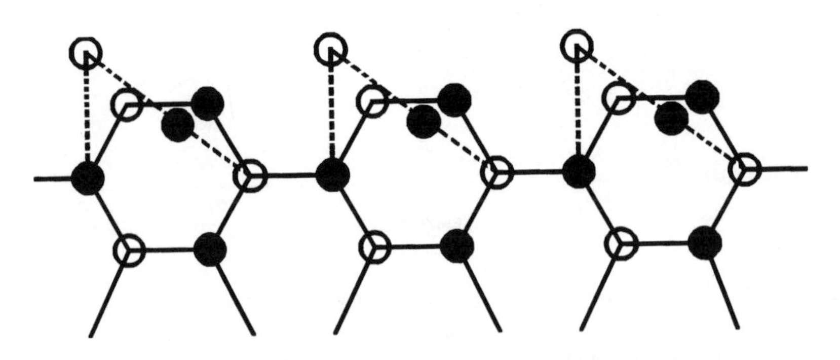

Figure 1.16 Reconstruction of the GaAs (110) surface. Solid line, ideal surface. Dotted line, reconstructed surface. (Source: C.B. Duke, *J. Vac. Sci. Technol.* 1976;13:761. Reprinted with permission.)

been the subject of widespread study by a range of diffraction techniques, involving both low- and high-energy electrons and helium atoms. The complexity of the reconstruction makes an unequivocal evaluation of I-V curves extremely difficult, although studies by Takayanagi [1985], using high-energy electron diffraction, had suggested the model eventually shown to be correct. An STM image of this surface is shown in Figure 1.18, along with the Takayanagi model of the reconstructed unit cell. In the micrograph, the rhombic shape of the unit cell is outlined. The adatom and hole positions can be seen with high contrast. The letters *F* and *U* indicate the faulted and unfaulted halves of the unit cell, as shown in the model.

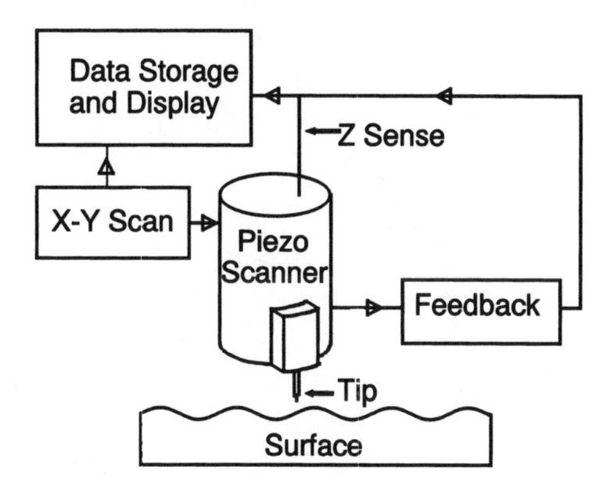

Figure 1.17 Schematic diagram of a scanning tunneling microscopy system.

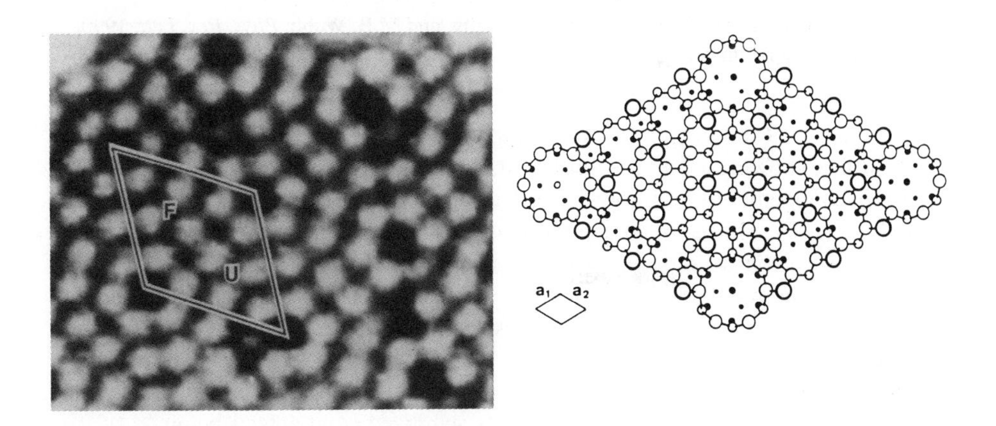

Figure 1.18 The Si (111) 7 × 7 surface as revealed by scanning tunneling microscopy. (a) STM image. The rhombic unit cell is outlined. The letters *F* and *U* refer to the faulted and unfaulted halves of the unit cell, as indicated in the accompanying model. (Source: R.J. Hamers, R.M. Tromp, and J.E. Demuth, *Phys. Rev. Lett.* 1986; 56:1972. Reprinted with permission.) (b) Schematic representation of surface. The adatoms are indicated by heavy circles; a stacking fault runs vertically across the center of the diagram. (Source: K. Takayanagi, Y. Tanashiro, S. Takahashi, and M. Takahashi, *Surface Sci.* 1985; 164:367. Reprinted with permission.)

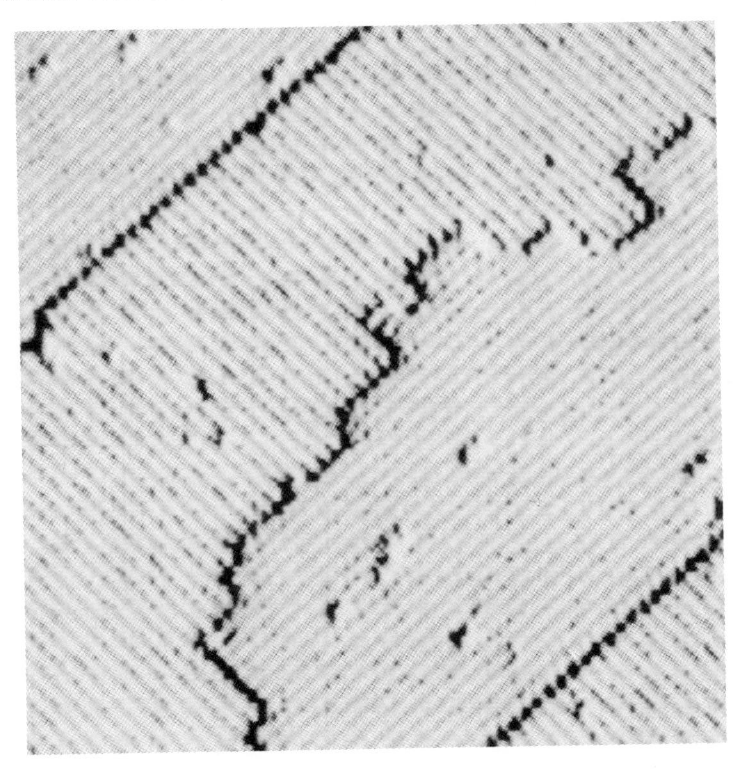

Figure 1.19 STM topograph of an Si (100) 2 × 1 reconstructed surface, showing monatomic height steps and the dimer rows formed in the reconstruction. (Source: B.S. Swartzentruber, Y-W. Mo, R. Kariotis, M.G. Lagally, and M.B. Webb, *Phys. Rev. Lett.* 1990; 65:1913. Reprinted with permission.)

Figure 1.19 shows an STM micrograph of a second silicon surface. This is an Si (100) surface showing a 2 × 1 reconstruction, in which adjacent rows of surface atoms have moved together to form rows of surface "dimers." This is a common reconstruction of the (100) surfaces of tetrahedrally bonded semiconductors, as the formation of the bonds between rows reduces the system energy associated with "dangling" bonds at the surface. The micrograph clearly shows the rows of dimers on the terraces. It also shows the presence of ledges and of kinks on the ledges. Note that there are two domains of the 2 × 1 reconstruction, at 90° to one another, and that the ledges that terminate the two different orientations have very different kink spacing. We will return to this difference when we consider thermally stabilized point defects below.

Surface Point Defects

Now consider the question of short-range nonidealities on crystal surfaces, or surface point defects. These nonidealities can be separated into two classes: those that are thermodynamically stable and those that are kinetically stable. The kinetically sta-

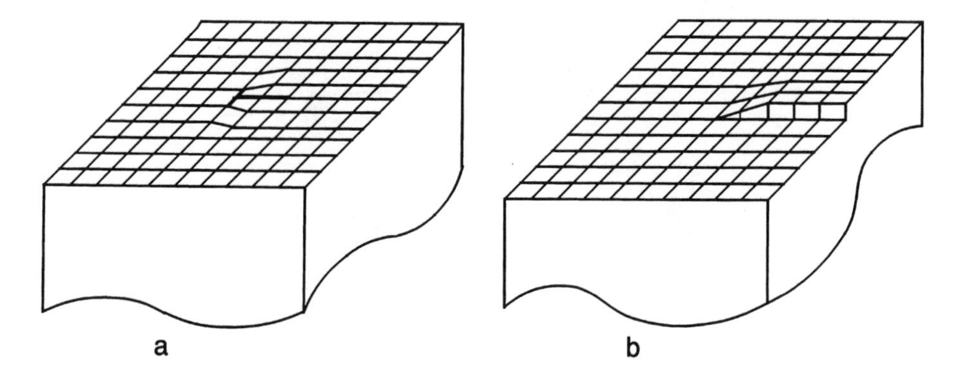

Figure 1.20 TLK model representation of the sites of emergence of dislocations onto a surface, showing (a) edge dislocation, and (b) screw dislocation.

ble class includes primarily points of dislocation emergence at the surface. Dislocations, which are line defects arising from mismatches in the bulk lattice of the crystal, are present in the bulk of practically all crystalline solids. They must be present either as closed loops or networks within the crystal or must terminate on the crystal surface. These points of termination disrupt the orderly array of surface atoms shown by the TLK model. The nature of the disruption is shown in Figure 1.20. Note that there are basically two types of dislocation, called edge dislocations and screw dislocations.

The edge dislocation is associated essentially with an extra half-plane of atoms in the bulk of the crystal, indicated by the $\perp$ in Figure 1.20a. The importance of the sites of emergence of edge dislocations is that they represent a different bonding configuration than normal surface atoms and as such may behave differently in gas-surface interactions such as adsorption or surface chemical reactions.

The screw dislocation is formed by shearing one-half of the crystal lattice with respect to the other half, in a portion of the crystal. The emergence of a screw dislocation creates a step on the surface of the crystal, one end of which is tied to the site of dislocation emergence. The presence of such steps can play a major role in the growth of crystals from the vapor or from solution. Because the step is tied to the site of dislocation emergence, it has the property of being continuously regenerated as crystal growth proceeds.

The second class of surface point defects consists of those defects that are present at equilibrium at any temperature above 0 K for thermodynamic reasons. These defects are similar in concept to the vacancies or interstitial atoms found at equilibrium in the bulk of crystals. Defects of this sort, formed in the ledges and terraces of crystal surfaces, all have a finite, positive free energy of formation from the structure given by the TLK model, but they are stable in finite quantities because of the favorable entropy of mixing term associated with the disorder produced by forming defects in an initially ideal system.

To consider the nature of the possible defects and the energetics of their forma-

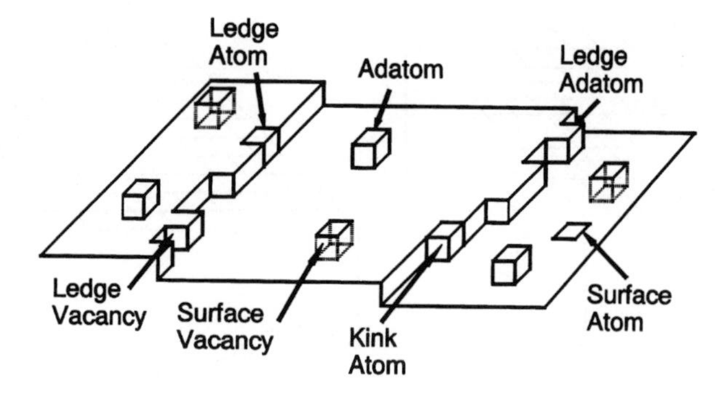

Figure 1.21 Typical surface sites and defects on a simple cubic (100) surface.

tion, begin with a surface vicinal to the (100) surface of a simple cubic crystal, as shown in Figure 1.21, and consider the process of forming the defects shown in the figure by removing an atom from a kink site and moving it to some other location, or vice versa. In determining energy changes, assume that the lattice energy can be represented by a nearest neighbor model, with each atom in the crystal bonded to its six nearest neighbors. The relative energies for other possible atomic sites are shown in Figure 1.22.

Consider now the process of forming a surface vacancy. We can do this in principle by removing an atom from a terrace site and replacing it at a kink site. In this process, one must break five bonds to remove the atom from the terrace, and reform three bonds by placing the atom in the kink site. Thus the net change is minus two bonds. That is,

$$\Delta G_v = W_K - W_T, \tag{1.15}$$

where W_K and W_T are defined in Figure 1.22. From a straightforward thermodynamic calculation, we can show that the equilibrium concentration of surface vacancies will be

$$\frac{n_v}{M} = \exp\frac{-\Delta G_v}{kT} \tag{1.16}$$

in which M is the total number of surface sites per unit area. Note that ΔG_v as defined in Equation 1.15 is inherently positive, so that the equilibrium concentration of surface vacancies will rise exponentially with increasing temperature.

In the same way, one can calculate energies of formation for other surface point defects. A surface adatom can be formed in principle by removing an atom from a kink site and placing it in the middle of a terrace. In this case, three bonds are broken in removing the kink atom, and one is reformed in putting it back on the surface, so that

$$\Delta G_a = W_A - W_L. \tag{1.17}$$

Again, the equilibrium concentration will be given by

$$\frac{n_a}{M} = \exp\frac{-\Delta G_a}{kT} \tag{1.18}$$

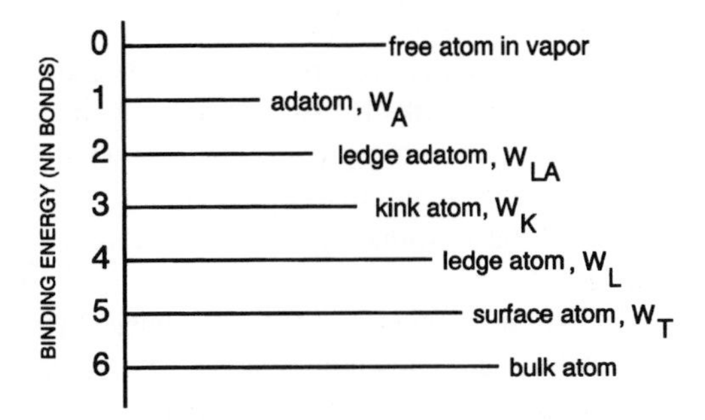

Figure 1.22 Relative binding energies for various surface sites and defect configurations on a simple cubic (100) surface, according to the nearest neighbor model.

Similar arguments hold for ledge adatoms and ledge vacancies:

$$\Delta G_{lv} = W_K - W_L,$$

$$\frac{n_{lv}}{N} = \exp\frac{-\Delta G_{lv}}{kT} \qquad 1.19$$

$$\Delta G_{la} = W_{LA} - W_K,$$

$$\frac{n_{la}}{N} = \exp\frac{-\Delta G_{la}}{kT} \qquad 1.20$$

where in this case N is the number of ledge sites per unit area of crystal surface.

In real crystals, the argument presented above is generally valid, but the values of the various energies of formation are not accurately given by the simple nearest neighbor assumption, as it neglects both relaxation around the defect and effects associated with second and more distant neighbors. The magnitude of the expected relaxation has been calculated theoretically and measured experimentally in great detail for the case of bulk defects. The theoretical calculations involve, for the case of a bulk vacancy, removing an atom from the lattice, placing it at a kink site on the surface, and then calculating the degree of relaxation of atoms around the vacancy by an energy minimization process. Experimental measurements of vacancy formation energy involve heating a sample to a high temperature, at which the vacancy concentration will be high, then rapidly quenching to freeze in this high concentration, and finally measuring the change in some property of the sample, such as density or resistivity, as the sample anneals. From the dependence of the property change on the temperature before quenching, ΔG_v can be obtained. As an indication of the degree of relaxation in various systems, a calculation neglecting relaxation gives $\Delta G_v \approx 3.8$ eV for bulk vacancies in both gold and germanium. Experimental values are about 1.1 eV for gold and 2.6 eV for germanium, indicating a much greater relaxation for the metallic lattice.

Such calculations and experiments are much more difficult for the case of surface defects. In order to go beyond a simple nearest neighbor model, one must understand the relaxation process at surfaces, where the constraints are quite different from those in the bulk. Experimentally, the quenching type experiments used to study bulk defects are not feasible, as there are many sinks for surface defects and a rapid rate of migration to these sinks during the quenching process. It is possible, however, to see the defects postulated above by field ion microscopy. Adatoms are especially easy to see, as they are above the level of the surrounding terrace and thus show up with high contrast, as do defects in ledges.

These defects can also be seen in STM micrographs. In the case of the STM image of the Si (100) –2 × 1 surface shown in Figure 1.19, it is possible to calculate the energy associated with the ledge and the formation energy for kinks on the ledges by counting the number of kinks per unit length of ledge as a function of the temperature at which the surface was equilibrated before the STM measurement. In the figure, the ledges that are parallel to the dimer rows have an energy of 0.028 eV per atom, while those that run perpendicular to the dimer rows have an energy of 0.09 eV per atom. An excess energy of 0.08 eV is associated with each kink site.

We will return to the subject of surface point defects in Chapter 17, as they are important factors in two major surface processes. The various surface defects represent stages of incorporation of atoms into the crystal lattice and, thus, are of significance in the process of crystal growth. We shall also see in Chapter 17 that ledges play a vital role in crystal growth kinetics. The various surface point defects also provide a mechanism for surface transport in the process of surface self diffusion.

Cooperative Effects among Defects

One final question on the subject of thermally induced surface defects remains to be discussed: What happens if the temperature is high enough to cause a high concentration of surface defects? If the defect concentration becomes very large, then the defects will begin to interact with one another, and their energy of formation may be affected. In the extreme case, if the formation energy decreases as the concentration increases, we may observe a catastrophic process leading to a general roughening or a so-called *surface-melting*.

Several attempts have been made to determine whether or not such a process occurs in practice. A rigorous calculation of this effect would require a knowledge of the formation energies of surface defects and of the way in which these formation energies change with increasing defect concentration. This information is lacking at present. It is possible, however, to carry out a first-order calculation, assuming that the possible defects involved are surface vacancies, adatoms, divacancies or diadatoms. This calculation involves the assumption of values for the formation energies of vacancies and adatoms, and for the energy difference between two single defects and a defect pair. These energies are balanced against an entropy of mixing term involving all four of the postulated species to determine the minimum surface free energy. If it is assumed that the energy to form a defect pair from two single defects is negative, the calculation does in fact lead to surface melting. The temperature at which this process is predicted to take place, however, is well above the bulk melting point of the crystals involved, at least for singular surfaces.

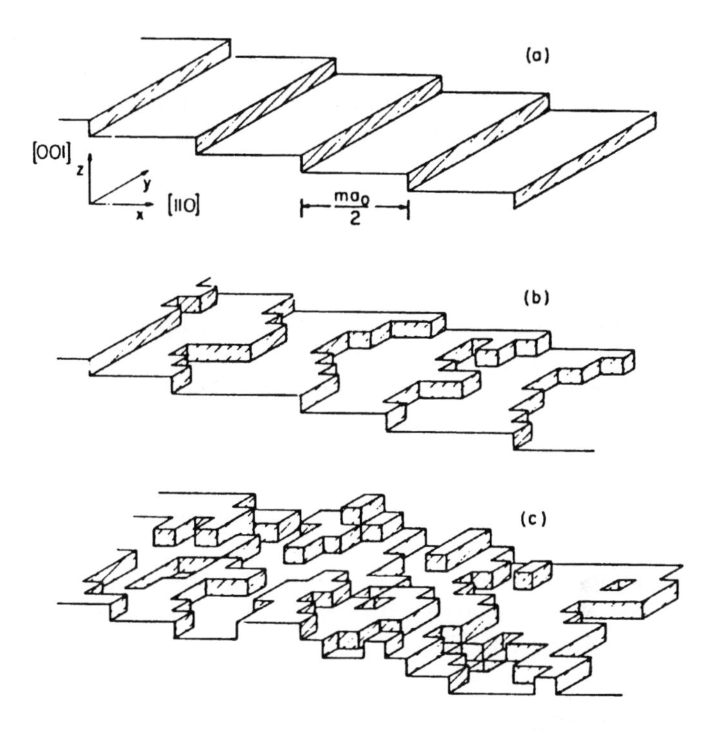

Figure 1.23 Schematic view of the surface roughening transition on an FCC (11m) surface, showing (a) $T = 0$ K, (b) meandering steps, and (c) above the roughening temperature. (Source: E.H. Conrad, R.M. Aten, D.S. Kaufman, L.R. Allen, T. Engel, M. den Nijs, and E.K. Riedel, *J. Chem. Phys.* 1986;84:1015. Reprinted with permission.)

The situation is quite different, however, for the case of vicinal surfaces. In recent years several studies have been carried out involving principally the elastic scattering of rare gas atoms from the surface under study, which indicate that, for metals, many of these vicinal surfaces are extensively disordered at temperatures well below the melting point of the bulk crystal. This process is shown schematically in Figure 1.23, which originally appeared with an article that reported the observance of surface melting on a Cu (115) surface. These measurements are discussed in more detail in Chapter 9.

Problems

1.1. Calculate the lattice vectors of the surface unit mesh for the (100) face of a metal having the face centered cubic structure, such as is shown in Figure 1.6a, assuming that the lattice parameter of the bulk crystal is 0.40 nm.

1.2. If an LEED experiment were carried out on the crystal of Problem 1.1, using 100 eV primary electrons, what would be the angle between the 0,0 and the 1,0 diffracted beams?

1.3. Construct the expected LEED pattern for a clean Ag (110) face for a primary electron energy of 150 eV and electron incidence normal to the silver surface.

1.4. Indicate the positions of the additional spots that would appear in the LEED pattern constructed in Problem 1.3 if oxygen were adsorbed on the surface in a $p(2 \times 2)$ structure.

1.5. Calculate the equilibrium surface adatom and surface vacancy concentrations for an Ag (111) surface, assuming that the bonding in the crystal can be described by a nearest neighbor bonding model, for T = 0 K, 100 K, and 1000 K.

1.6. How would the values calculated in Problem 1.5 change if relaxation around the defects reduced the energy of formation by a factor of three?

1.7. A Cu (111) surface is initially covered with 19 adsorbed copper atoms per square centimeter. If it can be assumed that the bonding in the system can be described by a nearest neighbor model;

a. Calculate the change in energy per square centimeter if these adatoms accrete to form hexagonal islands of 19 atoms each.

b. As the island size increases, how does the average number of bonds per atom change?

1.8. A platinum single crystal is cut to expose a surface that is misoriented from the (111) plane by 3.11° along a [110] direction. Describe the equilibrium structure of the surface at 0 K, using the TLK model.

Bibliography

Blakely, JM. *Introduction to the Properties of Crystal Surfaces.* New York: Pergamon Press, 1973.

Bringans, RD, Feenstra, RM, and Gibson, JM, eds. *Atomic Scale Structure of Interfaces.* Pittsburgh: Materials Research Society, 1990.

Burton, WK, Cabrera, N, and Frank, FC. The growth of crystals and the equilibrilium structure of their surfaces. *Phil. Trans. Roy. Soc.* 1951; A243:299.

Estrup, PJ, and McRae, EG. Surface studies by electron diffraction. *Surface Sci.* 1971; 25:1.

Garcia, N, ed. *STM '86.* Amsterdam: Elsevier, 1987.

Nicholas, JF. *An Atlas of Models of Crystal Surfaces.* New York: Gordon and Breach, 1965.

Somorjai, GA. *Chemistry in Two Dimensions: Surfaces.* Ithaca, N.Y.: Cornell University Press, 1981.

Strozier, JA, Jepsen, DW, and Jona, F. Surface crystalography. In: Blakely, J.M., ed. *Surface Physics of Materials.* New York: Academic Press, 1975; 1:1–77.

Takayanagi, K, Tanashiro, Y, Takahashi, S, and Takahashi, M. Structure analysis of Si(111)-7 × 7 reconstructed surface by transmission electron diffraction. *Surface Sci.* 1985; 164:367.

Van Hove, MA, and Tong, SY. *Surface Crystalography by LEED.* New York: Springer-Verlag, 1979.

Zangwill, A. *Physics at Surfaces.* New York: Cambridge University Press, 1988.

CHAPTER 2

Electronic Structure of Surfaces

This chapter addresses the question of the electronic structure of solid surfaces. It will be concerned with where the electrons are, both in terms of charge density near the surface and in terms of the allowed energy levels, both occupied and unoccupied. It will also present methods for measuring these properties, and in addition describe some surface analytical techniques whose principle of operation depends on surface electronic structure.

Metal Surfaces—the Jellium Model

Consider first the case of the electronic structure of clean metal surfaces. Conceptually, we can separate the behavior of the valence, or conduction, electrons, which are more-or-less free to move about within the metal, from the behavior of the so-called core electrons, which are bound to the ion cores in the metal structure and are not significantly involved in the binding energy of the metal. The energy terms that are involved in the binding energy of the crystal relative to a collection of free atoms are thus:

1. the coulomb interactions among the valence electrons.
2. the coulomb interactions among the ion cores.
3. the coulomb interactions between the valence electrons and the ion cores.
4. the kinetic energy of the valence electrons.
5. vibrations of the ion cores.
6. a negative contribution from the ionization energy of the free atoms.
7. electron correlation and exchange effects.

This last term includes quantum mechanical effects arising from the Pauli exclusion principal and the correlated motion of electrons due to coulomb repulsion.

The Infinite Crystal

Let us look first at how we can treat these terms for the case of an infinite crystal, ignoring the presence of surfaces for the moment. The simplest treatment is to use what is known as the *Jellium model*. In this model, the ion cores are replaced by a uniform sea of positive charge, having the same average charge density as the real crystal under consideration. For the case in which a surface is present, which we will consider in the sections on infinite surface barrier and finite surface barrier, the crystal surface is the edge of this area of positive charge. This model has been used extensively for calculations and can give reasonable agreement with some measured crystal properties, such as the work function. The major deficiency is that because it ignores the periodic nature of the crystal lattice, it washes out all effects associated with this periodicity, such as the variation in properties from one crystal face to another on a given crystal.

In applying the Jellium model to an infinite crystal, begin by considering a cube of length L, having a positive charge density $\rho_+(r) = ne$, in which n is the number of ions per unit volume. If we neglect all many body effects, which means essentially that we assume that the energy states available to a given electron are independent of how many other electrons are present, we can write the wave function of the whole system of electrons as a superposition of one electron wave functions of the type

$$-\left(\frac{\hbar^2}{2m}\right)\nabla^2\psi_i + V\psi_i = \varepsilon_i\psi_i, \qquad 2.1$$

in which $\hbar$ is Planck's constant divided by 2π, m the electron mass, V the potential a given electron sees as a result of all other particles in the system, and ϵ_i the total allowed electron energy.

For the infinite crystal, the allowed values of ψ_i (the eigenvalues of the equation) are those that are periodic in L in the x, y, and z directions (that is, parallel to the cube sides). For this case, we also have V = constant, and

$$\psi_{\mathbf{k}} = \frac{1}{L^{2/3}}\exp(i\mathbf{k}r)$$

$$= \left(\frac{1}{L^3}\right)^{1/2} \exp\left[i\left(\mathbf{k}_x x + \mathbf{k}_y y + \mathbf{k}_z z\right)\right] \qquad 2.2$$

for the electron kinetic energy term, where $\mathbf{k}$ is the electron wave vector. The fact that the charge density is uniform throughout, and thus that V is constant, means that V is indeterminate and can be neglected. The allowed electron kinetic energies are thus calculated to be

$$\varepsilon_{\mathbf{k}} = \left(\frac{h^2}{2m}\right)\mathbf{k}^2 = \left(\frac{h^2}{2m}\right)\left(\mathbf{k}_x^2 + \mathbf{k}_y^2 + \mathbf{k}_z^2\right), \qquad 2.3$$

where

$$\mathbf{k}_i = \frac{2\pi n_i}{L}, \quad n_i = 0, 1, 2, \ldots \qquad 2.4$$

Note that this result is simply the quantum mechanical result for the allowed kinetic energies of a particle in a box of uniform potential.

Because electrons have a spin of 1/2, and consequently follow Fermi–Dirac statistics, we can have no more than one electron per energy state or, because there are two spin states for each allowed energy, no more than two electrons per energy level. As a result, at 0 K, all available states are filled up to the value of $k = k_F$, the Fermi momentum, corresponding to an energy, the Fermi energy, ϵ_F, such that

$$k_F = (3\pi^2 n)^{1/3} \qquad 2.5$$

and

$$\varepsilon_F = \left(\frac{h^2}{2m}\right)\left(3\pi^2 n\right)^{2/3}, \qquad 2.6$$

in which n is the mean number density of valence electrons. Note that n is the only parameter that controls ϵ_F in this model. For future reference, we shall also define a parameter called the "effective radius" of valence electrons as

$$\frac{4\pi r_s^3}{3} = \frac{1}{n}. \qquad 2.7$$

Thus, large values of r_s imply a low density of valence electrons. Typically r_s is on the order of 0.1 nm.

Infinite Surface Barrier

Consider next the ways in which the picture developed in the previous section changes when we allow for the presence of a surface. The presence of the surface can be formally introduced into the quantum mechanical problem in several ways. The simplest method mathematically is to assume an infinitely high, square potential barrier at the surface by defining $V = O$ inside the metal and $V = \infty$ in the vacuum space outside the crystal at $x = O, L$, but to maintain the assumption of infinite extent in the y and z directions. For this set of assumptions, the eigenvalues of ψ_i are given by

$$\psi_{\mathbf{k}} = \left(\frac{2}{L^3}\right)^{1/2} \sin(\mathbf{k}_x x)\exp\left[i(\mathbf{k}_y y + \mathbf{k}_z z)\right]. \qquad 2.8$$

This expression differs from that for the infinite solid (Equation 2.2) in the form of the dependence on x, but the y and z behavior are unchanged. For this case, the allowed values of $\mathbf{k}_x$ are

$$\mathbf{k}_x = \frac{(n_x \pi)}{L}, \quad n = 1, 2, 3, \ldots \qquad 2.9$$

One consequence of this result is that the electron charge density is no longer uniform throughout the crystal. The charge density in general is defined by

$$\rho_- = -e\sum_k (\psi_k)^2. \qquad 2.10$$

Using the k-values appropriate to the present case yields

$$\rho_- = -e\sum_{\mathbf{k}}\left[\left(\frac{4}{L^3}\right)\sin^2(\mathbf{k}_x x)\right]g(x), \tag{2.11}$$

where $g(x)$ is the number of states having a given $\mathbf{k}_x$, but different values of $\mathbf{k}_y$ and $\mathbf{k}_z$. Evaluation of the sum leads to

$$\rho_- = -ne\left[1+\frac{3\cos\chi}{2\chi}-\frac{3\sin\chi}{3\chi}\right], \tag{2.12}$$

where $\chi = 2k_F x$.

This result is shown graphically in Figure 2.1. The x-axis in this figure is in terms of Fermi wavelengths, a term which is related to k_F by

$$\lambda_F = \frac{2\pi}{k_F} = \frac{4\pi x}{\chi} = \pi^{2/3}\left(\frac{32}{9}\right)^{1/3} r_s. \tag{2.13}$$

Typical values of λ_F are in the range of 0.5 nm.

Consider the form of this result. Because the barrier is assumed to be infinitely high, $\rho_- = 0$ outside the barrier. Physically, this causes ρ_- to drop off much too rapidly away from the surface. It is observed that there are oscillations in the charge density extending back into the crystal, a general feature of most treatments. The charge density has a finite value beyond the edge of the positive sea. This is also a generally observed

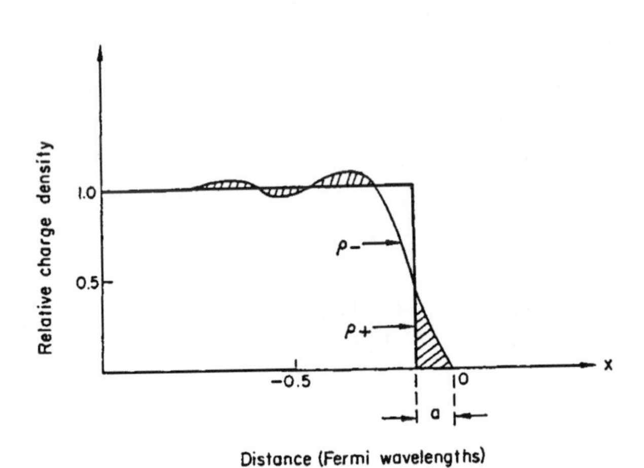

Figure 2.1 Charge distribution near the surface of a Jellium model solid with an infinite surface barrier at $x = 0$. Distances are in units of the Fermi wavelength $\lambda_F = \pi^{2/3}(32/9)^{1/3} r_s$. (Adapted from W. Swiatecki, *Proc. Phy. Soc. London* 1951;A64:227; by J.M. Blakely, *Introduction to the Properties of Crystal Surfaces*. New York, Pergamon Press, 1973. Reprinted with permission.)

feature and gives rise to a surface dipole layer, with the negative end of the dipole outermost. Note that the edge of the positive sea has had to be moved relative to the barrier in order to conserve charge. This model also neglects electron correlation effects.

Finite Surface Barrier

Now consider a second set of assumptions that is more realistic. In this case, assume that the potential barrier at the metal surface is finite in height, with the height being defined in terms of a parameter called the work function, Φ. The defining relation is

$$\Phi = (-eV - \mu), \tag{2.14}$$

in which μ is the electron chemical potential. Φ is thus the difference in potential energy between an electron just outside the surface and the chemical potential just inside the metal. The chemical potential, in turn, is defined by the energy of the electron in the highest energy filled state at T = 0 K. These relative potentials are shown schematically in Figure 2.2.

By convention, we choose the zero of potential energy as the energy of an electron "just outside" the surface. Equation 2.14 thus reduces to

$$\Phi = -\mu. \tag{2.15}$$

There is one problem with this assignment that we will deal with in the later section on inner potential. In the present context, *just outside* means a distance on the order of 10^{-4} cm. This may not be the same as the potential energy of an electron at infinite distance because of the field of the surface dipole layer. The value of the reference potential may thus change from one crystal face to another. This is observed in practice as a difference in work function from one face of a crystal to another. For the calculation of charge density, however, this convention is the most convenient and will be used for now.

Once the choice of reference potential is made, we can define μ in terms of the energy change an electron undergoes in the process of moving from "just outside" the surface to "just inside," that is, the energy change when an electron at rest "just out-

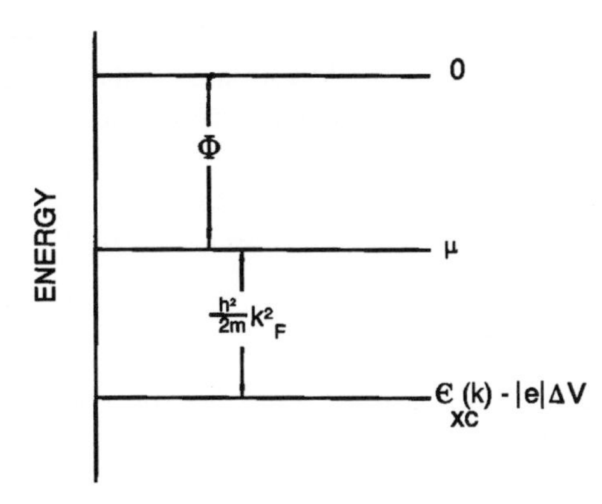

Figure 2.2 Contributions to the energy of electrons at the Fermi level in a Jellium model.

side" the surface is put "just inside" the surface with a kinetic energy equal to ϵ_F. In calculating that energy change, we must consider three types of contribution:

1. the potential energy change due to the potential difference across the surface dipole layer, ΔV.
2. exchange and correlation energy, defined as $\epsilon_{xc}(k_F)$.
3. the electron kinetic energy at k_F.

The form of Φ is then

$$\Phi = -\mu = e\Delta V - \varepsilon_{xc}(k_F) - \left(\frac{h^2}{2m}\right)k_F^2. \qquad 2.16$$

The ΔV due to the dipole layer can be characterized by the dipole moment per unit area, P, which in turn can be characterized in terms of the total charge distribution across the layer. That is,

$$\Delta V = -4\pi P = -4\pi \int_{-\infty}^{\infty} x\rho(x)dx . \qquad 2.17$$

If we stick to the Jellium model, with the positive charge density dropping abruptly to zero at $x = 0$, then we have

$$\rho(x) = n(e) - \rho\_(x) \text{ at } x < 0 \text{ (inside the metal)}, \qquad 2.18$$

$$\rho(x) = -\rho\_(x) \text{ at } x > 0 \text{ (outside the metal)}. \qquad 2.19$$

A number of investigators have treated this problem, using several different calculational techniques, with varying degrees of sophistication. The general features of the results obtained are shown in Figure 2.3. All treatments show features similar to those seen in the figure. There is a dipole layer at the surface, with $\rho\_$ dropping to zero at 0.1 to 0.3 nm from the surface. Oscillations in charge density extend back into the crystal roughly twice as far. Note that, as mentioned earlier, this treatment still does not tell us anything about the variation of the properties of the system with crystallographic orientation. A general empirical observation is that for a given crystal, the work function is greatest for the most closely packed surfaces. This result is explained on the basis that surfaces that are rougher on an atomic scale (less closely packed) have ion cores "sticking out" into the double layer. This is shown schematically in Figure 2.4. These protruding, positively charged ion cores reduce the dipole moment and, consequently, reduce the field at the surface, leading to a lower value of ΔV and consequently Φ.

The Inner Potential

Before leaving the subject of surface potentials, consider briefly one additional term, the so-called *inner potential*. This term is often used in surface theory and is especially important in theoretical treatments of low-energy electron diffraction. By

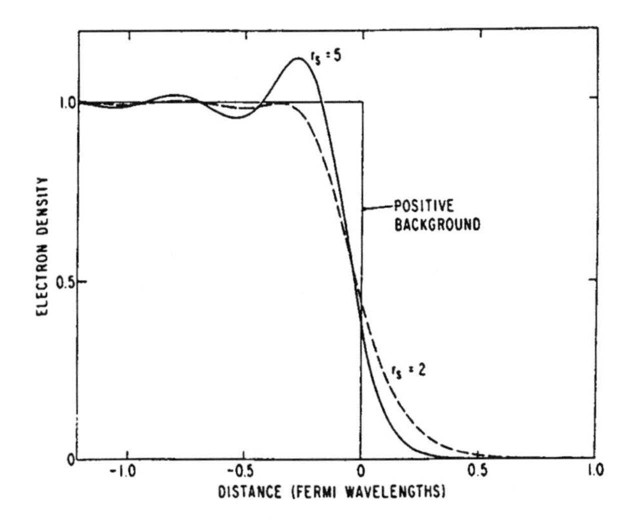

Figure 2.3 Charge distributions near the surface of a Jellium model solid with a finite surface barrier at $x = 0$. Solid curve: $r_s = 5$ (low total charge density). Dashed curve: $r_s = 2$ (high total charge density). (Source: N.D. Lang and W. Kohn, *Phys. Rev.* 1970;B1:4555. Reprinted with permission.)

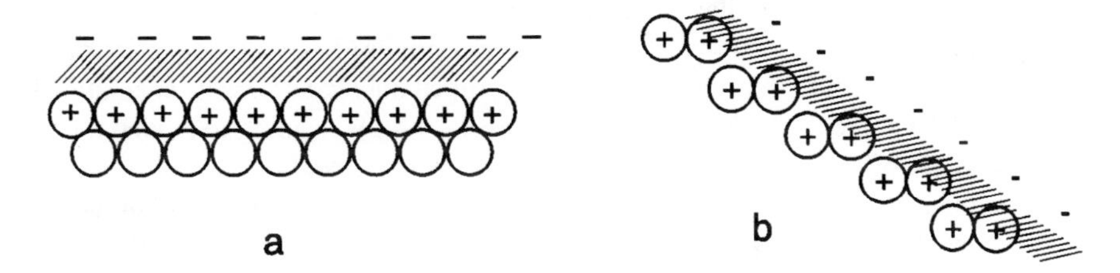

Figure 2.4 Schematic picture of surface electron charge densities for closely packed (a) and non-closely-packed (b) surfaces. Part (a) shows large surface dipole and high work function, while (b) shows small surface dipole and low work function.

analogy to optics, it is essentially a refractive index for electron waves. It is defined as the difference in potential energy of an electron of given kinetic energy in a metal relative to the energy of an electron at a large distance from the metal. For the Jellium model, this difference is given by

$$\Phi_{\text{inner}} = -e\Delta V' + \epsilon_{xc}(\mathbf{k}). \tag{2.20}$$

Here $\Delta V'$ is the electrostatic potential inside the crystal relative to that far away from the crystal, and the ϵ_{xc} is calculated for the value of $\mathbf{k}$ appropriate to the electron in question. That is, as the energy of the incident electron is changed, the value of $\mathbf{k}$ changes, as shown in our discussion of LEED. Since ϵ_{xc} is a function of $\mathbf{k}$, the value of the inner potential will depend on the energy of the electron in question, much as the refractive index of a material, in optical terms, depends on the wavelength of the optical radiation. The inner potential is related to the work function by

$$\Phi_{\text{inner}} = -\Phi - e(\Delta V' - \Delta V) + [\varepsilon_{xc}(\mathbf{k}) - \varepsilon_{xc}(k_F)] - \left(\frac{h^2}{2m}\right)k_F^2. \tag{2.21}$$

Cluster Treatments of Surface Bonding

One more approach to the problem of the theoretical calculation of surface properties will be considered. In what has been discussed so far, the comminution approach has been used to describe the surface. That is, the discussion began with the Jellium model for a bulk crystal, then looked at what happened when a surface potential barrier was imposed on the system. One can equally well take the opposite approach and essentially synthesize a surface by building up successively larger clusters of a given kind of atom. In this case, one is essentially carrying out molecular orbital calculations on successively larger molecules. This approach has been used both to deduce the properties of clean surfaces and to study the adsorption of gases on surfaces.

The basic approach can be illustrated most simply by looking at the case of the hydrogen molecule. Quantum mechanical calculations of the available energy states of the individual hydrogen atom lead to a value of –13.6 eV for the ground state, in agreement with experiment. A similar calculation for a system composed of two hydrogen nuclei (protons) and two electrons leads to a stable configuration, with both electrons in a single molecular orbital that surrounds both hydrogen nuclei, as shown schematically in Figure 2.5. The energy of this orbital is roughly –15.8 eV, leading to a bond energy for the molecule of 4.45 eV, again in agreement with experiment.

A similar procedure can be applied to systems containing increasing numbers of atoms and atoms of more complex electronic structure. Of course, the calculations

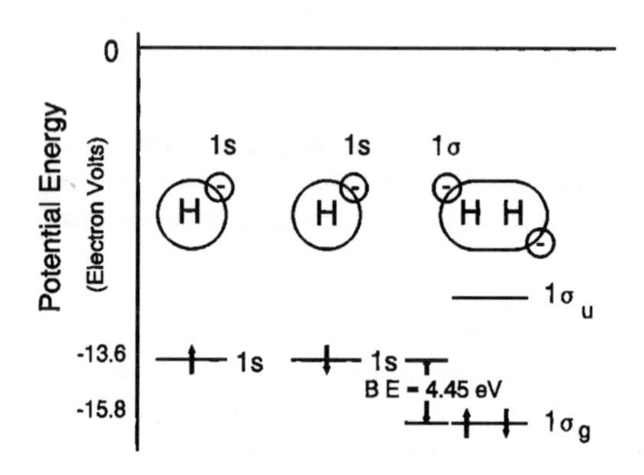

Figure 2.5 Potential energy diagram for electron orbitals for hydrogen atoms and hydrogen molecule.

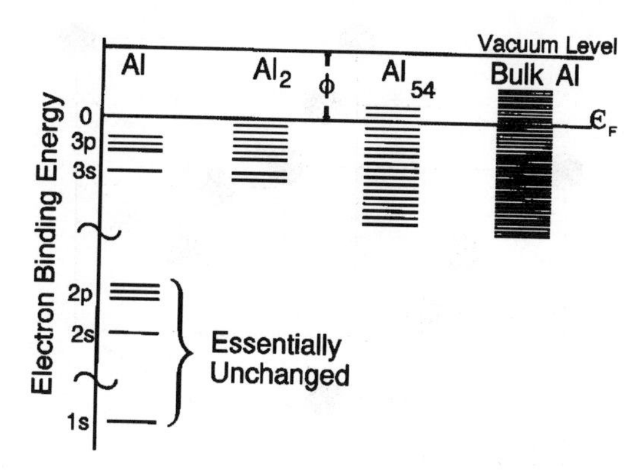

Figure 2.6 Potential energy diagram for aluminum clusters of various sizes, as determined from cluster calculations (schematic).

become much more difficult as the complexity of the system increases, but it has proved possible to treat clusters containing in excess of 50 metal atoms. In these calculations, it is customary to fix the positions of the ion cores and calculate the equilibrium configurations and energies of the orbitals occupied by the valence electrons. The configuration of the cluster is usually chosen to give at least one atom the same environment, in terms of nearest and next-nearest neighbors, as it would have in the bulk solid. The results of this procedure are shown schematically in Figure 2.6 for the case of aluminum. As the number of atoms in the cluster increases, so do both the number of valence electrons and the number of possible orbitals. The general trend with increasing cluster size is that the separation between adjacent levels decreases and that the energy of the highest occupied level approaches the Fermi level of the bulk crystal.

Results of a practical example of such a calculation, for the case of a ten-atom platinum cluster configured to represent the Pt (100) surface, are shown in Figures 2.7–2.9. Figure 2.7 shows the cluster model. Figure 2.8 shows the calculated orbital energies, with the highest occupied orbital, the $7b_2$, labeled with an arrow. Note that the calculated energy of this orbital is –5.3 eV, which is close to the value of –5.7 eV measured for the work function of the Pt (100) surface, and is considerably less than the ionization potential of an individual platinum atom, which is 9.0 eV. Figure 2.9 shows an electron density map of the 12e orbital, showing a strong lobe of high electron density extending away from the surface (in the *z* direction). This orbital is thought to be important in the chemisorption of molecules to the Pt (100) surface.

Work Function Measurement

As has been shown, the surface electron charge density can be described in terms of the work function. Let us now look at ways of measuring this parameter. We will consider both absolute measurements and relative measurements, for as we shall see, both types of measurement have practical utility.

Considering absolute measurements first, recall the energy picture of the surface

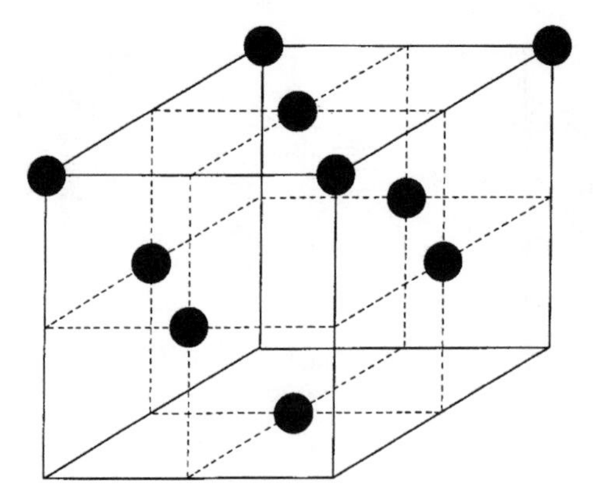

Figure 2.7 Ten atom cluster model of the Pt (100) surface. (Courtesy of Keith Johnson.)

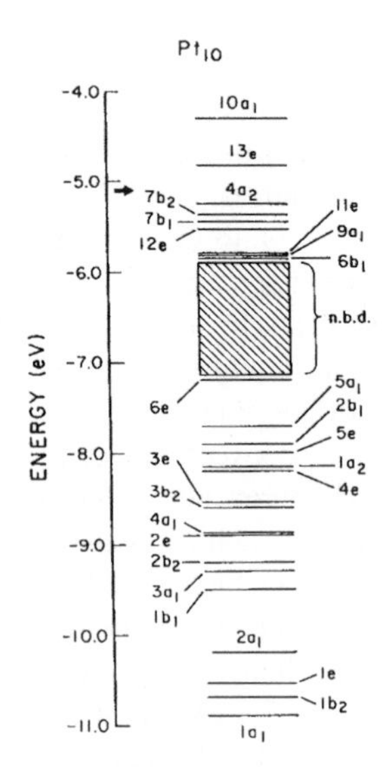

Figure 2.8 Calculated energy levels of the ten atom Pt cluster shown in Figure 2.7. (Courtesy of Keith Johnson.)

shown in Figure 2.2. The electrons in the conduction band, because they follow Fermi–Dirac statistics, will occupy a wide range of energy states, with the probability of occupancy being given by

$$f(\varepsilon)=\left\{\exp\left[\frac{\varepsilon-\mu}{kT}\right]+1\right\}^{-1}. \tag{2.22}$$

The typical density of states curve for a metal, in which we plot the number of allowed

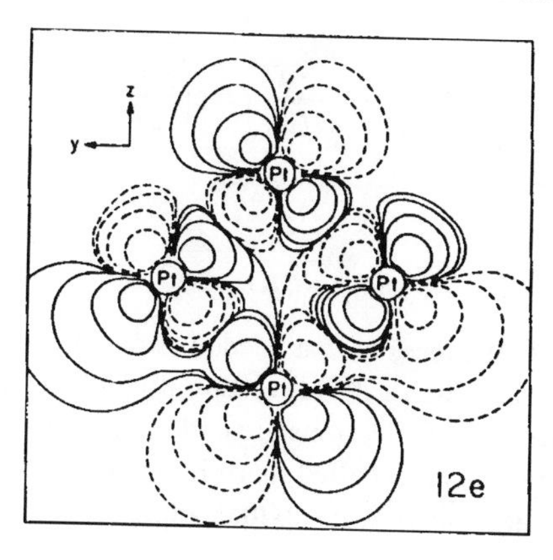

Figure 2.9 Electron charge density map of the 12e orbital of the ten atom Pt cluster, in the plane perpendicular to the surface. (Courtesy of Keith Johnson.)

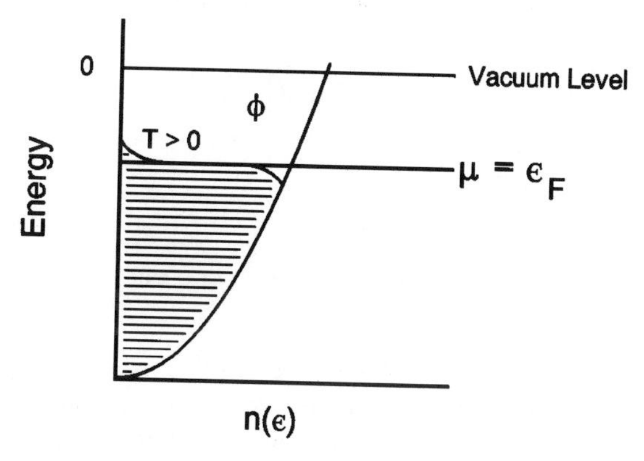

Figure 2.10 Density of states curve for the conduction band of a free electron metal.

states, for both filled and unfilled states, looks qualitatively as shown in Figure 2.10. At $T = 0$ K, all of the states for which $\epsilon \leq \mu$ will be filled and all states for which $\epsilon > \mu$ will be empty. The value of ϵ corresponding to $\epsilon = \mu$ is called the Fermi energy, ϵ_F. At temperatures above 0 K, the probability of occupancy of states just above ϵ_F becomes finite, and the probability of occupancy of states just below ϵ_F becomes correspondingly less than unity. The various methods for measuring the work function can be explained in terms of this diagram.

Thermionic Emission

At temperatures above 0 K, states with $\epsilon > \epsilon_F$ will have a finite probability of occupancy. At high enough temperatures, states with $\epsilon - \epsilon_F > \Phi$ will have a probability of occupancy given by

$$f(\varepsilon) \approx \exp\left(\frac{-\Phi}{kT}\right)\exp\left(\frac{-\varepsilon'}{kT}\right), \qquad 2.23$$

where ϵ' is given by

$$\epsilon' = \epsilon - (\mu + \Phi) \qquad 2.24$$

and represents the energy in excess of that needed to reach the vacuum level. Any electron that reaches a state for which $\epsilon' > 0$ is essentially outside of the crystal and can be collected as a free electron. Such electrons are called thermionic electrons, and the process by which they are produced is called thermionic emission. Note that this is a thermodynamic effect, similar in some ways to the evaporation of atoms from a condensed phase into the vapor.

Because the probability of occupancy of a state above the vacuum level depends on the work function, the thermionic electron current observed at a given temperature can be used to measure the work function. The expression relating thermionic current, j, to temperature is the Richardson–Dushman equation:

$$j = AT^2 \exp\left(\frac{-\Phi}{kT}\right), \qquad 2.25$$

in which

$$A = \frac{4\pi mk^2 e}{h^3} \approx 120\ A/\text{cm}^2\,\text{deg}^2. \qquad 2.26$$

Note that all of the parameters in A are fixed physical constants: m, the electron mass; k, Boltzmann's constant; e, the electronic charge; and h, Planck's constant. To obtain Φ from a thermionic emission measurement, one plots ln (j/T^2) as a function of $1/T$. The value of Φ is obtained from the slope of this curve. A typical example is shown in Figure 2.11. Note that if the work function is not uniform over the surface being studied, because the sample is polycrystalline, for example, or has areas that are covered by a contaminant species, one will see patchwise emission. Because of the exponential dependence of j on Φ, the areas of low work function will give rise to the bulk of the emitted current. This effect has been used in practice in the development of the thermionic emission microscope, in which contrast is generated by the difference in work function from point to point on the sample.

Photoelectron Emission

A great deal can be learned about the occupied electron energy states in a solid, and especially near the surface of the solid, by measuring the electron current excited by bombarding the surface with photons in the energy range from 1 to 100 eV, (corresponding to wavelengths from 100 to 10 nm in the ultraviolet region of the spectrum). This extensively used technique is known as ultraviolet photoelectron spectroscopy, or UPS.

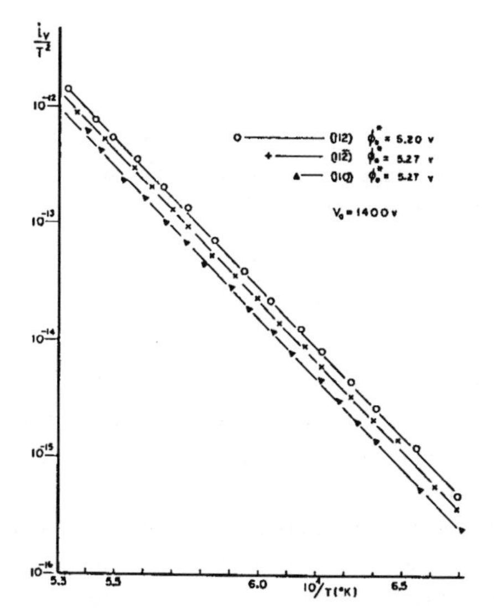

Figure 2.11 Plot of the Richardson–Dushman equation for different planes of tungsten. (Source: F.L. Hughes, H. Levin, and R. Kaplan, *Phys. Rev.* 1959; 113:1023. Reprinted with permission.)

In order to understand the information that can be obtained in a UPS experiment, look again at a typical density of states curve, as is shown in Figure 2.12. Any photon incident on the sample surface can interact with an electron in the solid to transfer its energy to the electron. If the photon energy is greater than the difference between the initial energy of the electron and the vacuum level, then the electron may escape from the solid, assuming that its velocity is directed toward the surface and that it does not lose energy by colliding with another electron in the solid before it reaches the surface. This process is called photoelectron emission, or simply photoemission. The ejected electrons are called photoelectrons. At very low temperature, when only those states at or

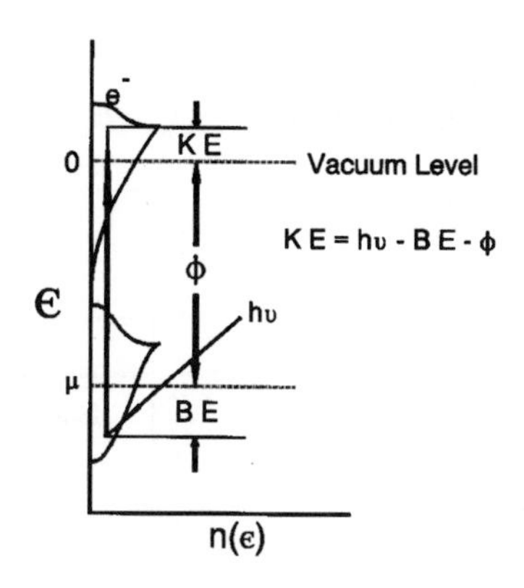

Figure 2.12 Illustration of the energy relationships in the photoelectron emission process.

below ϵ_F are filled, the minimum photon energy, $h\nu$, that can cause electron ejection is

$$h\nu_0 = \Phi . \qquad 2.27$$

At higher temperatures, when some electrons occupy states above ϵ_F, photons of slightly lower energy can cause photoemission. The total electron current generated in this process is given by the Fowler equation:

$$j = BT^2 f\left[\frac{h(\nu - \nu_0)}{kT}\right], \qquad 2.28$$

where B is a parameter that depends on the material involved. If measurements are made of the photoemitted current as a function of photon energy, extrapolation allows the determination of ν_0, and thus of Φ, providing a second means of determining the absolute value of the work function.

Much additional information about the density of electron energy states can be obtained by measuring the energy distribution of the photoemitted electrons. The kinetic energy of an electron that is excited by a photon of energy $h\nu$, and emitted from the sample without suffering an inelastic collision, is given by

$$KE = h\nu - BE - \Phi. \qquad 2.29$$

In this equation BE is the electron binding energy in the solid relative to the Fermi energy, as shown in Figure 2.12.

Figure 2.13 shows the experimental setup for a photoemission experiment. Photons from a monochromatic source, having a fixed frequency or very narrow frequency range, impinge on the sample, which is maintained in ultrahigh vacuum. The emitted photoelectrons pass through an electron energy analyzer, which separates them into an energy spectrum, and are detected at the output of the spectrometer. The output current as a function of electron kinetic energy is used to construct the photoelectron spectrum. Note that the higher the incident photon energy, the "deeper" in binding energy one can see into the sample.

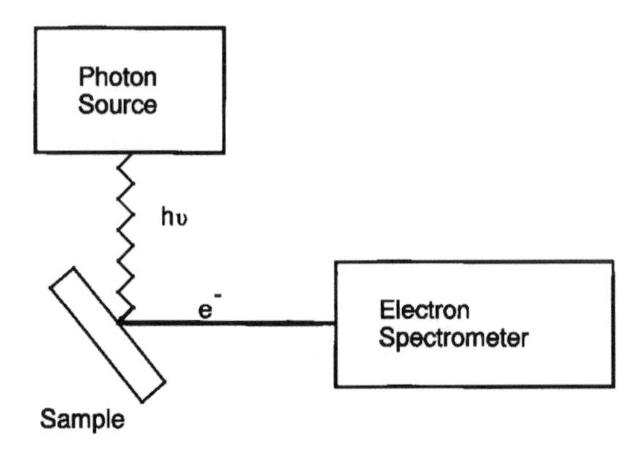

Figure 2.13 Schematic diagram of the experimental setup for a photoemission experiment.

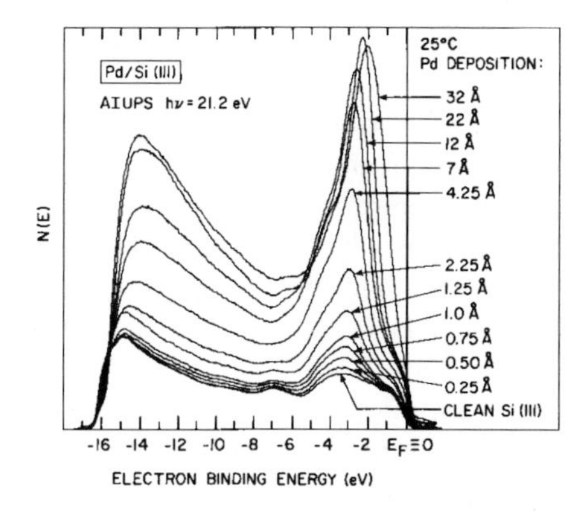

Figure 2.14 Ultraviolet photoemission spectra for the clean Si (111) surface and for various coverages of Pd deposited at room temperature, using 21.2 eV photons. (Source: G.W. Rubloff, P.S. Ho, J.F. Freeouf, and J.E. Lewis, *Phys. Rev.* 1981; B23:4183. Reprinted with permission.)

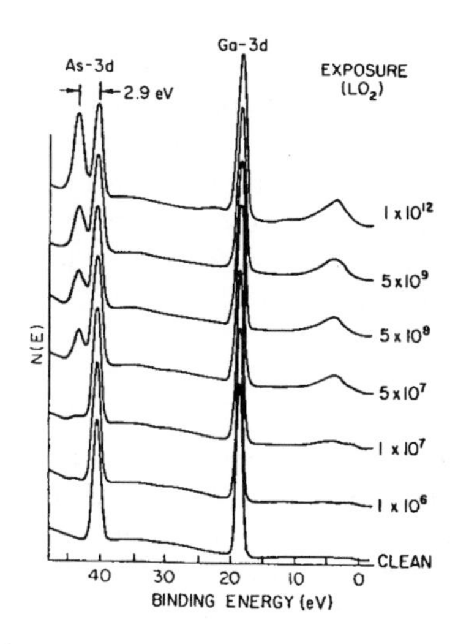

Figure 2.15 Photoelectron spectra for a GaAs surface, using 100 eV photons, showing changes associated with increasing exposure to oxygen. (Source: W.E. Spicer, in: L. Fiermans, J. Vennik and W. Dekeyser, eds., *Electron and Ion Spectroscopy of Solids*. New York: Plenum Publishers, 1978, p. 68. Reprinted with permission.)

Typical results of UPS measurements are shown in Figures 2.14 and 2.15. Figure 2.14 shows a series of spectra, taken with 21.2 eV photons, of an Si (111) surface, both clean and after deposition of increasing amounts of palladium. The clean surface shows the typical bulk silicon peaks at –7 and –3.5 eV. With the addition of palladium, the emission intensity increases rapidly, as a result of emission from the palladium 4d states (the large peak at –2 eV). As the palladium coverage increases, metallic states appear at the Fermi level. The position of the palladium 4d feature in the coverage range from 2 to 12 Å is typical of Pd_2Si. At higher coverages, this peak shifts to a value typical of bulk palladium metal.

Figure 2.15 shows a photoelectron spectrum obtained using 100 eV photons, generated as synchrotron radiation in an electron storage ring. In this case, the higher photon energy permits us to obtain information on more tightly bound electrons than in the previous case. The example shown is a spectrum of GaAs, showing photoemission from the 3d orbitals of As and Ga and the energy shifts associated with increasing exposure of the surface to oxygen. Note that for this semiconductor, there is very little emission just below ϵ_F.

In principle, one can obtain information about the density of filled states in the solid at equilibrium from such UPS measurements. In practice, this is often possible, although there are several complications. The penetration depth of the photons into the solid is finite, though this is not usually a limiting factor. The escape depth of the photoelectrons is also limited, due to inelastic scattering from other electrons. This is the same effect that is responsible for the surface sensitivity of LEED and means in practice that all of the photoelectrons that are detected without scattering arise in the top 1 to 10 nm of the sample. As in the case of LEED, this means that UPS will be a very surface sensitive technique. Finally, the excitation probability, that is, the probability that a photon of given energy will interact with an electron in a particular state, is a complicated function of both the photon energy and the energy structure of the solid and must be worked out in detail for each system studied. In spite of these limitations, UPS is a powerful tool both for the study of the electronic structure of clean surfaces and for the study of adsorbed layers on top of these surfaces. This latter application will be discussed in more detail in Chapter 16.

Field Emission

A third method for obtaining electron emission from a solid material, and in the process obtaining a measure of the work function, is known as field emission. The physical situation involved in this process is shown in Figure 2.16. Here we see again a potential energy diagram for the near-surface region of a metal. In this case we have indicated the shape of the surface energy barrier, on the assumption that the shape of

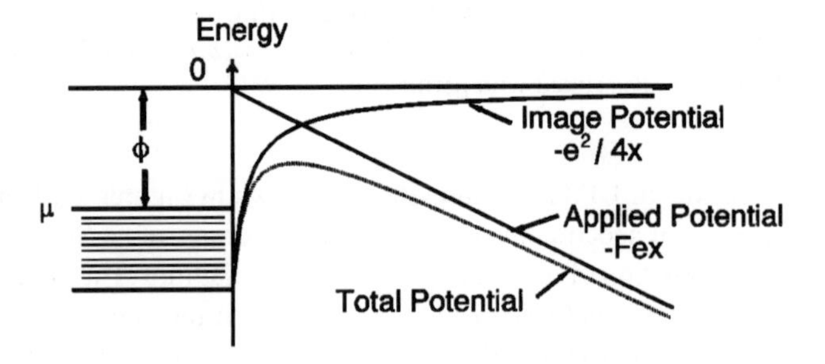

Figure 2.16 Potential energy curves for an electron near a metal surface. "Image Potential" curve: no applied field. "Total Potential" curve: applied external field = $-(Fex)$.

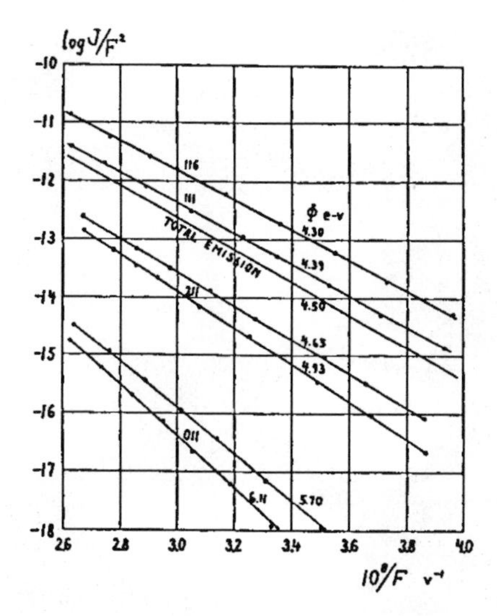

Figure 2.17 Fowler–Nordheim plots of field emission from various planes of tungsten. (Source: E.W. Muller, *J. Appl. Phys.* 1955;26:732. Reprinted with permission.)

the barrier is dominated by image potential effects. If no imposed electric field exists at the surface, the potential rises to the vacuum level and remains constant at that level with increasing distance from the surface. If a strong electric field is present at the surface, however, with the surface electrically negative with respect to the surroundings, the effective surface barrier will be the superposition of the zero-field barrier and the applied field, and is shown dotted in the figure. In this case, there is still a finite barrier to electron emission from the surface, but the barrier height is decreased and the width of the barrier is now finite. In such a case, a quantum mechanical calculation indicates that electrons can escape through the barrier by quantum mechanical tunneling and appear in the space outside the metal as field-emitted electrons. The current arising from this process is given by the Fowler–Nordheim equation as

$$j=\left(6\times10^{6}\right)\left(\frac{\mu}{\Phi}\right)\left(\mu+\Phi\right)^{-1}\left(\frac{F}{\alpha}\right)^{2}\exp\left[\frac{-\left(6.8\times10^{7}\alpha\Phi^{3/2}\right)}{F}\right], \tag{2.30}$$

where F is the electric field strength and α is a slowly varying function of F, of the order of unity. A plot of this relation, in terms of $\ln(j/F^2)$ vs. $(1/F)$ can be used to determine Φ, as shown in Figure 2.17.

This effect is also the basis of the field emission microscope, or FEM. The FEM has exactly the same physical setup as the FIM mentioned in Chapter 1 (Figure 1.3). In the present case, however, the polarity of the field is reversed, with the sample negative, and the space above the sample is held at ultrahigh vacuum. The tip radius used is typically on the order of 100 nm. As in the FIM case, the magnification is given by the ratio of screen radius to tip radius. The contrast observed in the FEM arises from the

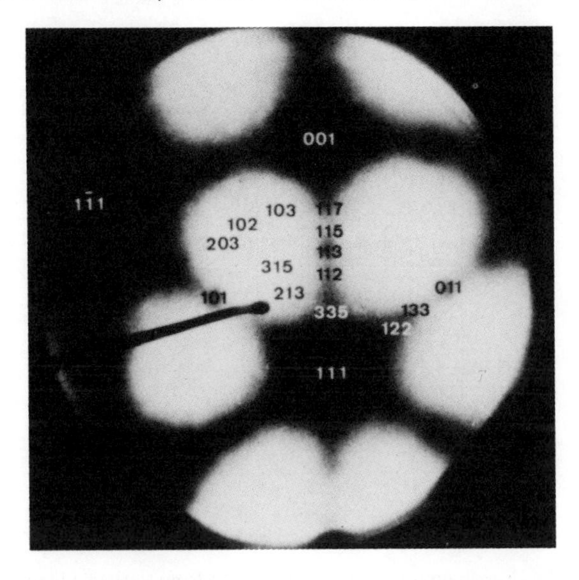

Figure 2.18 Field emission microscopy pattern of a clean (111) oriented rhodium surface, showing the location of various crystal planes on the surface. (Source: V.V. Gorodetskii and B.E. Nieuwenhuys, *Surface Sci.* 1981; 105:299. Reprinted with permission.)

variation in Φ with crystallographic orientation over the tip. Recall that closely-packed faces usually have higher work functions and thus appear dark. A typical FEM pattern, shown in Figure 2.18, demonstrates this effect.

The spatial resolution of this technique is on the order of 2 nm and is limited by the momentum of the emitted electrons parallel to the tip surface. It is possible to set up an FEM with a probe hole in the phosphor screen, or with a Faraday cup inside the bulb, to collect the current emitted from a single plane. This technique, along with the Fowler–Nordheim equation, allows the measurement of the variation of Φ with orientation for a wide variety of orientations on a single sample. The field emission microscope has also been used to some extent to measure adsorption and surface diffusion processes, making use of the work function change associated with the adsorption process. The fine structure of the current-voltage curve can also be used to provide information on the density of states near ϵ_F.

Now that we have introduced the concept of field emission, we can explain in more detail the physical basis of the field ion microscope. In this case, again, the presence of a strong electric field is critical. If an atom is in the gas phase, in field-free space, the potential energy for electrons is as shown in curve *a* on Figure 2.19. A potential well exists at the atom core, with a finite barrier extending away from the core. If the atom is placed in a strong electric field, such as that formed at the sample tip in the FIM, the barrier is finite as shown in curve *b* of Figure 2.19. Thus, there is a probability that an electron can escape from the atom by quantum mechanical tunneling, leaving the atom as a positive ion. In the case of the field ion microscope, the ionization takes place close to the tip, where the field is strongest, as shown in Figure 2.20. The

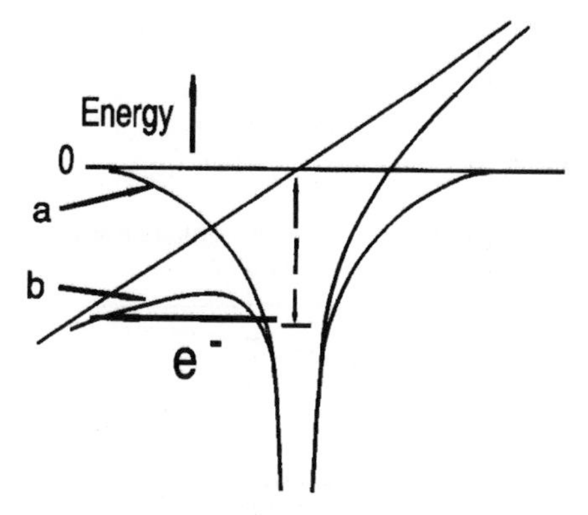

Figure 2.19 Potential energy for an electron in a free atom in (a) field-free space, and (b) strong electric field.

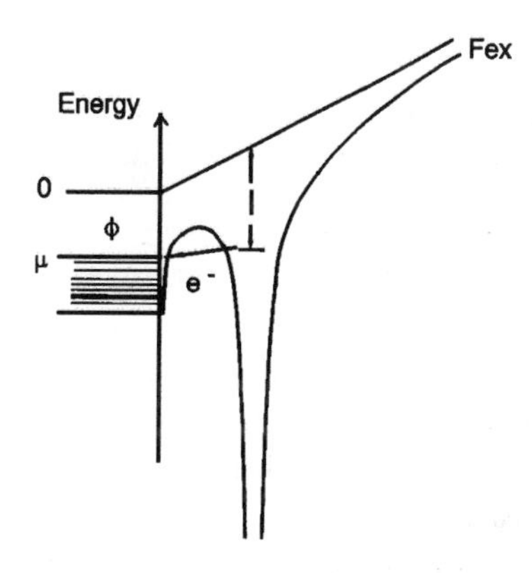

Figure 2.20 Potential energy for an electron in an atom near a field ion microscope tip.

electron that tunnels from the atom is taken up by the tip. If one looks at the theory of this process in detail, one finds that there is a critical distance, x_c, at which the tunneling probability is a maximum. This distance is typically about 0.4 nm. The very high spatial resolution and high contrast for features on the atomic scale arises from the fact that the electric field is enhanced in the vicinity of surface atoms, because of the higher local curvature.

A final field-induced process is known as field evaporation. This process involves the removal of atoms from the surface itself at very high field strengths. Figure 2.21 shows the energy diagram that is appropriate to this process. The effect of the field in this case is to reduce the effective binding energy of the atom to the surface and to give, in effect, a greatly increased evaporation rate relative to that expected at that temperature at zero field. In this process, atoms that are at positions of high local curvature, such as adatoms or ledge atoms, are removed preferentially. This process thus

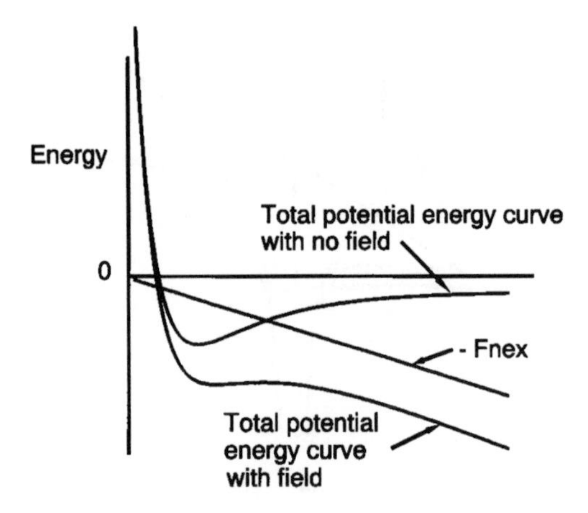

Figure 2.21 Potential energy curve for an atom at a surface. Upper solid curve: no field. Lower solid curve: applied external field = –(Fnex).

leads to a smoothing of the tip to a more spherical shape and is used routinely to produce regular tip surfaces such as that shown in Figure 1.4.

Relative Measurement of the Work Function

A number of techniques have been developed that permit measurement of relative values of the work function. Two such techniques are described here. These techniques are of use primarily because they do not involve heating the sample to high temperature or exposing it to a high electric field. In many studies of surface properties, especially those involving the adsorption of gas on a surface, the change in the work function as the process takes place provides valuable information on the course of the process. In these situations, it is not necessary to know the absolute value of the work function but only the way in which it changes in response to changes at the surface.

One way of measuring these changes is the so-called Kelvin, or vibrating capacitor method, shown schematically in Figure 2.22. In this method, a second surface is placed close to the surface whose work function is to be measured, and electrical connections are made to bias this reference surface relative to the sample and to measure any current that flows in the external circuit between the two surfaces. In the configuration shown in Figure 2.22, when an electrical connection is made between the two surfaces, charge will flow until the chemical potential of electrons is uniform throughout the system. The total charge flow will be

$$Q = C\Delta V, \ \ \Delta V = \Phi_R - \Phi_S, \tag{2.31}$$

where C is the interelectrode capacitance and R and S refer to reference and sample, respectively. If one now moves the reference surface relative to the sample at some constant frequency, the capacitance will change as the interelectrode separation changes.

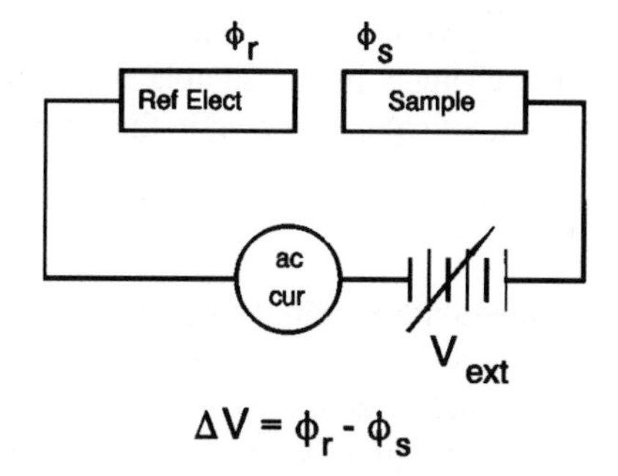

Figure 2.22 Work function measurement using the vibrating capacitor method.

An ac current will thus flow in the external circuit. If the reference electrode is then biased relative to the sample, to null out the voltage difference due to the difference in work function, the ac signal will go to zero at the point where $\Delta V = -V_{ext}$, allowing measurement of the work function of the sample relative to the reference surface.

A second means of relative work function measurement, the retarding potential difference method, is often used in systems where an LEED apparatus is available or, in general, where one has available a low energy electron source. In this method, one bombards the sample surface with a current of electrons of known, low energy. The electron current collected by the sample is measured as the sample potential is slowly increased in the negative sense. The resulting curve of sample current vs. retarding voltage is shown in Figure 2.23. For small values of the retarding voltage, practically all of the electrons from the gun that strike the surface are collected. As the retarding voltage is increased, eventually a voltage is reached such that the electrons from the gun have insufficient energy to reach the surface and are reflected. The voltage at which this occurs depends on the work functions of the sample and the electron gun

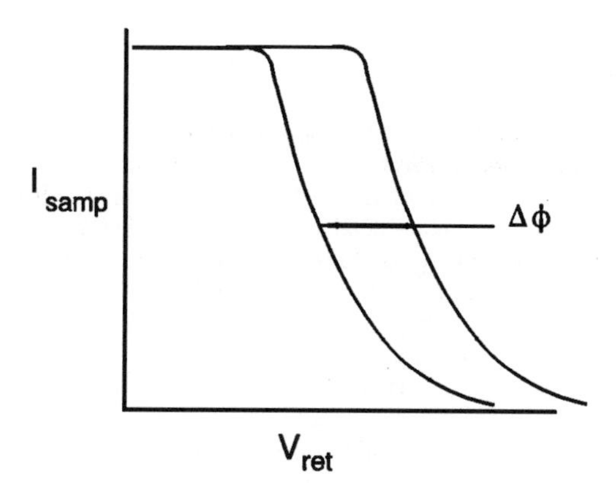

Figure 2.23 Current-voltage curves observed in the retarding potential difference method of work function measurement.

filament and on the accelerating voltage of the electron gun. If these last two parameters are held constant, then changes in the sample work function are indicated by changes in the cutoff retarding voltage, as shown in the figure.

Problems

2.1. Calculate the temperature required to obtain a thermionic electron current density of 10^{-2} A/cm^2 from polycrystalline wires of aluminum, cesium, iron, and tungsten. Which of these metals would make practical thermionic emitters?

2.2. Describe as many techniques as you can think of for obtaining an electron emission current of 10^{-3} A/cm^2 from an iron surface. The electron work function of iron is 4.5 eV, and its melting point is 1808 K.

2.3. Sketch the expected result of a UPS measurement for an iron surface for a photon energy of 20 eV. Assume that the energy bands in the solid have the same occupancy as the corresponding atomic levels in the free atom.

2.4. Sodium metal can be well characterized using the Jellium model. The electronic configuration of the atom is $1s^2$, $2s^2$, $2p^6$, $3s^1$. The work function of the solid is 2.5 eV. The electron binding energies in the sodium atom are given below:

Level	1s	2s	2p	3s
Binding energy (eV)	1072	63	31	1

a. Sketch the density of states curve [E vs. n(E)] for sodium, showing filled and unfilled states at a finite temperature.

b. Sketch the photoelectron energy distribution that would be obtained if a sodium surface were bombarded by 1000 eV photons.

2.5. In order for field ionization of a helium atom to take place in a field ion microscope, the applied field must be strong enough that the 1s level of the helium atom is at a higher energy than the Fermi level of the surface when the atom is 0.2 nm from the surface. Calculate the required field strength for the case of a tungsten surface having a work function of 4.5 eV. The ionization potential of helium is 24.1 eV.

Bibliography

Bardeen, J. Theory of the work function: II, The surface double layer. *Phys. Rev.* 1936;49:653.

Bauer, RS. *Surfaces and Interfaces: Physics and Electronics*. Amsterdam: Elsevier, 1983.

Cottrell, A. *Introduction to the Modern Theory of Metals*. London: The Institute of metals, 1988.

Duke, CB. Electronic structure of clean-metal surfaces. *J. Vac. Sci. Technol.* 1969;6:152.

Johnson, KH, Yang, CY, Vvedensky, D, Messmer, RP, Salahub, DR. Electronic structure of metal and alloy surfaces. 1977 ASM Materials Science Seminar on Interfacial Segregation. Chicago, Ill., 1977.

Lang, ND. The density functional formalism and the electronic structure of metal surfaces. In: F. Seitz, and D. Turnbull eds. *Reviews of Solid State Physics*. Academic Press, New York, 1973.

Lang, ND, and Kohn, W. Theory of metal surfaces; charge density and surface energy. *Phys. Rev. B* 1970;1:4555.

Smith, JR. Self-consistent many-electron theory of electron work functions and surface potential characteristics for selected metals. *Phys. Rev.* 1969;181:522.

CHAPTER **3**

Surface Tension

This chapter and the two following chapters will consider the conditions required for equilibrium in capillary systems, that is, in systems in which the overall behavior of the system is significantly affected by the presence of surfaces or interfaces. This chapter will present definitions of the parameter surface tension, both for the case of fluid–fluid and fluid–solid interfaces, and show how this parameter is involved in determining the criteria for mechanical equilibrium in capillary systems. The two following chapters will consider overall thermodynamic equilibrium, for single component and multicomponent systems, respectively.

Surface Tension—The Young Model

A model of the interface called the Young model can be used to define the conditions for mechanical equilibrium in capillary systems. This model assumes that, as far as mechanical behavior is concerned, one can replace the actual interface, which has finite thickness, with an infinitesimally thin elastic membrane, called the surface of tension, stretched between the two bulk phases. (This model is valid for any interface between two fluid phases but not for solids).

Using this model, we can define the surface tension as follows: Imagine an interface of arbitrary shape, between two bulk phases, such as is shown in Figure 3.1. If we cut this surface with a plane, such as plane *ABCD* in the figure, we will find that a force must be applied along each unit length of the cut surface, $d\ell$, tangential to the surface, in order to maintain the system in mechanical equilibrium. This force per unit length, represented by the vector γ in Figure 3.1, is defined as the surface tension at the element of length $d\ell$. Note that this is a purely mechanical definition and holds strictly true only for cases in which both the bulk phases are fluids. In addition, if it is found that γ at every point on the interface is independent of the direction of the element $d\ell$ and if γ has the same value for every point of the interface, then the surface is said to be in a state of uniform tension and γ is the surface (or interfacial) tension of the surface. This parameter has units of force per unit length, dynes per centimeter in c.g.s. units.

One may compare the Young model with the real state of affairs that must exist at an interface in a real system, in order to better understand the concept of surface tension. To do this, consider the interfacial region in a real system, as represented in Fig-

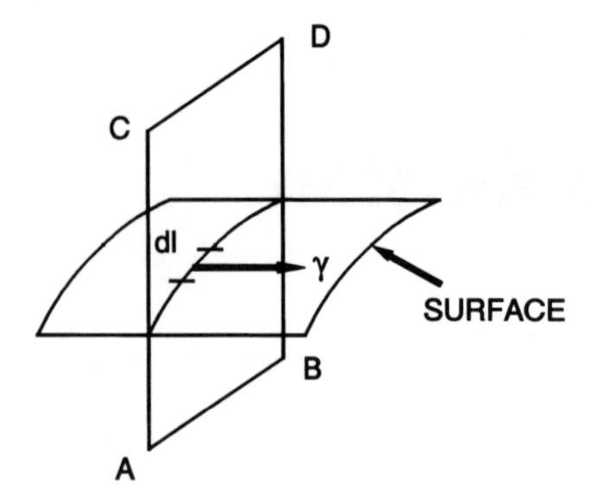

Figure 3.1 Definition of surface tension.

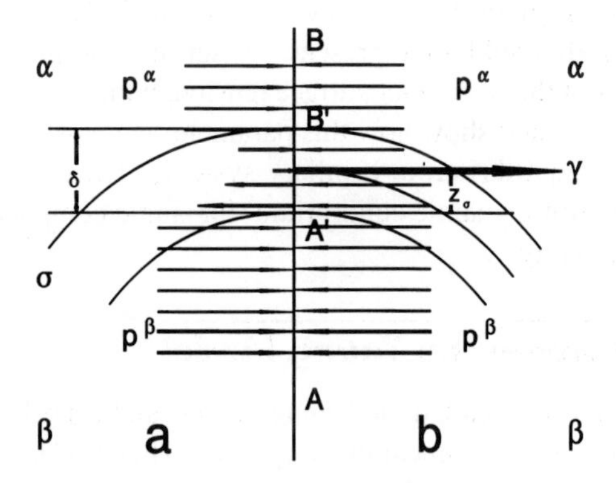

Figure 3.2 The Young model of surface tension, showing pressures and tensions at the interface. (a) Real system. (b) Young model.

ure 3.2. This system has two uniform phases, α and β, separated by an interface of finite thickness, δ. Let us pass a plane, AB, through this system, perpendicular to the interface. We can represent the pressure on this plane, as a function of position along the z-axis, by vectors, assuming that pressures are negative and that tensions are positive (this is simply a convention and has no physical significance) and representing the magnitude of the pressure or tension at a given position along the z-axis by the length of the vector.

For the case of a real system, shown on side *a* of Figure 3.2, the pressure will be uniform in the bulk phases, α and β, but not necessarily equal in the two phases. In the interfacial region between A' and B', the pressure will vary continuously in a generally unmeasurable way. The Young model representation of this real system is shown on side *b* of Figure 3.2. In this case, the pressures in the bulk phases are assumed to be uniform, but not necessarily equal, in the two phases, right up to an infinitesimally thin surface of tension located at z_σ. The surface tension, γ, is presumed to act at this surface. Note that by an appropriate choice of the location of z_σ and the magnitude of γ, the model can be made to represent the real system exactly, as far as mechanical equilibrium is concerned.

Let us make a calculation to determine the appropriate values of γ and z_σ for the case shown in Figure 3.2. Consider the forces normal to the plane *AB* for the two cases, the real system and the model. In the region between *A* and *A'*, there is a uniform pressure, p^β, in both the real system and the model. In the region between *B'* and *B*, there is a uniform pressure, p^α, in both the real system and the model. For the real system, the total applied force along the distance from *A'* to *B'* is given by

$$F_{real} = d\ell \int_0^\delta t\, dz \tag{3.1}$$

per unit length of the plane *A'B'*, where ℓ is the extent of this plane in the direction perpendicular to the figure and δ is the interface thickness. For the case of the model, in this same region from *A'* to *B'* the force is given by

$$F_{\text{model}} = \gamma d\ell - p^\beta z_\sigma d\ell - p^\alpha(\delta - z_\sigma)d\ell. \tag{3.2}$$

In order for the two cases to be mechanically equivalent, we must have

$$F_{\text{real}} = F_{\text{model}} \tag{3.3}$$

$$\int_0^\delta t\, dz = \gamma - p^\beta z_\sigma - p^\alpha(\delta - z_\sigma) \tag{3.4}$$

or

$$\gamma = \int_0^{z_\sigma} \left(p^\beta + t\right) dz + \int_{z_\sigma}^\delta \left(p^\alpha + t\right) dz. \tag{3.5}$$

In addition, the first moments of these forces about an axis through *A'* must also be equal. That is,

$$M_{\text{real}} = M_{\text{model}} \tag{3.6}$$

$$\int_0^\delta tz\, dz = \gamma z_\sigma - p^\beta z_\sigma\left(\frac{z_\sigma}{2}\right) - p^\alpha(\delta - z_\sigma)\left(\frac{\delta + z_\sigma}{2}\right) \tag{3.7}$$

or

$$\gamma z_\sigma = \int_0^{z_\sigma} \left(p^\beta + t\right) z\, dz + \int_{z_\sigma}^\delta \left(p^\alpha + t\right) z\, dz. \tag{3.8}$$

We thus have two equations in two unknowns, γ and z_σ, which is sufficient to specify these two parameters of the Young model and to ensure that the model is indeed mechanically equivalent to the real system.

Criteria for Equilibrium

Now that we have shown that the Young model can be used to represent mechanical equilibrium in real systems, let us use the model to develop criteria for this mechanical equilibrium. For simplicity, consider the case of a spherically curved surface, having a radius of a curvature, r, as shown in Figure 3.3a. Consider the balance of forces in the direction perpendicular to a plane passed though such a surface, as shown in Figure 3.3b. Any element of area dA on the surface will have a projected area on the plane of $dA' = dA \cos \theta$. The force perpendicular to the plane associated with this element dA', in the z direction, will be

$$(p^\beta - p^\alpha)\, dA' = (p^\beta - p^\alpha) \cos \theta\, dA\,. \tag{3.9}$$

The sum of these forces over the whole area of the surface will be

$$(p^\beta - p^\alpha)\, A' = (p^\beta - p^\alpha)\, \pi\rho^2. \tag{3.10}$$

The surface tension associated with the surface will contribute a force $\gamma d\ell$ along each element of the intersection between the surface and the plane. The component of this force perpendicular to the plane, in the z direction, will be $-\gamma \cos \Phi\, d\ell$ per unit length. The total contribution of these forces, summed around the perimeter of the intersection, will be

$$-\gamma \cos \Phi \int_0^{2\pi\rho} d\ell = -\gamma \cos \Phi (2\pi\rho) \tag{3.11}$$

$$= -\gamma\left(\frac{\rho}{r}\right)(2\pi\rho) \tag{3.12}$$

$$= -\left(\frac{2\pi\rho^2}{r}\right)\gamma, \tag{3.13}$$

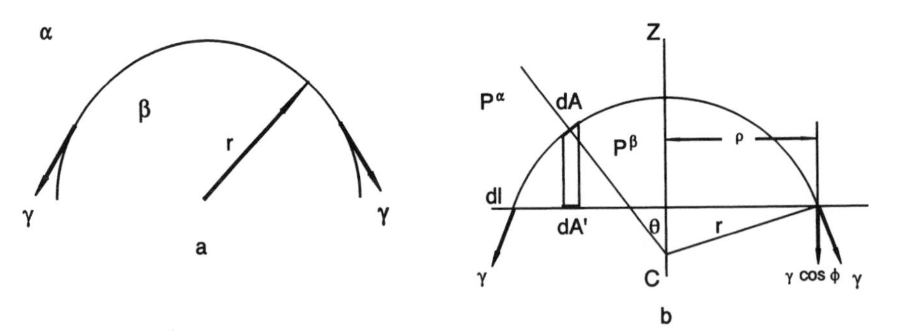

Figure 3.3 (a) Spherically curved surface. (b) Force balance on a plane passed through a spherically curved surface.

since $\cos \Phi = \rho/r$.

Because the system has rotational symmetry about the axis *CZ*, all forces perpendicular to this axis must cancel. Thus, for mechanical equilibrium to be assured, the forces parallel to *CZ* must also sum to zero, or

$$\left(p^{\beta} - p^{\alpha}\right)\pi\rho^2 - \left(\frac{2\pi\rho^2}{r}\right)\gamma = 0 \qquad 3.14$$

or

$$\left(p^{\beta} - p^{\alpha}\right) = \frac{2\gamma}{r}. \qquad 3.15$$

That is, the pressure in the interior phase, β, must exceed the pressure in the exterior phase, α, by the amount $2\gamma/r$. Moreover, if $r = \infty$ (in other words, if we have a flat surface), then $2\gamma/r = 0$ and $p^{\alpha} = p^{\beta}$. That is, no pressure difference exists across a flat interface.

For the more general case of a curved but nonspherical interface, a similar treatment would yield

$$\left(p^{\beta} - p^{\alpha}\right) = \left(\frac{1}{r_1} + \frac{1}{r_2}\right)\gamma, \qquad 3.16$$

in which r_1 and r_2 are the principle radii of curvature of the surface. This relation is known as LaPlace's formula.

A consequence of this relation is that the mean curvature of any interface between two fluid phases, defined by

$$\frac{1}{r_m} = \frac{1}{2}\left(\frac{1}{r_1} + \frac{1}{r_2}\right), \qquad 3.17$$

is always constant over the extent of the surface. This will be true in any case in which gravitational effects can be neglected. In the event that these effects must be accounted for, they can be represented by the changes in p^{α} and p^{β} with height. This subject will be discussed in more detail later in this chapter, in the section on measurement of surface tensions of liquids, when the techniques for determining γ by measuring the shapes of liquid drops in a gravitational field are considered.

Three-Phase Intersections

One other case of mechanical equilibrium in capillary systems containing only fluid interfaces must be considered, namely, the condition for equilibrium in the case where three bulk phases coexist, with interfaces between each of the two bulk phases intersecting along a line of contact, as shown in cross section in Figure 3.4. This situation is realized physically when a lens of one liquid is floated upon the surface of another liquid that which it is immiscible, or when a lens is formed in the interface

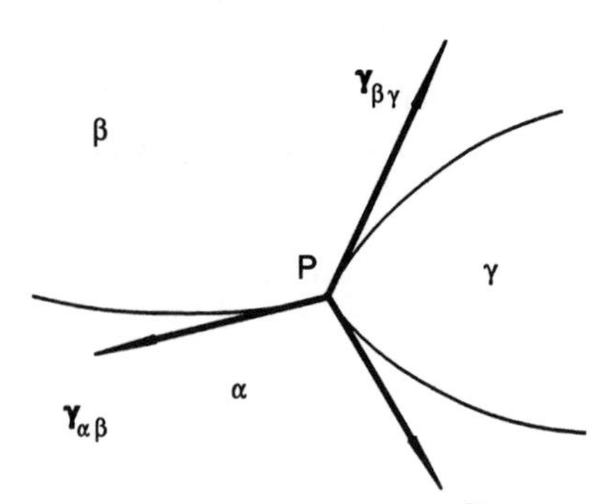

Figure 3.4 Forces acting at a three-phase intersection.

between two immiscible liquids by a third liquid which is immiscible in both of the other liquids. In such a system, any element of length $d\ell$ along the line of intersection of the interfaces, such as is shown at P in Figure 3.4, will be subjected to forces $\boldsymbol{\gamma}_{\alpha\beta}d\ell$, $\boldsymbol{\gamma}_{\beta\gamma}d\ell$, and $\boldsymbol{\gamma}_{\alpha\gamma}d\ell$, represented by the vectors shown in the figure. In order to have mechanical equilibrium, the vector sum of these forces must equal zero. That is,

$$\boldsymbol{\gamma}_{\alpha\beta} + \boldsymbol{\gamma}_{\beta\gamma} + \boldsymbol{\gamma}_{\alpha\gamma} = 0 . \qquad 3.18$$

This may be shown diagrammatically by drawing the three surface tension vectors as shown in Figure 3.5. Here one starts at an arbitrarily chosen origin and proceeds in the direction associated with $\boldsymbol{\gamma}_{\alpha\beta}$ for a distance proportional to the magnitude of $\boldsymbol{\gamma}_{\alpha\beta}$. One then proceeds in the direction associated with $\boldsymbol{\gamma}_{\beta\gamma}$ a distance proportional to the magnitude of $\boldsymbol{\gamma}_{\beta\gamma}$, and similarly for $\boldsymbol{\gamma}_{\alpha\gamma}$. In order to have mechanical equilibrium in the system, the triangle thus formed must close. This construction is known as Neumann's triangle.

Mechanical Equilibrium at Fluid–Solid Interfaces

Now consider briefly the complications that arise in systems involving interfaces between fluid and solid phases. Here one cannot proceed directly from purely mechan-

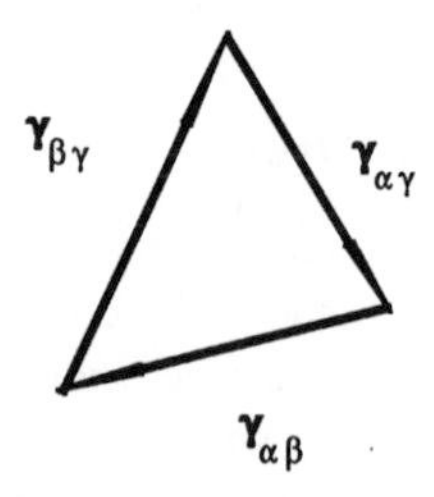

Figure 3.5 Neumann's triangle construction.

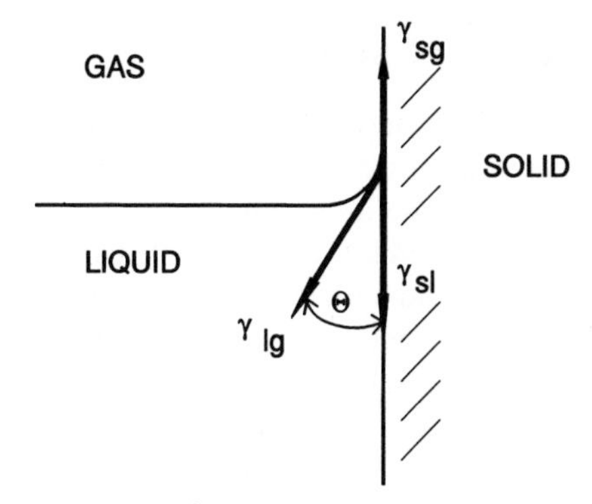

Figure 3.6 Forces acting at a solid–liquid–vapor contact line.

ical considerations, as the solid is not mobile like a fluid and thus is not in general susceptible to measurement of a surface tension. This is a point we will discuss in much more detail later in this chapter, in the section on the surface tension of solids. For the present we will just consider the indirect evidence for solid surface tensions obtained from the interactions between solid and fluid interfaces.

For example, consider the line of contact between three phases, a solid, a liquid, and a gas, as shown schematically in Figure 3.6. Recall from the discussion of three-phase intersections in fluids that the surface tension in each surface exerts a force on the line of intersection proportional to γ for that interface. If no similar forces were present in the solid–gas and solid–liquid interfaces, then no point of equilibrium would be possible.

Since equilibrium does occur, with the liquid–gas interface coming to some equilibrium contact angle, θ, with the solid surface, then similar forces must be involved in this case. If we call these γ_{sl} and γ_{sg}, then at equilibrium when all forces are balanced, we must have

$$\gamma_{lg}\cos\theta = \gamma_{sg} - \gamma_{sl}\,. \qquad 3.19$$

This is known as Young's equation. It is not rigorous, as it neglects forces in the direction perpendicular to the solid surface, which if large enough to deform the solid would destroy the validity of Young's equation. It is usually a good approximation in practice, however.

Depending on the relative magnitudes of γ_{lg} and $(\gamma_{sg}—\gamma_{sl})$, we can distinguish three cases of behavior:

1. $\gamma_{sg} > (\gamma_{sl} + \gamma_{lg})$. In this case, no value of θ will satisfy Young's equation. The liquid will cover the whole surface, and complete wetting results.
2. $\gamma_{sl} > (\gamma_{sg} + \gamma_{lg})$. Again, no equilibrium value of θ can be found. In this case, no contact area is possible and the liquid will not wet the solid at all.
3. $-\gamma_{lg} < (\gamma_{sg} - \gamma_{sl}) < \gamma_{lg}$. In this case, θ is finite and the Young equation can be balanced. Here we get partial wetting.

The Surface Tension of Solids

Consider now a detailed description of the surface tension of solid surfaces, or solid–fluid interfaces. We have already shown that in order to have mechanical equilibrium in a system containing both solid and fluid phases, we must assume that the solid–fluid interface behaves as though it had a surface tension associated with it.

Looking at the question of solid surface tension in detail, however, we see that there is a fundamental difference in behavior between solid and liquid surfaces. This difference manifests itself when we consider the process of changing the area of an existing surface. In the case of a liquid, for example, the surface of a liquid in a vessel as shown in Figure 3.7, the surface area may be increased simply by tilting the vessel, as shown in the figure. In this case, the incremental surface area is formed by material flowing from the bulk of the liquid to form additional surface having identical properties to the preexisting surface. That is, the state of the surface is not changed as its extent is changed.

The situation in a solid is quite different. Because the solid has a finite resistance to shear, one must stress the solid in order to cause an increase in its surface area. In the stressing process, bond angles and bond distances are changed. The state of the surface is no longer that associated with the unstressed surface.

We can describe the changes in the state of the surface associated with extending the surface area of a solid in terms of a parameter called the surface stress. This is a parameter that is related to, but not in general equal to, the surface tension. To understand the differences between the two parameters, and to develop the relations between them, consider the work done in the two following processes:

1. Take a unit cube of a solid material, as shown in Figure 3.8a, and stretch it along the x-direction, at constant y and unconstrained z. The work done in this case is

$$W_0 = \Sigma f dx = f_{\text{bulk}}\, dx, \tag{3.20}$$

where f_{bulk} is the force required to strain the bulk material of the cube.

2. Take the same unit cube, split it along the x–y plane as shown in Figure 3.8b, and then stretch the two halves along the x-direction as before, at constant y and unconstrained z. In this case, the work done is

$$W_1 = \Sigma f dx = f_{\text{bulk}}\, dx + 2f_{xx} dx, \tag{3.21}$$

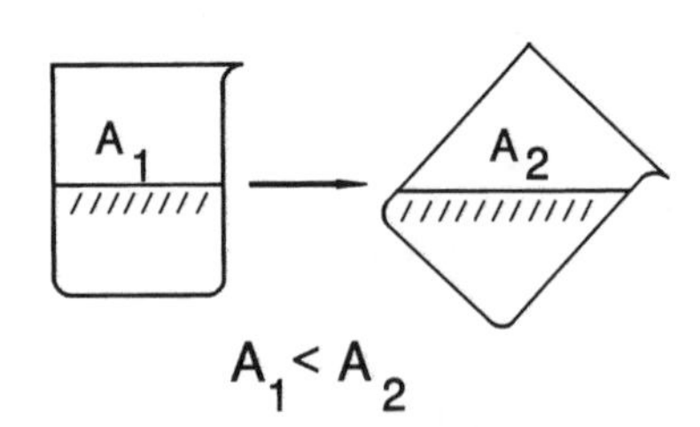

Figure 3.7 Illustration of the process of changing the surface area of a liquid.

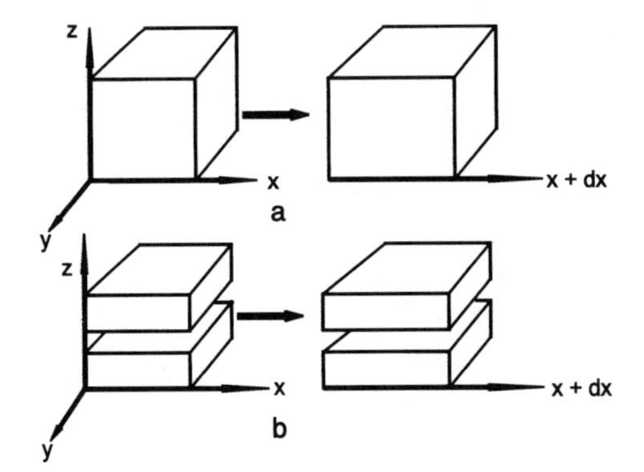

Figure 3.8 Operations carried out to define the surface stress in a solid, showing (a) stretch unit cube along x at constant y; and (b) cut cube along x–y plane, then stretch along x at constant y.

in which f_{xx} is the surface stress in the x-direction. From Equations 3.20 and 3.21

$$f_{xx} = \frac{W_1 - W_0}{2dx} = \frac{W_1 - W_0}{2\varepsilon_{xx}}, \quad 3.22$$

where ϵ_{xx} is the strain along the x-axis. That is, $(W_1 - W_0)$ is the work required to deform the surface along the x-direction and f_{xx} is the force (or stress) required to do the work that results in a strain ϵ_{xx}.

We could carry out similar experiments in which the unit cube was stretched in the y-direction at constant x, or in which the cube was sheared in the x–y plane. These exercises would yield

$$f_{yy} = \frac{W_1' - W_0'}{2\varepsilon_{yy}}$$

and

$$f_{xy} = \frac{W_1'' - W_0''}{2\varepsilon_{xy}}. \quad 3.23$$

What we have developed in this exercise are the parameters f_{xx}, f_{yy}, and f_{xy}, which are the components of a second-order tensor describing the state of stress in the surface. This demonstrates one initial significant difference between surface stress and surface tension: surface stress is a tensor, surface tension is a scalar.

Let us go on to develop the relation between the surface stress components and the surface tension. To do this, we will consider the work done in performing a given overall process following two different paths.

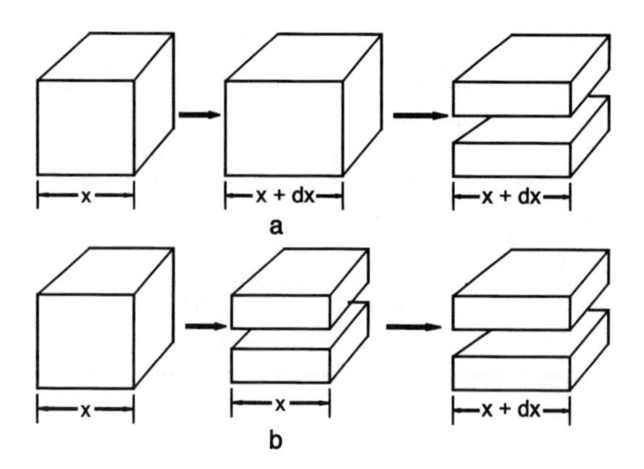

Figure 3.9 Operations carried out to show the relation of surface tension to surface stress, showing (a) stretch unit cube along x, then split, and (b) split unit cube, then stretch along x.

Starting again with a unit cube of a solid material, first stretch the cube along the x direction, then split the cube along the x–y plane, as shown in Figure 3.9a. The work done in this process is

$$W_{\mathrm{I}} = W_0 + 2(\gamma + \Delta\gamma)(1 + \epsilon_{xx}) \approx W_0 + 2\gamma + 2\Delta\gamma + 2\gamma\epsilon_{xx}, \quad 3.24$$

neglecting the second-order term $2\Delta\epsilon_{xx}$. As an alternative, first split the cube along the x–y plane, then stretch the two halves along the x direction, as shown in Figure 3.9b. For this case the work done is:

$$W_{\mathrm{II}} = 2\gamma + W_1. \quad 3.25$$

Note that in both cases the terms containing γ are associated with the work done in forming the new surfaces.

Since both of the above processes are assumed to have been carried out reversibly, and with no heat transfer from the surroundings,

$$W_I = W_{II} \quad 3.26$$

$$W_0 + 2\gamma + 2\Delta\gamma + 2\gamma\epsilon_{\mathrm{xx}} = W_1 + 2\gamma, \quad 3.27$$

or

$$\frac{W_1 - W_0}{2\varepsilon_{xx}} = \gamma + \frac{\Delta\gamma}{\varepsilon_{xx}} \approx \gamma + \frac{d\gamma}{d\varepsilon_{xx}}, \quad 3.28$$

or

$$f_{xx} = \gamma + \frac{d\gamma}{d\varepsilon_{xx}}. \quad 3.29$$

By a similar set of steps we could deduce that

$$f_{yy} = \gamma + \frac{d\gamma}{d\varepsilon_{yy}}, \quad f_{xy} = \frac{d\gamma}{d\varepsilon_{xy}}. \qquad 3.30$$

Thus, we can define the components of the surface stress tensor in terms of the surface tension and the changes in the surface tension with the respective strain terms.

The relations just developed will be applied to two practical cases, in order to see the explicit relationship between the stress components and the surface tension. Consider first the case of a cube of liquid. In this case, as discussed previously, the change in surface tension associated with stretching the surface will be zero, as material can flow from the bulk to the surface as extension proceeds. Since γ is unchanged, $d\gamma/d\epsilon = 0$ and we have

$$f_{xx} = \gamma, \quad f_{yy} = \gamma, \quad f_{xy} = 0 . \qquad 3.31$$

That is, the surface stress in a liquid surface is isotropic and numerically equal to the surface tension.

Alternatively, one may look at the relation between surface stress and surface tension in a simple solid. To do this, one must develop a model of the solid and postulate its response to stress. This will be done by assuming a simple cubic crystal structure and describing the bonding in the crystal by a nearest neighbor model. That is, the binding energy of the crystal is simply the sum of the energies of all of the nearest neighbor bonding interactions.

In such a solid, the surface tension will be just the sum of the energies associated with the nearest neighbor bonds that are broken in the process of creating the surface by splitting the crystal along the desired crystallographic plane. That is,

$$\gamma = \frac{\Phi \nu}{2}, \qquad 3.32$$

where Φ is the energy per nearest neighbor bond, ν is the number of bonds per unit area, and the factor of 2 accounts for the fact that two surfaces are formed in the splitting process.

For the case of the (100) surface of a simple cubic crystal, as shown in Figure 3.10,

$$\Phi = \frac{\Delta H_s}{3}, \qquad \nu = \frac{nm}{nma^2} = \frac{1}{a^2},$$

or

$$\gamma = \frac{\Phi}{2a^2} = \frac{\Delta H_s}{6a^2}. \qquad 3.33$$

For the case of a surface vicinal to this (100) surface, with a misorientation angle θ, as shown in Figure 3.11,

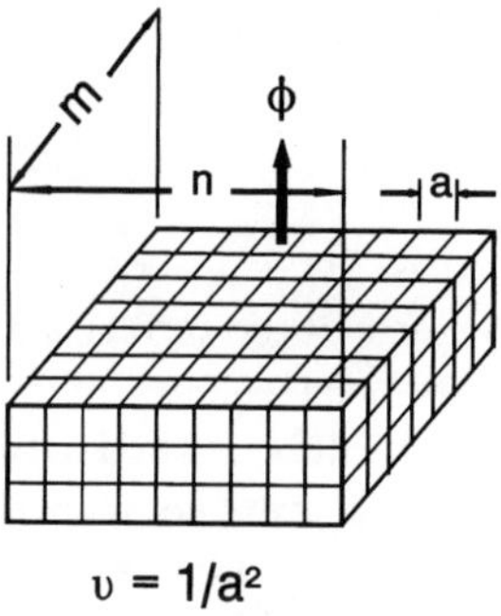

m
φ
n
a
υ = 1/a²

Figure 3.10 Nearest neighbor model for determining the surface tension of a solid for the case of a singular surface.

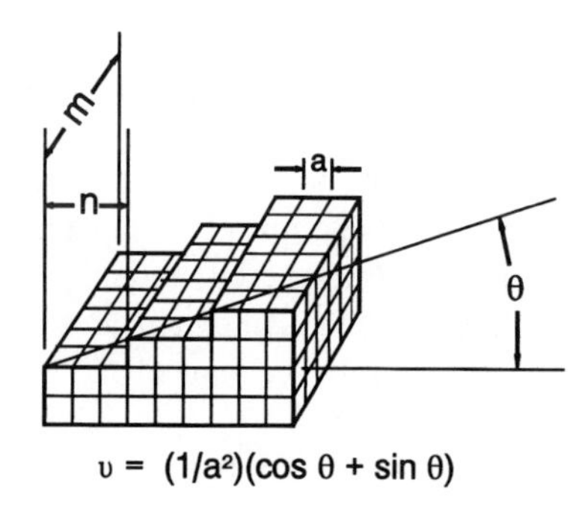

Figure 3.11 Nearest neighbor model for determining the surface tension of a solid for the case of a vicinal surface.

$$\Phi = \frac{\Delta H_s}{3}, \quad \nu = \frac{1 + \tan\theta}{a^2/\cos\theta},$$

or

$$\gamma = \left(\frac{\Phi}{2a^2}\right)(\cos\theta + \sin\theta)$$

$$= \left(\frac{\Delta H_s}{6a^2}\right)(\cos\theta + \sin\theta). \qquad 3.34$$

Note that since ($\cos\theta + \sin\theta$) is always greater than unity, the γ of the vicinal surface will always exceed that of the corresponding singular surface.

What, then, is the surface stress for such a surface? The nearest neighbor bond distance in the crystal at equilibrium is set by the condition that this equilibrium length minimizes the energy of the system. Consequently, any change in position of the surface atoms relative to their positions in the bulk of the crystal will cause an unfavorable increase in system energy. Thus, the surface as formed will be unstrained. Moreover, since the surface tension is associated only with bonds broken perpendicular to the surface (in this simple cubic model), the contribution to the surface energy per bond is unchanged when the surface is extended. However, the surface tension per unit area (the conventional way of measuring surface tension) will decrease as stretching proceeds, as the number of bonds per unit area will decrease. The change in surface tension with strain will be given by

$$\gamma + \Delta\gamma = \left(\frac{\Phi\nu}{2}\right)\left(\frac{1}{1+\varepsilon_{xx}}\right)$$

$$\approx \left(\frac{\Phi\nu}{2}\right)(1-\varepsilon_{xx}). \qquad 3.35$$

Thus,

$$\frac{\Delta\gamma}{\varepsilon_{xx}} \approx \frac{d\gamma}{d\varepsilon_{xx}}$$

$$= -\frac{\Phi\nu}{2} = -\gamma, \qquad 3.36$$

and, from Equation 3.29,

$$f_{xx} = \gamma + \frac{d\gamma}{d\varepsilon_{xx}} = \gamma - \gamma \qquad 3.37$$

or

$$f_{xx} = 0. \qquad 3.38$$

Thus, no surface stress is present, even though the surface tension is finite. In any real solid, the surface stress components would not in general be equal to zero, but neither would they be equal to the surface tension.

The critical point of the foregoing discussion is that the relation between the surface stress components and the surface tension is determined by whether or not the nature of the surface is changed in the process of extending the surface. In a liquid, with no resistance to shear, the surface tension is unchanged by an extension process. The new surface formed is identical to the old, and surface stress is equal to surface tension (assuming only that the extension process is slow enough to permit adsorptive equilibrium to be maintained). In a solid, except at very high temperatures (a case which we will consider later in this chapter, in the section on measurement of solid surface tensions), because of the limits on atomic mobility, the state of the surface does change in the extension process and the surface stress is no longer identical to the surface tension.

The γ Plot

The orientation dependence of the surface tension of a solid surface can be described graphically using a construction known as the γ plot. In this construction, the surface tension of a crystal face of orientation S is described in terms of $\boldsymbol{\sigma}$, a unit vector normal to the orientation S, and the magnitude of γ associated with this orientation, $\gamma(\sigma)$. A polar plot is then constructed in which the radius vector is

$$\mathbf{R} = \gamma(\sigma)(\boldsymbol{\sigma}) \,. \tag{3.39}$$

The surface generated by the tip of $\mathbf{R}$ is the γ plot and is in general a closed surface of some arbitrary shape.

Consider the shape of the γ plot for a few simple cases. The simplest case is that of a liquid, for which γ is independent of orientation. For this case, $\mathbf{R}$ will have the same magnitude for all directions of $\boldsymbol{\sigma}$, and the γ plot will be a sphere.

As a second case, consider the model discussed above, that of a simple cubic crystal held together with nearest neighbor bonds. For this case, we determined that, for faces vicinal to (100),

$$\gamma(\theta) = (\Phi/2a^2)(\cos\theta + \sin\theta) \,. \tag{3.40}$$

The two-dimensional γ plot for this case is shown in Figure 3.12. Note that the plot is made up of the outer envelope of four circles, each having a diameter $\Phi/\sqrt{2}a^2$, all tangent to the origin of the plot.

This plot represents a simple model, but it illustrates a number of features that are common to all such plots. The presence of cusps at low index faces is a common feature of all plots based on pair bonding models. The nearest neighbor model shows cusps only at singular orientations; more sophisticated models would show additional cusps. Extension of the model to three dimensions, for the same model assumptions discussed previously, would lead to a γ plot that was the outer envelope of eight spheres, each having a radius $\sqrt{3}\Phi/2a^2$, all tangent to the origin of the plot and centered on the corners of a cube. This feature that the γ plot is made up of segments of spheres

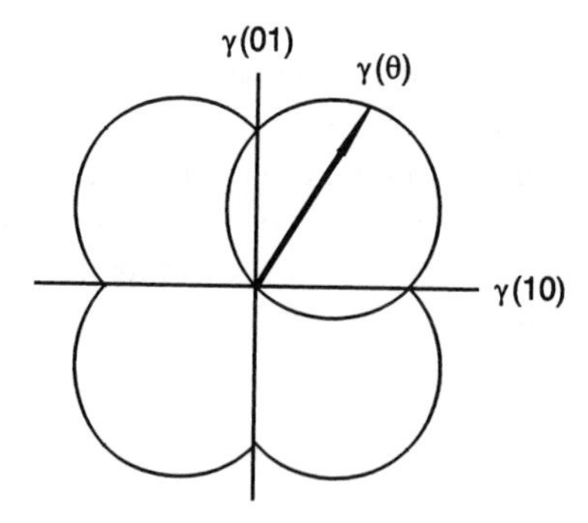

Figure 3.12 γ plot for a two-dimensional simple cubic solid, assuming nearest neighbor bonding.

passing through the origin is a general feature of all plots based on pair bonding models. Consideration of second nearest and farther neighbors merely complicates the plot by adding more spheres to the construction and, consequently, more cusps to the plot. The general expression for γ in the two-dimensional case of a many neighbor model is

$$\gamma(\theta) = \frac{\gamma_0 + \left(\frac{\varepsilon}{a}\right)\tan\theta}{1/\cos\theta}$$

$$= \gamma_0 \cos\theta + \left(\frac{\varepsilon}{a}\right)\tan\theta, \qquad 3.41$$

where $\epsilon = f(\theta)$. For the case of nearest neighbors only, $\epsilon/a = \gamma_0$, and we revert to Equation 3.40.

Finally, it must be noted that an assumption in these calculations is that the surface is defect free. Thermally induced adatom-vacancy pairs, or other defect pairs that would be present in finite concentrations at finite temperatures, will modify the treatment above, as γ is associated with the change in Helmholz free energy involved in the formation of new surface. Since $F = E - TS$, any process that changes E (e.g., vacancy formation energy) or S (e.g., mixing defect sites with perfect sites) will change γ and hence the γ plot. The major effect of these changes is to blunt the cusps of the γ plot, so that $d\gamma/d\theta$ will not be discontinuous at singular orientations. The trend in all cases is toward a more nearly spherical γ plot as the melting temperature is approached.

Measurement of Surface Tension of Liquids

The two final sections of this chapter will consider practical methods for measuring the surface tension, first for the case of liquid surfaces and then for solids. A number of techniques have been developed for the measurement of liquid surface tension. Of the two most common, one makes use of the application of LaPlace's equation to liquid in a capillary tube, and the other uses the balance between surface tension and gravitational forces.

The so-called capillary rise method of surface tension measurement can be explained with reference to Figure 3.13. Here we have a small diameter capillary tube, immersed in a large body of liquid. If the liquid wets the walls of the capillary tube, (that is, if the contact angle is zero), then the initial situation, when the tube is first immersed in the liquid, will be as shown in Figure 3.13a. If we consider the balance of forces in the vertical direction in the tube, just below the surface of the meniscus, we see that p^α is less than p^β by the amount

$$\left(p^\beta - p^\alpha\right) = \frac{2\gamma}{r}, \qquad 3.42$$

from LaPlace's equation. To restore the balance of forces, liquid must rise up the capillary until the weight of the liquid column balances this difference in pressure.

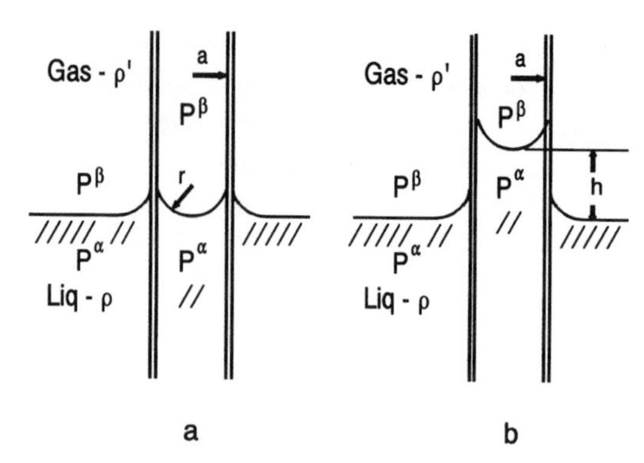

Figure 3.13 Capillary rise method of measuring the surface tension of a liquid, showing (a) initial state, and (b) final equilibrium state.

Figure 3.13b shows this situation. The weight change associated with the rise of the liquid column is

$$hg\ (\rho - \rho')\ , \tag{3.43}$$

in which ρ is the density of the liquid and ρ' the density of the gas phase. Thus, at equilibrium

$$p^{\beta} = p^{\alpha} + hg\ (\rho - \rho') \tag{3.44}$$

or

$$\left(p^{\beta} - p^{\alpha}\right) = \frac{2\gamma}{r} = hg(\rho - \rho'). \tag{3.45}$$

If a small enough tube is used that the radius of curvature of the liquid surface, r, is equal to the radius of the tube, a, and assuming that $\rho' \approx 0$, Equation 3.45 reduces to

$$\gamma = \tfrac{1}{2}\, hg\ \rho\ a, \tag{3.46}$$

thus expressing γ in terms of the gravitational constant, g, and the experimentally observable variables h, ρ, and a. In general, if the contact angle θ is not zero, a term $r = a/\cos\ \theta$ will appear in the expression, yielding

$$\gamma = \tfrac{1}{2}\, hg\ \rho\ (a/\cos\ \theta)\ . \tag{3.47}$$

This relation gives γ in terms of constants and observables.

A similar argument holds for systems in which the liquid does not wet the tube. Figure 3.14 shows the situation in this case. Because the liquid surface is convex, the meniscus must be depressed to reach equilibrium.

The second method of determining γ is a little less obvious than the first one, but has been used often to measure the surface tension of liquid metals. This technique depends on the combined effects of gravity and surface tension on the shape of a drop hanging from a tip (pendant drop) or resting on a flat surface (sessile drop). The situation is shown in Figure 3.15.

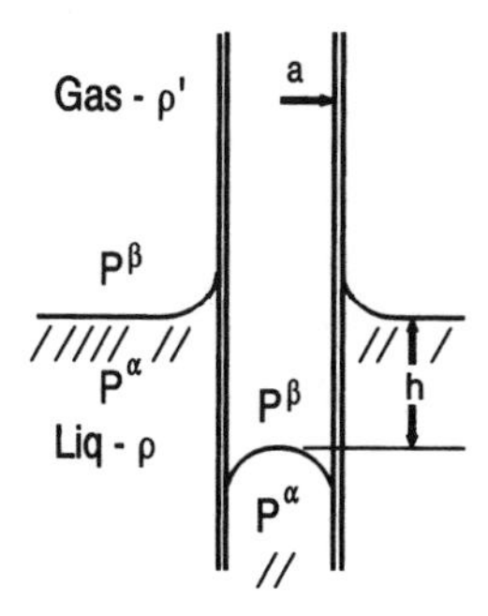

Figure 3.14 Capillary depression in a system in which the liquid does not wet the solid.

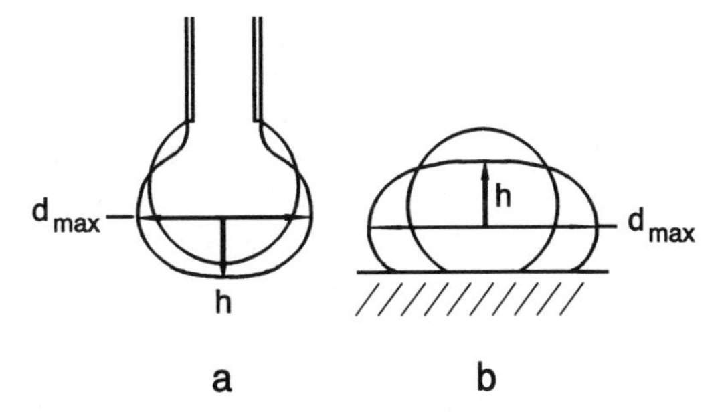

Figure 3.15 Drop shape methods of measuring the surface tension of a liquid for (a) pendant drop and (b) sessile drop.

In the absence of gravitational forces, the drop would be a sphere. Gravitational forces distort the drop to a shape that is complex but that can be rigorously defined. These shapes have been calculated to high precision in terms of γ and the gravitational constant, g. These shapes can be characterized in terms of the maximum diameter of the drop and the height of the maximum diameter above the bottom of the drop, for the case of the pendant drop, or in terms of the height of the top of the drop above the maximum diameter and maximum diameter for the case of the sessile drop. That is, by measurement of these parameters (usually from a photograph of the drop or using a telescope to look at a drop in an oven), one can determine the surface tension from published tables and a knowledge of the density of liquid.

Measurement of Solid Surface Tensions

Although, as we have discussed previously, there are fundamental differences between the concept of surface tensions in solids and those in liquids, it is possible in some circumstances to make a direct measurement of the surface tension of a solid material. The classic measurement of this kind is based on a technique developed by Udin, Shaler, and Wulff [1952], which allows determination of the surface tension of a solid from a purely mechanical test.

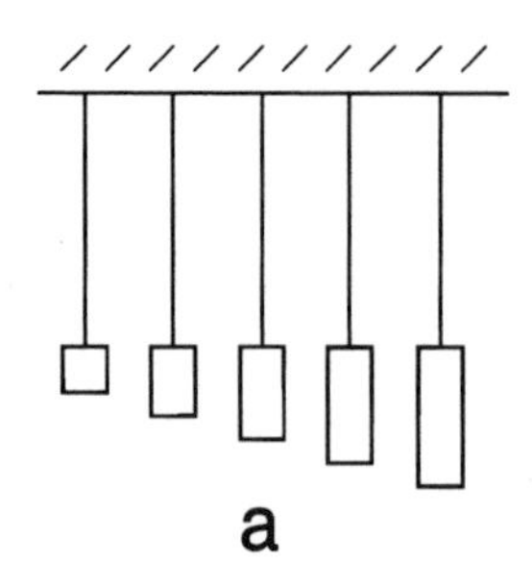

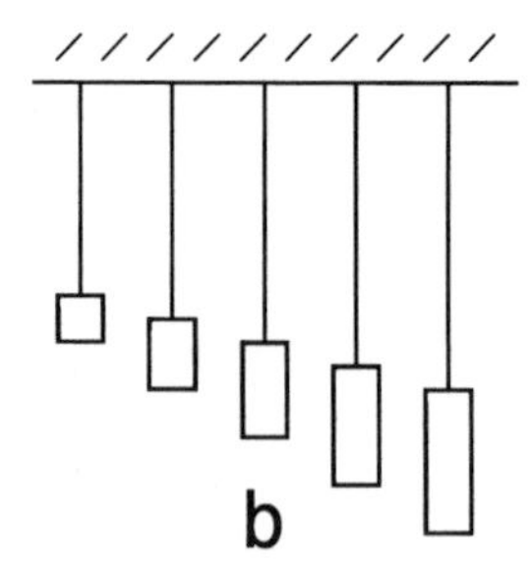

Figure 3.16 Measurement of the surface tension of solid wires by the zero creep method of Udin, Shaler and Wulff. Part (a) shows wire lengths before heating and part (b) shows wire lengths after creep has occurred at high temperature.

The basis of this measurement lies in the difference between the work of formation of a surface, and the stretching of a solid surface. The stresses associate with stretching arise because the solid cannot, in general, respond to the stretching force by atom movements at a rate fast enough to relieve the applied stress, as a liquid can. If the force is small enough, however, and is applied to the surface of a solid at a high enough temperature that diffusion is rapid, and in a configuration which supplies large numbers of the defects required for diffusive transport, then this limitation can be circumvented.

In practice, one begins with fine polycrystalline wires, loads them with a range of small loads, and holds them at temperatures close to the melting point for long periods of time. The initial and final conditions observed in such a study are shown in Figure 3.16. It can be seen from the figure that wires with small loads contract, while those with large loads extend.

Mechanical considerations dictate that in a situation such as this, net diffusive transport will go on unless and until the pressure tensor within the solid is isotropic. Thus, for the no-motion case, the at equilibrium

$$\mathbf{p}_{xx} = \mathbf{p}_{yy} = \mathbf{p}_{zz} = p^{\beta}. \tag{3.48}$$

If one looks at a cross-section of the wire subjected to the load P that corresponds to the no-motion case, as shown in Figure 3.17, the result for the balance of forces in the vertical direction is

$$2\pi r\gamma - \pi r^2 p^{\beta} = P - \pi r^2 p^{\alpha}, \tag{3.49}$$

in which $p^{\beta} \neq p^{\alpha}$ because of the considerations which led to LaPlace's equation. Application of LaPlace's equation to this case leads to

$$\left(p^{\beta} - p^{\alpha}\right) = \frac{\gamma}{r} \tag{3.50}$$

(Note that the result is γ/r, rather than $2\gamma/r$ because the curved surface is cylindrical rather than spherical.) Substitution of this expression into the force balance equation yields

$$2\pi r\gamma - \pi r^2 p^{\alpha} - \pi r\gamma = P - \pi r^2 p^{\alpha}, \tag{3.51}$$

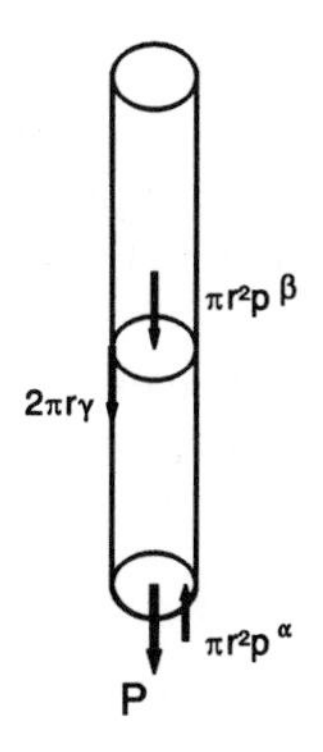

Figure 3.17 Force balance in the zero-creep method for a single crystal wire.

or

$$P = \pi r \gamma, \tag{3.52}$$

allowing the surface tension to be calculated from the applied load for zero motion and the wire diameter.

In practice, one must also allow for the presence of grain boundaries in the wire. Generally the specimens used in this type of experiment have the configuration shown in Figure 3.18. Consideration of the grain boundary surfaces and the surface tension associated with them modifies the above equation for γ of the solid to

$$P = \pi r \gamma_{surf} - \left(\frac{n}{\ell}\right)\left(\pi r \gamma_{gb}\right), \tag{3.53}$$

where n/ℓ is the number of grain boundaries per unit length of wire and γ_{gb} is the interfacial tension of the grain boundary.

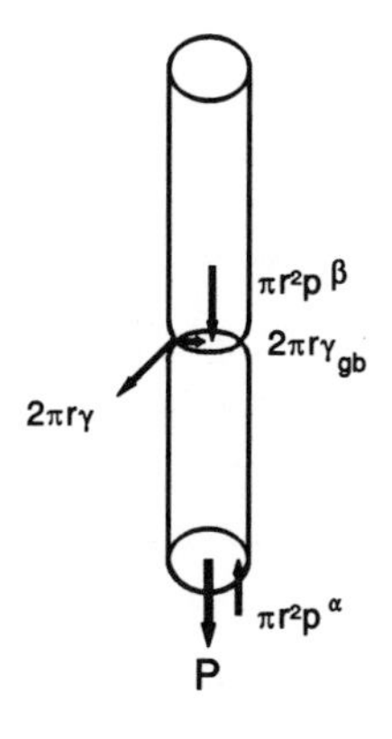

Figure 3.18 Force balance in the zero-creep method for a polycrystal wire.

Problems

3.1. Calculate the work done when a simple cubic crystal having a cross sectional area of 1 cm^2 in the (100) plane is fractured to produce a surface misoriented by 14° from the (100) orientation.

3.2. Sketch the results of capillary rise experiments carried out using glass capillary tubes having diameters of 0.01, 0.1, and 1.0 mm, respectively for the following situations (all at 300 K):

a. H_2O in a terrestrial laboratory

b. Hg in a terrestrial laboratory

c. H_2O in a zero gravity environment

The relevant surface energies and densities are: $\gamma_{H_2O} = 80$ dyn/cm^2, $\gamma_{Hg} = 500$ dyn/cm^2, $\rho_{H_2O} = 1$ g/cm^3, $\rho_{Hg} = 13.6$ g/cm^3.

3.3. A sphere of copper, 5 mm in diameter, is floated on a liquid lead surface inside a sealed quartz container at 1300 K. The surface tension of the lead is 1000 dyn/cm and that of copper is 1300 dyn/cm. The copper–lead interfacial tension is 900 dyn/cm. What is the equilibrium shape of the copper particle? (You may neglect any effects associated with adsorption or gravity.)

3.4. Consider the system shown in Figure 3.19, consisting of a solid material, a condensible vapor, A, and an inert gas, B, in a closed, isothermal container. There is a drop of liquid A 2 cm in diameter on the surface of the solid and another drop of A in a crevice in the solid. The crevice is a conical hole with a cone angle of 20° and is filled to a depth of 1 cm. The partial pressure of B is 760 Torr. The contact angle of A on the solid is $\theta = 30°$. The solid–liquid interfacial tension is 75 dyn/cm, and the surface tension of the liquid is 80 dyn/cm.

a. What is the solid–gas interfacial tension?

b. What are the total pressures inside the two drops?

3.5. Construct the γ plot for a two-dimensional hexagonal close packed crystal, using the nearest neighbor bonding model. Give both a graphical and an analytical answer. Qualitatively, how would the plot differ if second nearest neighbor terms were also included in the calculation?

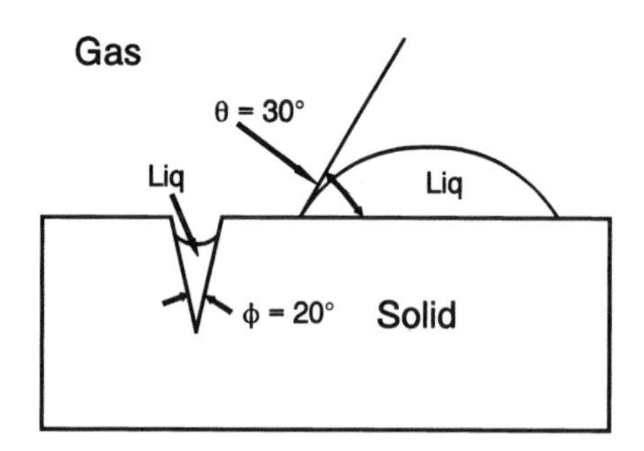

Figure 3.19 Capillary forces in a hypothetical gas–solid–liquid system.

Bibliography

Adamson, AW. *Physical Chemistry of Surfaces*. New York: John Wiley and Sons, 1967. Second Edition.

Blakely, JM. *Introduction to the Properties of Crystal Surfaces*. New York: Pergamon Press, 1973.

Defay, R, Prigogine, I, Bellemans, A, and Everett, DH. *Surface Tension and Adsorption*. London: Longmans, Green & Co., 1966.

Gilman, JJ. Direct measurements of the surface energies of crystals. *J. Appl. Phys.* 1960; 31:2208–2218.

Mullins, WW. Solid surface morphologies governed by capillarity. In: W.D. Robertson, and N.A. Gjostein, eds. *Metal Surfaces*. Metals Park, Ohio: American Society for Metals, 1963; 17–66

Shewmon, PG, and Robertson, WM. Variation of surface tension with orientation. In: WD Robertson, and NA Gjostein, eds. *Metal Surfaces*. Metals Park, Ohio: American Society for Metals, 1963; 67–98.

Udin, H. Measurement of solid:gas and solid:liquid interfacial energies. In: *Metal Interfaces*. Cleveland: American Society for Metals, 1952; 114.

Udin, H, Shaler, AJ, and Wulff, J. *Trans. AIME* 1952; 185:186.

CHAPTER **4**

Thermodynamics of One-component Systems

This chapter continues the discussion of the equilibrium behavior of capillary systems; considering thermodynamic processes and the conditions required for thermodynamic equilibrium in one-component systems.

Work in Capillary Systems

Consider first the process of performing mechanical work in a capillary system, as expressions for this work will be needed to develop the connections between mechanical equilibrium and thermodynamic equilibrium. The general thermodynamic definition of work is

$$đW = \Sigma F dx, \tag{4.1}$$

in which the F represent all of the forces acting on the system and the dx all of the displacements of the system in response to these forces. By convention, work done by the system is a positive quantity.

The general expression for work for the particular case of a two-phase capillary system may be evaluated using as a model a system composed of two fluid phases, an α phase, which is continuous, and a β phase, which is present as a spherical drop of radius r within the bulk of the α phase. The whole system is contained in a chamber at a constant external pressure p^α. Figure 4.1 illustrates this system. The β phase is a pure liquid drop and the α phase is made up of the vapor of the same chemical species. The total volume of the system is

$$V = V^\alpha + V^\beta. \tag{4.2}$$

Now suppose that a small amount of the liquid in the drop evaporates into the vapor phase, pushing back the piston with a constant pressure, p^α. The work done is (note that since work is being done by the system, $đW$ will be positive):

$$đW = p^\alpha dV \tag{4.3}$$

$$= p^\alpha (dV^\alpha + dV^\beta) \tag{4.4}$$

$$đW = p^\alpha dV^\alpha + p^\beta dV^\beta - (p^\beta - p^\alpha) dV^\beta. \tag{4.5}$$

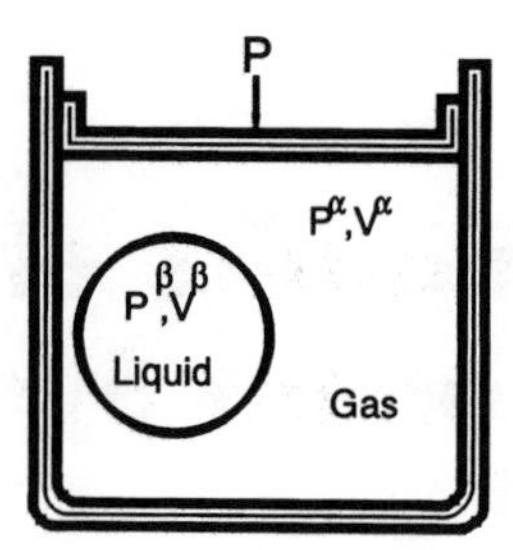

Figure 4.1 Conceptual system for determining the work done in a capillary system having fluid–fluid interfaces.

Using LaPlace's relation

$$p^{\beta} - p^{\alpha} = \frac{2\gamma}{r}, \tag{4.6}$$

leads to

$$đW = p^{\alpha}\,dV^{\alpha} + p^{\beta}\,dV^{\beta} - \left(\frac{2\gamma}{r}\right)dV^{\beta}. \tag{4.7}$$

But, since the β phase is a spherical drop,

$$V^{\beta} = \frac{4}{3}\pi r^{3}, \quad \mathrm{d}V^{\beta} = 4\pi r^{2}\mathrm{dr}, \tag{4.8}$$

and

$$A^{\beta} = 4\pi r^{2}, \; dA^{\beta} = 8\pi r dr. \tag{4.9}$$

Combining Equations 4.7, 4.8, and 4.9 yields

$$đW = p^{\alpha}dV^{\alpha} + p^{\beta}dV^{\beta} - \gamma dA. \tag{4.10}$$

Thus, γ is the reversible work done in extending the surface (or forming new surface) at constant surface configuration, pressure, and temperature. Note that in this development one must use the surface of tension defined in the Young model as the dividing surface between the two bulk phases in defining V^{α} and V^{β}. Otherwise the final expression for the work done will be in error, as it is the surface of tension that defines the value of the radius of curvature in LaPlace's equation. Note that in the case of a system containing only plane interfaces, in which $r = \infty$, and thus $p^{\alpha} = p^{\beta}$, LaPlace's equation may no longer be used to define the location of the surface of tension. In this case, the expression for the work done reduces to

$$đW = p(dV^{\alpha} + dV^{\beta}) - \gamma dA = pdV - \gamma dA, \tag{4.11}$$

and the location of the surface of tension need not be known in order to calculate the work done.

Capillary Work in Liquid–Solid Systems

Consider now the work done in a capillary system in which a solid surface is involved. Take, for example, the system shown in Figure 4.2, which is a solid plate dipping into a liquid. There is a contact angle, θ, between the liquid and the solid. P is the perimeter length and S is the cross-sectional area of the plate. The force required to hold the plate in position is

$$F = P\gamma_{lg}\cos\theta + mg - S(\rho_b - \rho_a), \qquad 4.12$$

in which the first term on the right is associated with capillary forces, the second with gravitational forces, and the third with buoyancy. In this situation, the work that must be done to raise the plate a distance dh will be $F\,dh$. This work can be split into two terms, one gravitational and the other capillary. The capillary term is

$$đW_{cap} = P\gamma_{lg}\cos\theta\, dh \qquad 4.13$$

or, using Young's equation

$$đW_{cap} = (\gamma_{sg} - \gamma_{sl})\, P\, dh. \qquad 4.14$$

This is called the work of emersion. If the areas of solid in contact with liquid and gas are A_{sl} and A_{sg}, respectively, then one can rewrite the equation above as

$$đW_{cap} = \gamma_{sl}\, dA_{sl} + \gamma_{sg}\, dA_{sg} \qquad 4.15$$

(since $P\,dh = dA_{sg} = -dA_{sl}$). That is, γ_{sl} is the work of formation of unit area of surface A_{sl}, and γ_{sg} is the work formation of unit area of surface A_{sg}.

Since it was assumed that the work was done so that Young's equation, which defines equilibrium, was maintained throughout, this is the reversible work of formation (provided that adsorptive equilibrium is maintained). Moreover, the process was implicitly assumed to have been carried out isothermally. Consequently, the work done

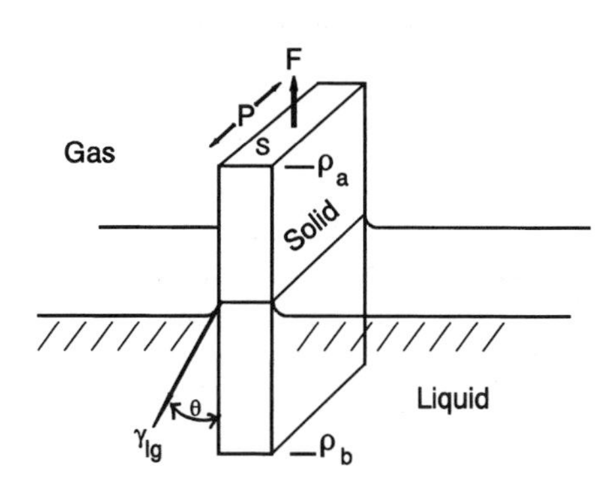

Figure 4.2 Conceptual system for determining the work done in a capillary system having a solid–fluid interface.

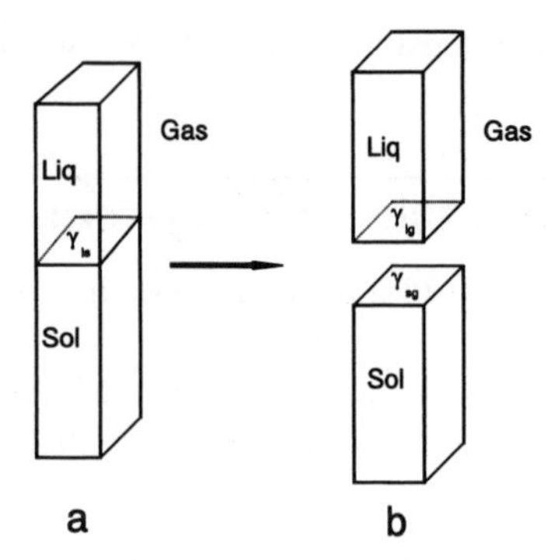

Figure 4.3 Illustration of the work done in separating a liquid–solid interface (a) to produce a liquid–gas and a solid–gas interface (b). $W_s = \gamma_{lg} + \gamma_{sg} - \gamma_{sl}$.

will be equal to the Gibbs free energy change during the process. Thus, γ_{sg} is the Gibbs free energy of formation of unit area of surface A_{sg}, and similarly for γ_{sl}. Note that the work must be done isothermally and in such a way that adsorptive equilibrium is maintained on all newly formed surfaces.

Consider one final mechanical process in capillary systems, namely the work involved in the separation of two surfaces initially in contact, as shown in Figure 4.3. This is called the reversible work of separation and will be equal to the difference in free energy between the cases shown in Figure 4.3a and Figure 4.3b. From the preceding definitions of γ_{sg}, and so on, as the free energy of formation of the surface in question, one may write

$$W_s = \gamma_{lg} + \gamma_{sg} - \gamma_{sl}. \tag{4.17}$$

This is called Dupre's equation. It may be combined with Young's equation to yield

$$W_s = \gamma_{lg} (1 + \cos \theta), \tag{4.18}$$

the Young–Dupre equation.

One could also consider the process of separating two parts of the same material, as is shown in Figure 4.4, where the material separated could be either a liquid or a solid. In this case

$$W_s = 2\gamma_{sg} - \gamma_{ss} \tag{4.19}$$

or

$$W_s = 2\gamma_{sg}. \tag{4.20}$$

This technique is used as a practical method for measuring the surface energy of crystals.

Thermodynamics of Curved Interfaces

The following section will develop some thermodynamic expressions for systems of one component, in thermodynamic equilibrium, containing curved interfaces. Expressions will be developed for the effect of the presence of the curved interface on

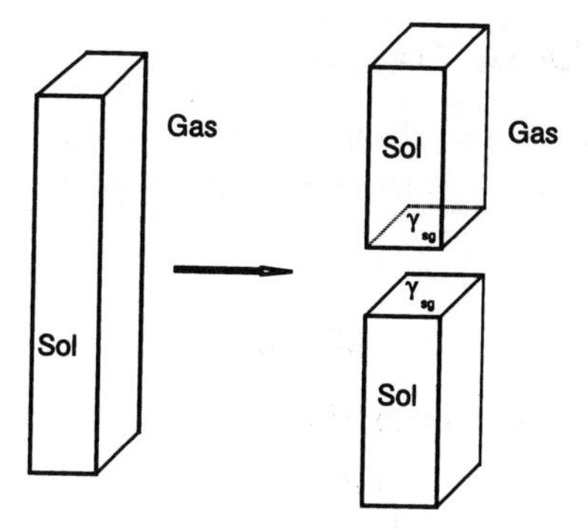

Figure 4.4 Illustration of the work done in cleaving a crystal to produce new solid surface. $W_s = \gamma_{sg}$.

the properties of the system, such as the equilibrium vapor pressure of a liquid drop or the chemical potential of material in the drop relative to the values of the same variables in an infinite bulk phase of the same substance.

To do this, one begins with LaPlace's equation

$$p^{\beta} - p^{\alpha} = \frac{2\gamma}{r}, \tag{4.21}$$

in which the β phase is the interior phase and α the exterior phase, and the condition for distributive equilibrium in a one-component system

$$\mu^{\alpha} = \mu^{\beta}, \tag{4.22}$$

where μ is chemical potential. This latter relation is a completely general thermodynamic equation.

For any displacement of a system of this sort in equilibrium, about a point of equilibrium, the differentials of the above expressions are

$$dp^{\beta} - dp^{\alpha} = d\left(\frac{2\gamma}{r}\right) \tag{4.23}$$

and

$$d\mu^{\alpha} = d\mu^{\beta} \tag{4.24}$$

Moreover, for each bulk phase there is a Gibbs–Duhem equation of the type

$$\mathcal{S}^{\alpha}dT - \mathcal{V}^{\alpha}dp^{\alpha} + d\mu^{\alpha} = 0 \tag{4.25}$$

$$\mathcal{S}^{\beta}dT - \mathcal{V}^{\beta}dp^{\beta} + d\mu^{\beta} = 0, \tag{4.26}$$

in which $\mathcal{S}^{i}$ is the molar entropy of the appropriate bulk phase and $\mathcal{V}^{i}$ is the appropriate molar volume.

These equations will be used to treat a number of situations.

Equilibrium Vapor Pressure

Consider first the effect that changing r, the radius of curvature, at constant temperature, has on the equilibrium vapor pressure of the system. For this case, in general,

$$-\mathcal{V}^{\alpha}dp^{\alpha} + d\mu^{\alpha} = 0 \quad (4.25)$$

$$-\mathcal{V}^{\beta}dp^{\beta} + d\mu^{\beta} = 0 \quad (4.26)$$

for changes at constant temperature or, since

$$d\mu^{\alpha} = d\mu^{\beta} \quad (4.27)$$

one has

$$\mathcal{V}^{\alpha}dp^{\alpha} = \mathcal{V}^{\beta}dp^{\beta}, \quad (4.28)$$

or

$$dp^{\beta} = \left(\frac{\mathcal{V}^{\alpha}}{\mathcal{V}^{\beta}}\right)dp^{\alpha}, \quad (4.29)$$

or

$$dp^{\alpha} = \left(\frac{\mathcal{V}^{\beta}}{\mathcal{V}^{\alpha}}\right)dp^{\beta}. \quad (4.30)$$

The differential of LaPlace's equation can thus be written as

$$d\left(\frac{2\gamma}{r}\right) = \left(\frac{\mathcal{V}^{\alpha}}{\mathcal{V}^{\beta}}\right)dp^{\alpha} - dp^{\alpha} \quad (4.31)$$

or

$$d\left(\frac{2\gamma}{r}\right) = \left(\frac{\mathcal{V}^{\alpha} - \mathcal{V}^{\beta}}{\mathcal{V}^{\beta}}\right)dp^{\alpha} \quad (4.32)$$

or, alternatively

$$d\left(\frac{2\gamma}{r}\right) = \left(\frac{\mathcal{V}^{\alpha} - \mathcal{V}^{\beta}}{\mathcal{V}^{\alpha}}\right)dp^{\beta}. \quad (4.33)$$

Applying the equation in terms of p^{α} to the case of a liquid drop in contact with its own vapor—for example a fog or mist—it can be assumed that $\mathcal{V}^{\beta}$ is much less than $\mathcal{V}^{\alpha}$ and is effectively zero. Moreover, if one assumes the vapor is an ideal gas,

$$\mathcal{V}^{\alpha} = \frac{RT}{p^{\alpha}}. \qquad 4.34$$

This leads to

$$d\left(\frac{2\gamma}{r}\right) = \left(\frac{RT}{\mathcal{V}^{\beta}}\right)\left(\frac{dp^{\alpha}}{p^{\alpha}}\right). \qquad 4.35$$

Integrating this expression from the flat surface ($r = \infty$) to finite r, and assuming that $\mathcal{V}^{\beta} \neq f(r,p^{\alpha})$, yields

$$\int_{p_0}^{p^{\alpha}} \left(\frac{RT}{\mathcal{V}^{\beta}}\right)\left(\frac{dp^{\alpha}}{p^{\alpha}}\right) = \int_{\infty}^{r} d\left(\frac{2\gamma}{r}\right), \qquad 4.36$$

or

$$\left(\frac{RT}{\mathcal{V}^{\beta}}\right)\ln\left(\frac{p^{\alpha}}{p_0}\right) = \frac{2\gamma}{r} \qquad 4.37$$

or

$$\ln\left(\frac{p^{\alpha}}{p_0}\right) = \left(\frac{2\gamma}{r}\right)\left(\frac{\mathcal{V}^{\beta}}{RT}\right) \qquad 4.38$$

for the droplet, in which p_o is the equilibrium vapor pressure over the flat surface.

This is Kelvin's equation. It says that the equilibrium vapor pressure over a small droplet (or collection of droplets) of radius r exceeds the equilibrium vapor pressure over a large volume of the bulk liquid by the amount given in the equation. This equation has two important consequences:

1. In a mist containing drops of various sizes, the small drops will have a higher vapor pressure than the larger drops, and the larger will tend to grow at the expense of the smaller; that is, the average droplet size will tend to increase with time.
2. If a drop satisfying the above equation exists in equilibrium in an infinite volume of vapor, this equilibrium is an unstable one—if the drop shrinks by evaporation, then r decreases, p^{α} for equilibrium increases above p^{α} existing in the vapor phase, and the droplet evaporates completely. If the drop grows slightly by condensation, r increases, p^{α} for equilibrium is less than the existing p^{α}, and the droplet will tend to grow further.

This concept of a droplet of given radius being in unstable equilibrium with a vapor at pressure $p^{\alpha} > p_0$ forms the basis of the capillarity theory of the nucleation process, which is discussed in detail in Chapter 17.

Consider next the similar but alternative case in which the vapor is the droplet phase, for example a small bubble within the body of a large bulk of liquid. This time the expression

$$d\left(\frac{2\gamma}{r}\right)=\left(\frac{\mathcal{V}^{\alpha}-\mathcal{V}^{\beta}}{\mathcal{V}^{\alpha}}\right)dp^{\beta} \qquad 4.39$$

is the starting point. Assuming the vapor to be ideal and the liquid incompressible and using

$$\mathcal{V}^{\beta}=\frac{RT}{p^{\beta}}, \qquad 4.40$$

one finds that

$$d\left(\frac{2\gamma}{r}\right)=-\left(\frac{RT}{\mathcal{V}^{\alpha}}\right)\left(\frac{dp^{\beta}}{p^{\beta}}\right). \qquad 4.41$$

Integrating as before

$$\ln\left(\frac{p^{\beta}}{p_{o}}\right)=-\left(\frac{2\gamma}{r}\right)\left(\frac{\mathcal{V}^{\alpha}}{RT}\right). \qquad 4.42$$

Thus, the smaller is r, the lower is the equilibrium vapor pressure. That is, the pressure inside the bubble is less than p_0 over the bulk liquid.

This explains why a liquid has to be superheated before it will boil. At the normal boiling point, $p_0 = 760$ Torr and is the same as the pressure above the bulk of the liquid (and in the liquid). However, the equilibrium p^{β} inside a small bubble would be less than p_0, and the bubble would collapse because of the unbalanced mechanical forces. Thus, bubbles will not form until p^{β} is equal to the applied pressure—this will require a temperature above the equilibrium boiling point for the flat free surface.

As an idea of the magnitude of these effects, consider the case of pure water shown in Table 4.1. From the table one can see that changes in equilibrium pressure are appreciable only for very small drops. One may question, too, the application of a macroscopic argument such as we have used to drops having a radius of 1 nm (about 3–4 atom distances).

Effect of Curvature on Chemical Potential

Consider another situation, this time the variation of chemical potential with radius of curvature at constant temperature. For this case

$$d\mu^{\alpha}=d\mu^{\beta}=d\mu\ . \qquad 4.43$$

If the temperature is constant, then the Gibbs–Duhem equations reduce to

$$-\mathcal{V}^{\alpha}\,dp^{\alpha}+d\mu=0 \qquad 4.44$$

Table 4.1 The effect of interface curvature on the equilibrium vapor pressure

r-cm	*Drop*	*Bubble*
	p/p_o	p/p_o
∞	1.000	1.000
10^{-4}	1.001	0.999
10^{-6}	1.115	0.897
10^{-7}	2.968	0.337

and

$$-\mathcal{V}^{\beta} dp^{\beta} + d\mu = 0 \tag{4.45}$$

or

$$d\mu = \mathcal{V}^{\alpha} dp^{\alpha} = \mathcal{V}^{\beta} dp^{\beta}, \tag{4.46}$$

an equation of general validity.

Applying this result to the case of a liquid droplet in a vapor, the differential of LaPlace's equation,

$$d\left(\frac{2\gamma}{r}\right) = dp^{\beta} - dp^{\alpha}, \tag{4.47}$$

becomes, again using

$$dp^{\alpha} = \left(\frac{\mathcal{V}^{\beta}}{\mathcal{V}^{\alpha}}\right) dp^{\beta} \tag{4.48}$$

and assuming $\mathcal{V}^{\beta} << \mathcal{V}^{\alpha}$,

$$d\left(\frac{2\gamma}{r}\right) = \left(\frac{\mathcal{V}^{\alpha} - \mathcal{V}^{\beta}}{\mathcal{V}^{\alpha}}\right) dp^{\beta} \approx dp^{\beta} = \frac{d\mu}{\mathcal{V}^{\beta}}, \tag{4.49}$$

or

$$d\left(\frac{2\gamma}{r}\right) = \frac{d\mu}{\mathcal{V}^{\beta}}. \tag{4.50}$$

Integrating this expression, as in previous situations, from $r = \infty$ to finite r, yields

$$\int_{\infty}^{r} d\left(\frac{2\gamma}{r}\right) = \int_{\mu^{\circ}}^{\mu} \frac{d\mu}{\mathcal{V}^{\beta}}, \tag{4.51}$$

or

$$(\mu - \mu^\circ) = \left(\frac{2\gamma}{r}\right)\mathcal{V}^\beta, \tag{4.52}$$

in which μ° is the chemical potential in the infinite bulk liquid or vapor phase. Since all of the terms on the right are inherently positive, $\mu > \mu^\circ$. That is, the chemical potential in the drop is greater than the chemical potential in the bulk liquid at the same temperature.

Alternatively, for bubbles in a liquid,

$$d\left(\frac{2\gamma}{r}\right) = \frac{-d\mu}{\mathcal{V}^\alpha}, \tag{4.53}$$

which leads to

$$(\mu - \mu^\circ) = -\left(\frac{2\gamma}{r}\right)\mathcal{V}^\alpha. \tag{4.54}$$

That is, the chemical potential of the vapor in the bubble is lower than that of vapor over bulk liquid having a plane surface.

Effect of Curvature on Equilibrium Temperature

Alternatively, consider processes at constant external pressure—such as the effect of curvature on the equilibrium temperature of a liquid drop.

In this case, since p^α is constant, the Gibbs–Duhem equations reduce to

$$\mathcal{S}^\alpha dT + d\mu^\alpha = 0 \tag{4.55}$$

$$\mathcal{S}^\beta dT - \mathcal{V}^\beta dp^\beta + d\mu^\beta = 0\,. \tag{4.56}$$

Using the relation

$$T(\mathcal{S}^\alpha - \mathcal{S}^\beta) = \Delta\mathcal{H}_e, \tag{4.57}$$

where $\Delta\mathcal{H}_e$ is the molar latent heat of evaporation, and realizing that $d\mu^\alpha = d\mu^\beta$ and dp^α is taken as zero, one obtains

$$(\mathcal{S}^\alpha - \mathcal{S}^\beta)\, dT + \mathcal{V}^\beta dp^\beta = 0 \tag{4.58}$$

or

$$\left(\frac{\Delta\mathcal{H}_e}{T}\right)dT + \mathcal{V}^\beta dp^\beta = 0. \tag{4.59}$$

Again, using $dp^\alpha = 0$ in the differential of LaPlace's equation,

$$dp^\beta = d\left(\frac{2\gamma}{r}\right). \tag{4.60}$$

Thus, substituting for dp^{β}, we have

$$\left(\frac{\Delta\mathcal{H}_e}{T}\right)dT + \mathcal{V}^{\beta}d\left(\frac{2\gamma}{r}\right) = 0, \qquad 4.61$$

or

$$\frac{dT}{T} = -\left(\frac{\mathcal{V}^{\beta}}{\Delta\mathcal{H}_e}\right)d\left(\frac{2\gamma}{r}\right). \qquad 4.62$$

Again integrating from $r = \infty$ to finite r, assuming $\Delta\mathcal{H}_e \neq f(r)$,

$$\int_{T^\circ}^{T}\frac{dT}{T} = -\left(\frac{\mathcal{V}^{\beta}}{\Delta\mathcal{H}_e}\right)\int_{\infty}^{r} d\left(\frac{2\gamma}{r}\right), \qquad 4.63$$

or

$$\ln\left(\frac{T}{T^\circ}\right) = -\left(\frac{2\gamma}{r}\right)\left(\frac{\mathcal{V}^{\beta}}{\Delta\mathcal{H}_e}\right), \qquad 4.64$$

which is known as the Thompson equation.

Thus, if the pressure in the vapor phase is held constant, then small drops must be cooler in order to be in equilibrium with bulk liquid. Note the relation to the vapor pressure equation developed previously: At a given temperature, the drop has a higher vapor pressure; at a given external pressure, the drop must be cooler to be in equilibrium with the bulk phase.

Effect of Curvature on the Surface Tension

Finally, consider the effect of curvature on the surface tension. The treatment to this point has included this change implicitly in the equations developed, as what has been integrated is $d(2\gamma/r)$. Thus, as long as the value of γ appropriate to the actual value of r is used, no error results. A rough calculation, of the change of γ with r will suffice to determine the general magnitude of the change.

For this case, begin with the equation previously developed from the differential of LaPlace's equation at constant temperature:

$$d\left(\frac{2\gamma}{r}\right) = \left(\frac{\mathcal{V}^{\alpha} - \mathcal{V}^{\beta}}{\mathcal{V}^{\alpha}}\right)dp^{\beta}, \qquad 4.65$$

and expand the left side to get

$$\left(\frac{2}{r}\right)d\gamma + 2\gamma d\left(\frac{1}{r}\right) = \left(\frac{\mathcal{V}^{\alpha} - \mathcal{V}^{\beta}}{\mathcal{V}^{\alpha}}\right)dp^{\beta}. \qquad 4.66$$

To procede further requires the use of another relation which has not yet been discussed but which will be treated in detail in Chapter 5. This is Gibbs equation, which for a one-component system is

$$d\gamma = -\Gamma d\mu \,, \tag{4.67}$$

in which Γ is the Gibbs adsorption at the surface of tension and is a measure of the amount of material per unit area associated with the presence of the interface.

At constant temperature for the bulk phase, the Gibbs–Duhem equation reduces to

$$d\mu^\beta = \mathcal{V}^\beta dp^\beta. \tag{4.68}$$

Thus, the Gibbs equation can be written

$$d\gamma = -\Gamma \mathcal{V}^\beta dp^\beta. \tag{4.69}$$

Using these two equations to eliminate dp^β leads to

$$\frac{d\gamma}{\gamma} = \frac{2\Gamma}{\frac{2\Gamma}{r} + c^\beta + c^\alpha} d\left(\frac{1}{r}\right) \tag{4.70}$$

in which $c^\beta = 1/\mathcal{V}^\beta$ and $c^\alpha = 1/\mathcal{V}^\alpha$ are the concentrations of the two bulk phases in moles per unit volume.

To obtain the variation in surface tension with radius of curvature, this equation must be integrated. This requires an estimate of Γ, which in turn involves the assumption of a location for the surface of tension. If one assumes that the surface of tension lies roughly half of a molecular diameter below the surface at which the density varies abruptly and, in addition, that the liquid is incompressible, one may integrate from $r = \infty$ to r finite and obtain

$$\frac{\gamma}{\gamma^\circ} = \frac{c^\beta}{\frac{2\Gamma}{r} + c^\beta}. \tag{4.71}$$

Since all terms on the right are inherently positive, $\gamma < \gamma^\circ$. An estimate of the magnitude of the difference for liquid drops of water having various radii is given in Table 4.2. Here again note that the variation is significant only for very small values of r, and again note the reservation about the applicability of macroscopic equations at $r = 10^{-7}$ cm.

Dependence of the Surface Tension on Temperature

As a final topic on liquid surfaces, consider briefly the qualitative effect of changing the temperature on the value of the surface tension. A quantitative treatment is beyond the scope of this book.

Basically, what one observes is that at low temperatures, the interface between a liquid and its own vapor is sharp, with an abrupt change in density at the interface. At high temperatures, approaching the critical temperature, T_c, the difference in properties between the liquid and the vapor becomes less sharp. Finally, at the critical tempera-

Table 4.2 Effect of interface curvature on the surface tension

r-cm	γ/γ^o
∞	1.000
10^{-4}	0.9996
10^{-6}	0.968
10^{-7}	0.755

ture, all difference between liquid and vapor disappears. Thus, no interface between them exists, and thus the surface tension must approach zero. That is, the surface tension must fall from a finite value at low temperature to zero at the critical temperature. This is indeed observed in practice.

This observation has led to the empirical equation of Eötvös that

$$\gamma \mathcal{V}^{2/3} = K_s (T_c - T) , \qquad 4.72$$

where $\mathcal{V}$ is the molar volume of the liquid and K_s is a constant for any given liquid, usually on the order of 2 erg/degree. This is a completely empirical equation but is valid over a wide range of temperature for many substances. It does have the desired property of making γ approach zero as T approaches T_c, but it is significantly in error close to T_c. A more exact treatment is difficult.

Curved Interfaces in Fluid–Solid Systems

As a final topic in the consideration of the behavior of systems containing curved interfaces, consider the complications that arise when one of the phases is a solid. In the prior discussion on the surface tension of solids in Chapter 3, it was determined that there was a fundamental difference relative to liquids because of the rigidity of the solid. A second fundamental difference, at least in the case of crystalline solids, is that the surface tension will differ for surfaces of different crystallographic orientation. This effect arises because the atomic configuration at the surface is different for different orientations.

This dependence of surface tension on orientation complicates the treatment of curved solid surfaces. Because the surface is curved, it must of necessity present a range of crystallographic orientations and, thus, a range of surface tensions. Consequently, the development of equations analogous to the Kelvin equation must take this variation into account. This discussion will not go through a rigorous development of this case but will reason from the case of the liquid drop to demonstrate the analogous expression for the dependence of chemical potential on curvature for the case of a solid surface. The resulting expression will be used when the measurement of surface self diffusion is discussed in Chapter 6. It was determined earlier that for the case of a spherical liquid drop

$$(\mu - \mu^\circ) = \left(\frac{2\gamma}{r}\right)\mathcal{V}^\beta . \qquad 4.73$$

A similar treatment for a nonspherical liquid drop would have lead to

$$(\mu - \mu^\circ) = \gamma \mathscr{V}^\beta \left(\frac{1}{r_1} + \frac{1}{r_2} \right), \tag{4.74}$$

or

$$(\mu - \mu^\circ) = \gamma \mathscr{V}^\beta (K_1 + K_2), \tag{4.75}$$

where K_1, K_2 are the principal curvatures of the drop. For a solid, in which the surface tension is a function of crystallographic orientation, one can express the orientation dependence in terms of $\gamma = f(\theta)$, in which θ is the crystallographic orientation relative to some chosen standard orientation, such as a close-packed plane. Using this terminology, one can develop an expression for the chemical potential difference having the form

$$(\mu - \mu^\circ) = \mathscr{V}^\beta \left[\left(\gamma + \frac{\partial^2 \gamma}{\partial \theta_1^2} \right) K_1 + \left(\gamma + \frac{\partial^2 \gamma}{\partial \theta_2^2} \right) K_2 \right], \tag{4.76}$$

where θ_1 and θ_2 represent the angles between each of the two principal curvatures and a reference orientation. For the case of a surface having a curvature in one direction only, that is, a cylindrical surface, this expression reduces to

$$(\mu - \mu^\circ) = \mathscr{V}^\beta \left(\gamma + \frac{\partial^2 \gamma}{\partial \theta^2} \right) K. \tag{4.77}$$

This result has a number of practical consequences. If a surface is formed that has a curvature that varies with distance, for example, an undulatory solid surface, then the chemical potential will vary with distance over the surface, providing a driving force for surface migration. (Moreover, as the following section will show, the equilibrium shape of a crystallite will be that which minimizes the total surface energy of the crystallite). Finally, just as in the case of the liquid droplet, the equilibrium temperature and vapor pressure of a very small crystallite will differ from those of the bulk crystal at a given temperature. That is,

$$\ln\left(\frac{T}{T^\circ} \right) = -\left(\frac{\mathscr{V}^\beta}{\Delta \mathscr{H}_e} \right) \left[\left(\gamma + \frac{\partial^2 \gamma}{\partial \theta_1^2} \right) K_1 + \left(\gamma + \frac{\partial^2 \gamma}{\partial \theta_2^2} \right) K_2 \right] \tag{4.78}$$

and

$$\ln\left(\frac{p}{p_0} \right) = \left(\frac{\mathscr{V}^\beta}{RT} \right) \left[\left(\gamma + \frac{\partial^2 \gamma}{\partial \theta_1^2} \right) K_1 + \left(\gamma + \frac{\partial^2 \gamma}{\partial \theta_2^2} \right) K_2 \right]. \tag{4.79}$$

The Equilibrium Shape of Small Crystals

As a final topic in the thermodynamics of one-component systems, consider the question of the equilibrium shape of a small crystallite and how this shape differs from the shape of a small liquid droplet.

The criterion for equilibrium in a small crystallite of constant volume is the Gibbs–Curie relation, which states that the equilibrium form is that which leads to a minimum in the function

$$\Phi = \sum_{a} \gamma^a A^a = \int_a \gamma^a dA\,, \qquad 4.80$$

where the a indicates summation over all facets of the crystallite. This theorem can be proven rigorously, but here will be taken as given. The theorem implies that the total surface energy of the system is minimized subject to the condition that the total volume is held constant.

This criterion will be applied to three cases, namely, a liquid drop, a solid for which γ is specified only for three mutually perpendicular orientations, and the two dimensional nearest neighbor model solid developed in Chapter 3. For the case of the liquid, γ is constant, independent of orientation, so that

$$\Phi = \gamma^a \int dA. \qquad 4.81$$

Thus, the shape that minimizes Φ will be the shape that minimizes A at constant volume; i.e., a sphere, as one would expect.

As a second case, consider a crystal that is a rectangular prism, having sides x, y, z and surface tensions γ_x, γ_y, γ_z, as shown in Figure 4.5. For this case

$$\begin{aligned} \Phi &= \sum_{a} \gamma^a A^a = \gamma_x(2yz) + \gamma_y(2xz) + \gamma_z(2xy) \\ V &= xyz. \end{aligned} \qquad 4.82$$

A minimum in Φ must be found subject to the condition that $dV = 0$. This process of finding a conditional extremum is one that is encountered often in thermodynamics. The classic case is in the development of the partition function in statistical thermodynamics. The process generally used in such cases is the LaGrange method of undetermined multipliers.

In the present case,

$$\begin{aligned} d\Phi &= d[\gamma_x(2yz) + \gamma_y(2xz) + \gamma_z(2xy)] \\ &= 2\gamma_x(ydz + zdy) + 2\gamma_y(xdz + zdx) + 2\gamma_z(xdy + ydx). \end{aligned} \qquad 4.83$$

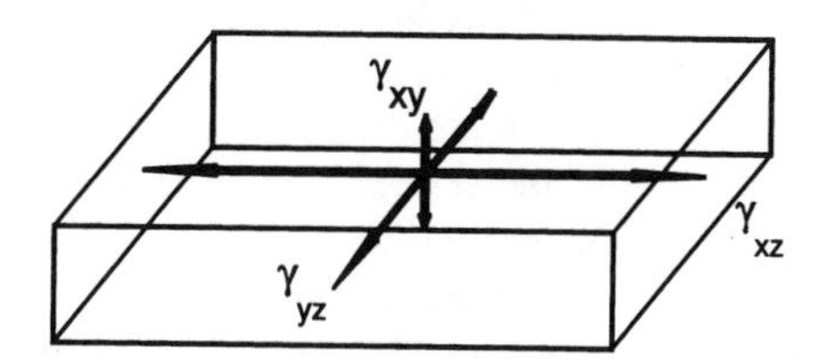

Figure 4.5 Determination of the equilibrium shape of a crystal having a γ plot defined only in three orthogonal directions.

The condition $dV = 0$ leads to

$$dV = xydz + zydx + xzdy = 0, \tag{4.84}$$

or, introducing the LaGrange multiplier λ,

$$\lambda xydz + \lambda zydx + \lambda xzdy = 0. \tag{4.85}$$

Summing Equations 4.83 and 4.85 yields

$$(2\gamma_x y + 2\gamma_y x + \lambda xy)dz + (2\gamma_x z + 2\gamma_z x + \lambda xz)dy + (2\gamma_z y + 2\gamma_y z + \lambda yz)dx = 0. \tag{4.86}$$

Since dx, dy, and dz are arbitrary and independent of one another, each of the terms in parentheses must be equal to zero, leading to

$$2\gamma_x y + 2\gamma_y x + \lambda xy = 0, \qquad \lambda = \frac{-2\gamma_x}{x} - \frac{2\gamma_y}{y}, \tag{4.87}$$

and equivalently for the other two terms. The final result is

$$\frac{\gamma_x}{(x/2)} = \frac{\gamma_y}{(y/2)} = \frac{\gamma_z}{(z/2)} = \frac{-\lambda}{2} = \frac{\Phi_{\min}}{3V}. \tag{4.88}$$

That is, the distance of each face from the center of the crystal is inversely proportional to its γ. The relations in equation 4.88 are known as Wulff's relations.

Using Wulff's relations and the concept of the γ plot, developed in Chapter 3, one can determine the expected equilibrium shape in the general case. To do this, one uses what is known as the Wulff construction. To carry out this construction, one begins with the γ plot. Lines are drawn outward from the origin of the plot, through its surface. At the point of intersection with the surface of the plot, a plane is passed perpendicular to the line. Such a plane, known as a Wulff plane, will have a γ equal to the γ for the crystal orientation perpendicular to the line drawn, as this is the way the γ plot is defined. The equilibrium shape of the crystal is simply the inner envelope of all such possible Wulff planes.

As an example of the Wulff construction, begin with the γ plot for the two-dimensional nearest neighbor model crystal considered in Chapter 3 and shown in Figure 4.6. As shown in the figure, the inner envelope of Wulff planes in this case is bounded by the planes at the cusp orientations, leading to a square equilibrium form, showing only the singular orientations. Similar reasoning holds in the extension of this concept to three dimensions. A more realistic crystal model would lead to more cusps in the γ plot and, consequently, a more complex equilibrium shape. It should also be noted that as the temperature of the crystal is increased and the γ plot becomes more nearly spherical, the equilibrium shape of the crystal becomes more rounded.

As a final point, the distinction must be drawn between the thermodynamic equilibrium shape of a crystal, as discussed here, and the shape that one would observe in a situation in which a crystal was being grown from the vapor or from solution at a finite

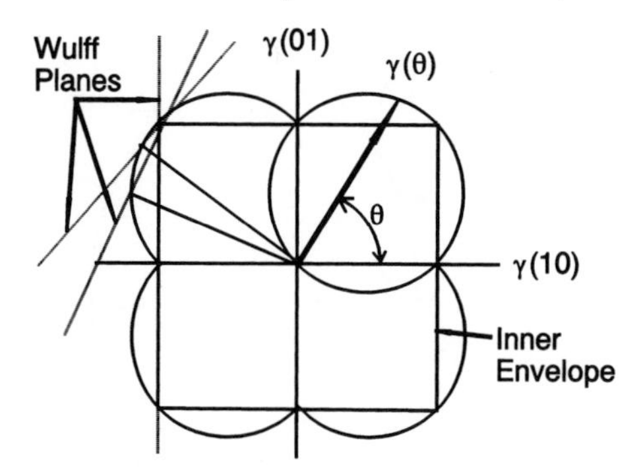

Figure 4.6 Equilibrium shape of a crystal having a two-dimensional simple cubic structure and nearest neighbor bonding.

growth rate. In this latter case, kinetic considerations may control the shape of the growing crystal, rather than thermodynamic equilibrium. The shape may be very different from that deduced from the considerations developed above.

Problems

4.1. Calculate the work done in blowing a hemispherical bubble of air at the end of a 1 mm inside diameter capillary tube immersed to a depth of 1 cm in water.

4.2. Calculate the temperature at which boiling will be observed in a beaker of water, subject to the following assumptions:

a. The equilibrium vapor pressure over bulk water is given by: $\log_{10}P = -2900/T - 4.565 \log_{10}T + 22.395$ Torr

b. The surface tension of liquid water is 80 dyn/cm, independent of temperature.

c. In order for boiling to be observed, a bubble of water vapor 50 nm in radius must be stable.

4.3. Consider a hollow bubble, with nitrogen gas (and water vapor) inside a spherical film of water, and nitrogen gas (and water vapor) outside the bubble. The nitrogen pressure outside the bubble is 1 atm, the temperature is 27°C, and the bubble diameter is 10 nm. Assume that the equilibrium vapor pressure of water over a flat surface is 25 Torr at 27°C.

a. What is the total pressure inside the bubble?

b. Is there a thermodynamic driving force for growth or shrinkage of the bubble? If so, what is it?

4.4. Calculate the melting point of a small, spherical indium crystal 10^{-4} cm in diameter. Assume that the surface tensions of both solid and liquid indium are 500 dyn/cm at the melting point, independent of crystallographic orientation. The melting point of bulk indium is 156.6°C. The densities of the two phases are $\rho_\ell = 7.02$ g/cm^3, $\rho_s = 7.31$ g/cm^3, and the atomic weight is 114.8 AMU. The heat of fusion is 859 cal/mol.

4.5. Determine the equilibrium shape of a small crystallite of a material having the γ plot shown in Figure 4.7.

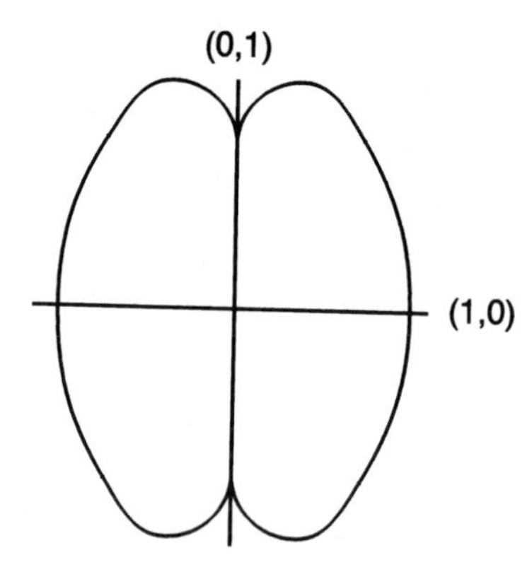

Figure 4.7 γ plot for a hypothetical two-dimensional crystal.

Bibliography

Adamson, AW. *Physical Chemistry of Surfaces*. 2nd ed. New York: John Wiley and Sons, 1967.

Blakely, JM. *Introduction to the Properties of Crystal Surfaces,* New York: Pergamon Press, 1973.

Cabrera, N. The Equilibrium of crystal surfaces. *Surface Sci.* 1964; 2:320.

Defay, R, Prigogine, I, Bellemans, A,and Everett, DH. *Surface Tension and Adsorption*. London: Longmans Green & Co, 1966.

J.W Gibbs, *The Collected Works of J. Willard Gibbs*. New York: Longmans, 1928.

Guggenheim, EA. *Thermodynamics*. Amsterdam: North Holland, 1959.

Herring, C. The use of classical macroscopic concepts in surface energy problems, In: R. Gomer and C.S. Smith, eds. *Structure and Properties of Solid Surfaces*. Chicago: University of Chicago Press, 1953; Chapter 1.

Hirth, JP. The kinetic and thermodynamic properties of surfaces, In: W.M. Meuller, ed. *Energetics in Metallurgical Phenomena*. New York: Gordon and Breach, 1965

Leamy, HJ, Gilmer, GH, and Jackson, KA. Statistical thermodynamics of clean surfaces. In: J.M. Blakely, ed. *Surface Physics of Metals. New* York: Academic Press, 1975; 1:121–188.

Mullins, WW. Solid surface morphologies governed by capillarity. In: W.D. Robertson and N.A. Gjostein, eds. *Metal Surfaces*. Metals Park, Ohio: American Society for Metals, 1963;17-66.

Shewmon, PG, and Robertson, WM. Variation of surface tension with orientation. In: W.D. Robertson and N.A. Gjostein, eds. *Metal Surfaces,* Metals Park, Ohio: American Society for Metals, 1963:67–98.

CHAPTER 5

Thermodynamics of Multicomponent Systems

This chapter extends the thermodynamic treatment of capillary systems to the case of systems of more than one component. In so doing it introduces the concept of adsorption, a process that will be dealt with at great length throughout this book. The first step is the development of general thermodynamic relations for multicomponent systems, using classical thermodynamics, followed by development of descriptions of adsorptive equilibrium and surface segregation equilibrium using statistical thermodynamics. The question of specific thermodynamic calculations for adsorption from the gas phase will be discussed again when the adsorption process is treated in detail in Chapters 10, 11, and 12.

The Gibbs Model

The process of developing a thermodynamic description of capillary systems containing more than one component begins with the development of a model of the surface known as the Gibbs model. The central problem in such a development is that, in any real system, the concentrations of the various components will change in some generally unknown way over some finite distance as one passes from one bulk phase, through the interface, into another bulk phase. In this respect, the problem is similar to the one faced in determining the conditions for mechanical equilibrium, in which the pressures and tensions parallel to the interface varied in some generally unknown way across a region of finite thickness. In the present case, one can circumvent the problem of unknown concentration variation in a manner similar to the earlier circumvention of the tension variation problem, namely, by constructing a model, the Gibbs model, which can describe the interface on a thermodynamic basis without requiring detailed knowledge of the composition variation.

The Gibbs model may be developed with reference to Figure 5.1. Figure 5.1a shows schematically the variation in concentration of two components of a two-phase system as a function of distance perpendicular to an interface between two bulk phases, α and β, and indicates that the region of nonuniform composition is finite in extent, as was the region of nonuniform tension in the case of the mechanical description of the interface. The Gibbs model treats this composition variation in a manner similar to that used to treat the tension variation in the Young model, namely, by assuming that the concentration in

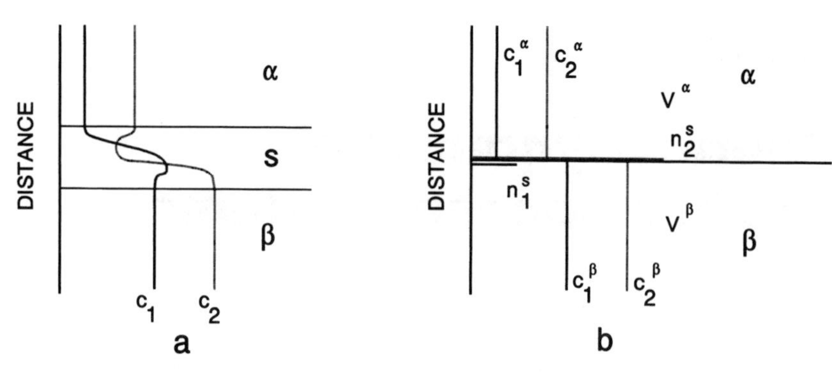

Figure 5.1 Composition variation across an interface in a two-component system. (a) The real system. (b) According to the Gibbs model.

each bulk phase is uniform up to some mathematical surface of zero thickness, called the Gibbs dividing surface, and that any excess material, energy, or entropy needed to make the model thermodynamically equivalent to the real system is associated with this dividing surface. This model is shown in Figure 5.1b. In the real system, the concentration of component i in the α phase is c_i^{α}. In the model, it is assumed that this concentration is maintained right up to the dividing surface. Thus, if V^{α} is the volume of the α phase in the model, then the number of moles of component i in the α phase is

$$n_i^{\alpha} = c_i^{\alpha} V^{\alpha} \tag{5.1}$$

and similarly for the β phase

$$n_i^{\beta} = c_i^{\beta} V^{\beta}. \tag{5.2}$$

Thus, if the real system contains a total of n_i moles of component i, in order for the model to be stoichiometrically equivalent to the real system, it must be the case that

$$n_i = n_i^{\alpha} + n_i^{\beta} + n_i^{s} \tag{5.3}$$

or

$$n_i^{s} = n_i - n_i^{\alpha} - n_i^{\beta} \tag{5.4}$$

is the amount of material associated with the dividing surface. Note that these n_i^s, n_i^{β}, and n_i^{α} are all artifacts of the model and have no physical significance in the real system. Moreover, the values of these terms will depend on where the dividing surface is located. This can be seen from consideration of Figure 5.2. The hatched area gives n_i^s if the dividing surface is located at Z_1. The dotted area gives n_i^s for a dividing surface located at Z_2. The two areas are clearly different.

Finally, one can express the surface excess per unit area in terms of the quantity

$$\Gamma_i = \frac{n_i^s}{A}, \tag{5.5}$$

where Γ_i is called the adsorption of component i.

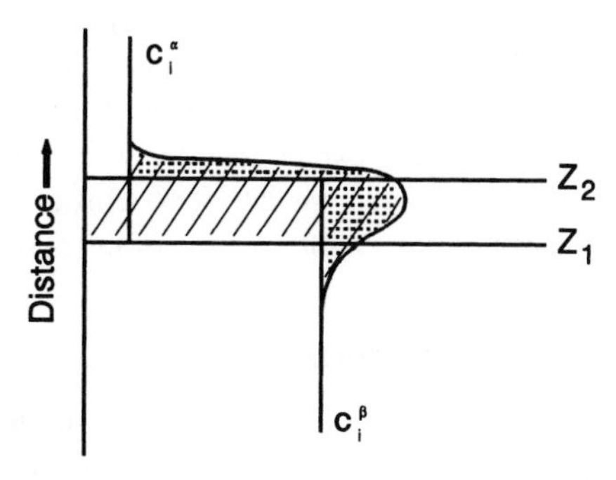

Figure 5.2 The effect of the choice of the location of the Gibbs dividing surface on the value of the surface excess of the component i.

By an argument similar to that presented previously for the material in the system, one can define the internal energy of the α and β phases (in the model) as

$$E^\alpha = e^\alpha V^\alpha \qquad 5.6$$

and

$$E^\beta = e^\beta V^\beta , \qquad 5.7$$

in which e^α is the internal energy per unit volume in the α phase of the real system at a point far from the interface and similarly for e^β. Using these parameters, one may write

$$E = E^\alpha + E^\beta + E^s \qquad 5.8$$

or

$$E^s = E - E^\alpha - E^\beta, \qquad 5.9$$

where E is the total internal energy of the real system and E^s is the internal energy associated with the dividing surface in the model.

Similarly one can write expressions such as

$$S^s = S - S^\alpha - S^\beta \qquad 5.10$$

and, from the defining relation $F = E - TS$,

$$F^s = E^s - TS^s. \qquad 5.11$$

Finally, since no volume is associated with the dividing surface,

$$V = V^\alpha + V^\beta. \qquad 5.12$$

Note that the statements made concerning n_i^s apply equally to E^s, S^s, and F^s. The magnitudes of these quantities depend on the choice of the location of the dividing surface; thus, they are all artifacts of the model and have no physical significance in the real system.

Two factors enter into the choice of location of the dividing surface, namely, the requirement that the Gibbs model be mechanically equivalent to the Young model and the desirability of choosing a location that will simplify the thermodynamic treatment of the system. In the case of systems containing curved interfaces, in order to make the Gibbs model and the Young model equivalent, it is necessary to make the dividing surface coincide with the surface of tension; otherwise, the volumes V^α and V^β associated with the two models would not be the same, and expressions for work done in the capillary system would not be consistent with the Gibbs model. In systems in which the effects of curved interfaces are absent, this restriction on the location of the dividing surface is not present.

One-Component Systems

One may relate the parameters of the Gibbs model to the surface tension by making use of the relation derived in Chapter 4 for work in capillary systems. It was determined that, for a two-phase capillary system, the work done by the system is

$$đW = p^\alpha dV^\alpha + p^\beta dV^\beta - \gamma dA. \tag{5.13}$$

From this relation one may write that, for a reversible process,

$$\gamma = -\left(\frac{đW}{dA}\right)_{n_i,V,T}. \tag{5.14}$$

The general thermodynamic expression for the work done in a reversible process is

$$đW = -dF\ . \tag{5.15}$$

Thus, for a one-component system, at constant total volume

$$\gamma dA = dF = d(F^\alpha + F^\beta) + dF^s\ . \tag{5.16}$$

If, in addition, the temperature is constant, one may write that

$$\gamma dA = \mu d(n^\alpha + n^\beta) + dF^s\ , \tag{5.17}$$

where

$$\mu_i = \left(\frac{\partial F}{\partial n_i}\right)_{T,V,n_j}$$

Moreover,

$$d(n^\alpha + n^\beta) = -dn^s = -\Gamma dA\ , \tag{5.18}$$

and finally,

$$F^s = f^s A, \qquad dF^s = f^s dA\ , \tag{5.19}$$

where f^s is the Helmholz free energy per unit area of dividing surface and, again, constant temperature and total volume have been assumed. Combining Equations 5.17 and 5.19,

$$\gamma dA = -\Gamma \mu dA + f^s dA\ , \tag{5.20}$$

or

$$\gamma = f^s - \Gamma \mu. \tag{5.21}$$

This relation states, in effect, that the surface tension, γ, is the surface density of (F – G). Note that both f^s and Γ will depend on the location of the dividing surface. However, since γ is independent of this choice, which has been shown to be an artifact of the model, the changes in f^s and Γ associated with changing the location of the dividing surface must be equivalent.

A common convention in working with one component systems is to choose $\Gamma = 0$, that is, to locate the dividing surface so that $\Gamma = 0$, so that

$$\gamma = f^s . \tag{5.22}$$

Having made this choice, if one separates f^s into

$$f^s = e^s - Ts^s, \tag{5.23}$$

in which

$$s^S = -\left(\frac{\partial f^S}{\partial T}\right)_{V,\Gamma} = -\left(\frac{\partial \gamma}{\partial T}\right)_V, \tag{5.24}$$

(the latter substitution is valid only because we have chosen $\Gamma = 0$), one has

$$e^S = \gamma - T\left(\frac{\partial \gamma}{\partial T}\right)_V . \tag{5.25}$$

Experimentally, $(\partial\gamma/\partial T)$ is found to be negative in all cases. This is as one would expect, as γ must approach zero at the critical point, as mentioned in Chapter 4. Because $(\partial\gamma/\partial T)$ is negative, s^s must, in all cases, be positive.

Multicomponent Systems

The Gibbs model may also be applied to multicomponent systems. For this case, Equation 5.21 becomes

$$\gamma = f^s - \sum_i \Gamma_i \mu_i . \tag{5.26}$$

One cannot now, in general, get rid of the $\Gamma_i\mu_i$ terms, as the choice of dividing surface location that gives $\Gamma_1 = 0$ for one component will not in general give $\Gamma_{i\neq 1} = 0$ for the other components.

Consider the question of the surface excess concentrations in terms of the experimentally observed phenomenon of adsorption. One can do this by considering the change in F for the whole system associated with a small displacement of a system initially at equilibrium. For such a process, in general, for a closed system

$$dF = -SdT - pdV \tag{5.27}$$

$$= dF^\alpha + dF^\beta + Adf^s \tag{5.28}$$

$$= dF^\alpha + dF^\beta + Ad\gamma + A\sum_i \mu_i d\Gamma_i + A\sum_i \Gamma_i d\mu_i . \tag{5.29}$$

Although the system as a whole is closed, material can move from one bulk phase to another or from one bulk phase to the interface. That is,

$$dF^{\alpha} = -S^{\alpha}\,dT - p^{\alpha}\,dV^{\alpha} + \sum_i \mu_i^{\alpha} dn_i^{\alpha} \qquad 5.30$$

$$dF^{\beta} = -S^{\beta}\,dT - p^{\beta}\,dV^{\beta} + \sum_i \mu_i^{\beta} dn_i^{\beta}. \qquad 5.31$$

Substituting these values for dF^{α} and dF^{β} into Equation 5.29, assuming that the interface is planar so that $p^{\alpha} = p^{\beta}$, and recalling that

$$S = S^{\alpha} + S^{\beta} + As^{s} \qquad 5.32$$

$$V = V^{\alpha} + V^{\beta} \qquad (\text{so that } dV^{\alpha} = -dV^{\beta}), \qquad 5.33$$

leads to

$$d_{\gamma} = -s^{s}\,dT - \sum_i \Gamma_i d\mu_i. \qquad 5.34$$

This is essentially a Gibbs–Duhem equation for the material associated with the dividing surface.

In the case of constant temperature, this yields

$$d_{\gamma} = -\sum_i \Gamma_i d\mu_i \qquad 5.35$$

or

$$\Gamma_i = -\left(\frac{\partial \gamma}{\partial \mu_i}\right)_{T,n_j,\mu_{j \neq i}}. \qquad 5.36$$

This relation is known as the Gibbs adsorption isotherm, and applies to any case of adsorptive equilibrium in a system at constant temperature and overall composition.

Consider next the form that the Gibbs adsorption isotherm takes in two common cases. Consider first the case of a solid–gas interface in a two-component system, as shown schematically in Figure 5.3. In this case it is convenient to locate the dividing surface right at the density discontinuity, so that $\Gamma_1 = 0$. This yields

$$d\gamma = -\Gamma_2 d\mu_2 \qquad 5.37$$

or

$$\Gamma_2 = -\left(\frac{\partial \gamma}{\partial \mu_2}\right)_{T,n_1,n_2}. \qquad 5.38$$

If the vapor phase is ideal, then

$$\mu_2 = \mu_2^{\circ} + RT \ln p_2, \qquad 5.39$$

and

$$\Gamma_2 = -\left(\frac{1}{RT}\right)\left(\frac{\partial \gamma}{\partial \ln p_2}\right)_T. \qquad 5.40$$

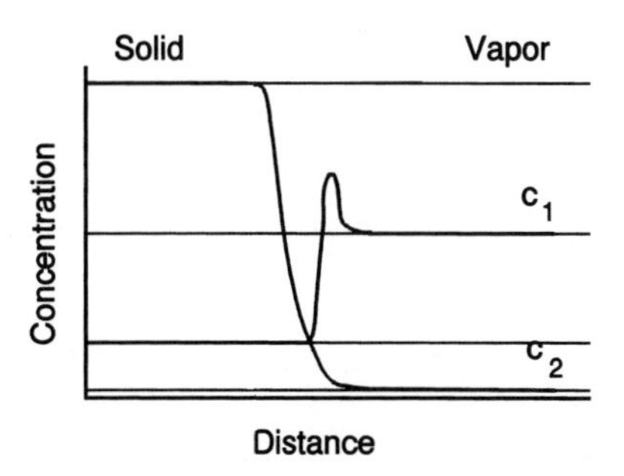

Figure 5.3 Possible concentration profile across a solid–gas interface, showing some solubility of the gas in the solid and an accumulation of gas in the interface region.

Note the consequences of this relation—in cases in which the gas species causes a large decrease in the surface tension of the solid, extensive adsorption will occur and vice versa.

The adsorption defined above can be measured experimentally in a wide range of systems, for example by carrying out a zero-creep experiment in an atmosphere of the gas of interest, or by doing a pendant drop experiment for the case of a liquid. The result of a typical experiment, for oxygen adsorption on solid silver, is shown in Figure 5.4. Note that Equation 5.40 is an equilibrium expression and applies only when there is no kinetic barrier to the attainment of equilibrium. This subject will be considered in more detail in Chapter 12.

A second case of interest is that of the segregation of one component of a two-component condensed phase system to the surface. In considering this case, assume a dilute solution of one component (component 2) in the bulk of a condensed phase composed primarily of the other component. In this case,

$$\mu_2 = \mu_2^\circ + RT \ln \gamma' X_2, \tag{5.41}$$

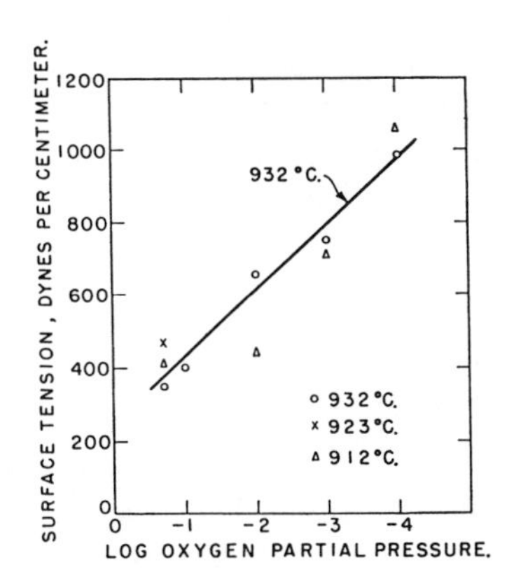

Figure 5.4 Variation of the surface tension of silver with oxygen partial pressure. (Reprinted with permission from F.H. Buttner, E.R. Funk, and H. Udin, *J. Phys. Chem.* 1952;56:657. Copyright 1952, American Chemical Society.)

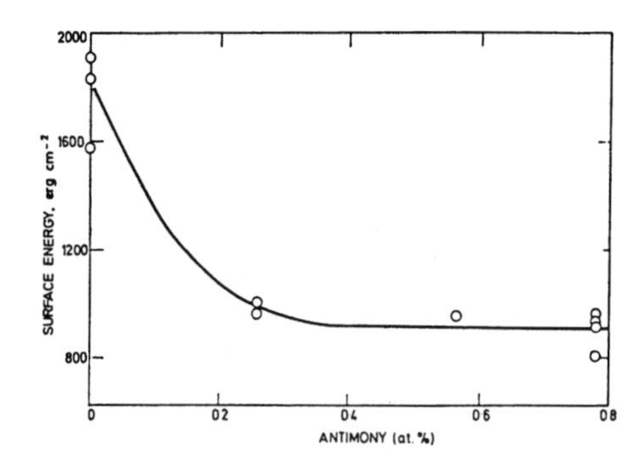

Figure 5.5 Variation of the surface tension of copper with the bulk concentration of dissolved antimony. (Source: E.D. Hondros and D. McLean, in: *Surface Phenomena of Metals:* Soc. Chem. Ind. Monograph #28. London: Soc. Chem. Ind., 1968, p. 39. Reproduced by permission of the SCI.)

in which X_2 is the mole fraction of component 2 in the system and γ' is the activity coefficient of component 2, which at low concentrations will be a constant. This leads to

$$\Gamma_2 = -\left(\frac{1}{RT}\right)\left(\frac{\partial \gamma}{\partial \ln X_2}\right). \tag{5.42}$$

Again, measurement of the surface tension as a function of the bulk-phase concentration of component 2 shows the extent of the adsorption process. Figure 5.5 shows the results of one such measurement, for the case of antimony in copper.

Statistical Thermodynamics of Adsorption

The methods of statistical thermodynamics can also be used to treat the processes of adsorption and surface segregation. Two such cases will be examined, namely, a simple model for adsorption from a gas phase known as the Langmuir isotherm and a model of segregation at the surface of a two-component solid phase. In both cases, the techniques of statistical thermodynamics will be applied to a model of the system in question that describes the various contributions to the system energy and entropy. The results of this approach parallel the classical treatment developed above.

The Langmuir Isotherm

This case can be treated using a model known as the ideal lattice gas. In this model, a number, N, of particles is assumed to be bound to a number, M, of adsorption sites, with the constraint $N \leq M$. Each adsorbed particle is assumed to be bound to the surface with an energy E_{00} (an inherently negative number), and to have three independent, orthogonal degrees of vibrational freedom.

The canonical ensemble partition function that is appropriate to this model is

$$Q = q^N \left[\frac{M!}{N!(M-N)!} \right]. \tag{5.43}$$

The term q is essentially a single particle partition function describing the sum over all of the allowed energy states of an adsorbed particle and is given by

$$q = \exp\left(\frac{-E_{00}}{kT}\right) \prod_i \exp\left(\frac{-h\nu_i}{2kT}\right) \prod_i \sum_n \exp\left(\frac{-nh\nu_i}{kT}\right), \tag{5.44}$$

in which the product is over the three vibrational modes and $n = 1, 2, 3, \ldots$ The factorial term is simply the number of ways of distributing the N particles over the M adsorption sites with no more than one particle per site.

The thermodynamic functions of this adsorbed phase may be obtained using the relation

$$F = -kT \ln Q. \tag{5.45}$$

Using Stirling's approximation to the factorial terms,

$$\ln N! = N \ln N - N, \tag{5.46}$$

one obtains

$$F = -kT[N \ln q + M \ln M - N \ln N - (M - N) \ln (M - N)], \tag{5.47}$$

and from this, using

$$\mu = \left(\frac{\partial F}{\partial N}\right)_{M,T}, \tag{5.48}$$

one has

$$\mu = kT\, \partial/\partial N[N \ln q - N \ln N - (M - N) \ln (M - N)] \tag{5.49}$$

or

$$\begin{aligned} \mu &= kT \ln\left[\frac{N}{(M-N)}\left(\frac{1}{q}\right)\right] \\ &= kT \ln\left[\frac{\frac{N}{M}}{\left(\frac{M}{M} - \frac{N}{M}\right)}\left(\frac{1}{q}\right)\right], \end{aligned} \tag{5.50}$$

or, if one defines the fractional occupancy of available adsorption sites as $\theta = N/M$ (the equivalent of Γ in the classical treatment), one has

$$\mu = kT \ln\left[\left(\frac{\theta}{1-\theta}\right)\left(\frac{1}{q}\right)\right]. \tag{5.51}$$

One may now determine the equilibrium amount of adsorption when the array of sites (the surface in question) is in contact with a gas phase at pressure, p, using the expression stated earlier for μ in a gas phase,

$$\mu = \mu^\circ + kT \ln p \,. \qquad 5.52$$

When the system is in equilibrium with respect to adsorption,

$$\mu_{ads} = \mu_{gas} \qquad 5.53$$

$$kT \ln\left[\left(\frac{\theta}{1-\theta}\right)\left(\frac{1}{q}\right)\right] = \mu^\circ + kT \ln p, \qquad 5.54$$

or

$$\frac{\theta}{1-\theta} = pq \exp\left(\frac{-\mu^\circ}{kT}\right). \qquad 5.55$$

or, defining

$$\chi_T = q \exp\left(\frac{-\mu^\circ}{kT}\right), \qquad 5.56$$

one has

$$\frac{\theta}{(1-\theta)} = \chi_T p, \qquad 5.57$$

or

$$\theta = \frac{\chi_T p}{1+\chi_T p} \qquad 5.58$$

as the desired equilibrium relation between the adlayer coverage, θ, and the gas phase pressure, p. The form of this relation is shown in Figure 5.6. Note that at low pressure, $\chi_T << 1$, $\theta = \chi_T p$. At high pressure, $\chi_T >> 1$, and $\theta \to 1$.

Chapter 10 will show that one can derive an equivalent set of equations on a purely kinetic basis. In addition, although the model chosen is very simple, it provides a good approximation to the behavior of many real systems.

Statistical Thermodynamic Treatment of Segregation

Alternatively, one may look at equilibrium segregation at a solid surface on a statistical thermodynamic basis. The argument in this case is similar to that used in Chapter 3 in the initial discussion of the concepts of surface tension and surface stress in solids. Again, a model will be used based on a nearest-neighbor quasi-chemical treat-

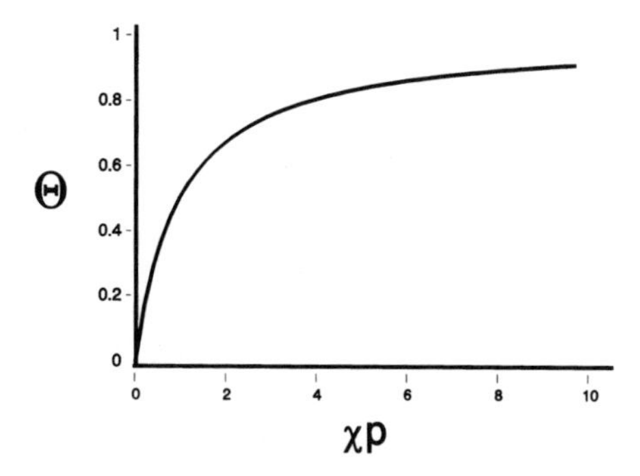

Figure 5.6 The Langmuir adsorption isotherm.

ment of the bonding in a crystalline solid phase, and the changes that take place when bonds are broken or rearranged to form a surface will be considered.

In this case, the argument will be presented on a somewhat more sophisticated basis than was done in Chapter 3. The model used will be the so-called *regular solution model,* for the case of a two-component system, with

$$\Delta\mathcal{H}_{mix} = \omega X_1 X_2 N_{Av} \tag{5.59}$$

$$\Delta\mathcal{S}_{mix} = -R(X_1 \ln X_1 + X_2 \ln X_2). \tag{5.60}$$

Here $\Delta\mathcal{H}_{\mathrm{mix}}$ and $\Delta\mathcal{S}_{\mathrm{mix}}$ are the molar enthalpy and molar entropy of mixing, ω is a constant describing the energetic interactions between the two components, N_{Av} is Avogadro's number, and X_1 and X_2 are the mole fractions of the two components. Assuming that the energy of the system can be described in terms of the energies of nearest neighbor bonds, one may write

$$\omega = \left(\frac{z}{2}\right)\Delta\varepsilon,$$

$$\Delta\epsilon = 2\epsilon_{12} - (\epsilon_{11} + \epsilon_{22}). \tag{5.61}$$

Here z is the number of nearest neighbor atoms in the crystal structure and ϵ_{11}, ϵ_{22}, and ϵ_{12} are the energies of 1-1, 2-2, and 1-2 bonds, respectively.

Recall now that from the preceding classical thermodynamic treatment

$$A\gamma = F^s - \sum_i n_i^s \mu_i. \tag{5.62}$$

One may also write that

$$\sum_i n_i^s = \sum_i n_i^t \mu_i - \sum_i \left(n_i^\alpha \mu_i + n_i^\beta \mu_i\right), \tag{5.63}$$

(where the superscript t refers to the total system) or, since

$$\sum_i n_i^{\alpha} \mu_i = G^{\alpha},$$
$$\sum_i n_i^{\beta} \mu_i = G^{\beta}, \qquad 5.64$$
$$\sum_i n_i^{t} \mu_i = G^{t},$$

and recalling that

$$F^s = F^t - F^{\alpha} - F^{\beta}, \qquad 5.65$$

one has

$$A\gamma = (F^t - G^t) - [(F^{\alpha} - G^{\alpha}) + (F^{\beta} - G^{\beta})] \qquad 5.66$$

$$A\gamma = \Omega^t + pV^{\alpha} + pV^{\beta} \qquad 5.67$$

$$A\gamma = \Omega^t + pV, \qquad 5.68$$

where $\Omega^t \equiv F^t - G^t$.

One can relate this expression to the results of statistical thermodynamics through the equation

$$\Omega^t = -kT \ln \Xi, \qquad 5.69$$

in which Ξ is the grand canonical ensemble partition function and is a function of μ, V, T. This expression is equivalent to the standard statistical thermodynamic expression

$$pV = kT \ln \Xi, \qquad 5.70$$

which was developed assuming that the only work term involved was pdV. Making this substitution, and using Equations 5.67 and 5.68,

$$A\gamma = -kT \ln \Xi + pV^{\alpha} + pV^{\beta} \qquad 5.71$$

or, recognizing that

$$pV^{\alpha} = kT \ln \Xi^{\alpha}, \quad pV^{\beta} = kT \ln \Xi^{\beta}, \qquad 5.72$$

one has

$$A\gamma = -kT \ln \left[\frac{\Xi}{\left(\Xi^{\alpha} . \Xi^{\beta}\right)} \right]. \qquad 5.73$$

In order to determine the value of γ from this statistical thermodynamic relation, all that remains now is to evaluate $\Xi(\mu, V, T,)$. The general form of Ξ, for a two-component, open system, is

$$\Xi = \sum_{N_1, N_2, j} \exp\left(\frac{N_1 \mu_1}{kT}\right) \exp\left(\frac{N_2 \mu_2}{kT}\right) \exp\left[\frac{-E_j(N_1, N_2, V)}{kT}\right], \qquad 5.74$$

where the E_j are the quantum-mechanically allowed values of the overall energy of the system. In order to evaluate Equation 5.74, one must assume a model of the system and perform the indicated summing operation over all of the allowed E_j. To do this, assume the following model:

1. The crystal consists of a set of atomic planes parallel to the crystal surface.
2. $X_1(r)$ and $X_2(r)$ are the atom fractions of components 1 and 2 in the rth plane of the crystal ($r = 1$ for the plane closest to the surface).
3. The density in the vapor phase above the topmost plane is effectively zero.
4. $X_1(\ell)$ and $X_2(\ell)$ are the mole fractions of components 1 and 2 far from the surface.
5. The total number of atoms is fixed. That is,

$$N_1 = nA\sum_{r=1}^{\ell} X_1(r), N_2 = nA\sum_{r=1}^{\ell} X_2(r), \qquad 5.75$$

   where n is the number of atom sites per unit area in the crystal planes.

6. Assume that one is dealing with the (111) face of a face centered cubic crystal. Thus, $z = 12$, and each atom has six in-plane neighbors, three in the plane above and three in the plane below.
7. The site occupancy within each plane is random.

In order to evaluate any term in the summation leading to Ξ, one must have a means of calculating E_j, the total energy associated with that term. For any arrangement of the two types of atoms within the layers, characterized by a set of values of $X_1(r), X_2(r)$,

$$E_j(N_1, N_2, V) = E_j[X_1(r), X_2(r)] = (M_{11}\epsilon_{11} + M_{22}\epsilon_{22} + M_{12}\epsilon_{12}), \qquad 5.76$$

where M_{11} is the number of 1–1 bonds and is given by

$$M_{11} = 3nA\sum_{r=1}^{\ell} X_1(r)\chi_1(r), \qquad 5.77$$

where

$$\chi_1(r) = X_1(r) + \tfrac{1}{2}\,[X_1(r+1) + X_1(r-1)]. \qquad 5.78$$

Similar expressions obtain for M_{22} and M_{12}.

The expressions for the E_j set up in Equation 5.76 must now be plugged into equation 5.74 and the sum evaluated. This evaluation involves the use of the maximum term method, a mathematical device that says, in effect, that in a sum of this type, in which N_1 and N_2 are very large numbers, one term of the sum, the maximum term, will be much larger than all of the rest of the terms put together. In the present case one must find the maximum in the sum Equation 5.74 subject to the condition that μ_1 and μ_2 are uniform throughout the solution. The result of such a calculation is

$$(\mu_1-\mu_2)+6(\varepsilon_{11}-\varepsilon_{12})=3[\chi_2(r)-\chi_1(r)]\Delta\varepsilon+kT\ln\left[\frac{X_1(r)}{X_2(r)}\right],\ \text{for } r=1,2,3\ldots\ell. \quad 5.79$$

That is, there will be a set of equations of the form of Equation 5.79, with one equation for each value of r. Solution of this set of equations defines the values of $X_1(r)$, $X_2(r)$ for all r. This set of $X_1(r)$, $X_2(r)$, inserted into Equation 5.76, using equations of the form of Equation 5.77, define the value of $E_j(N_1,N_2,V)$ associated with the maximum term in the sum of Equation 5.74. This maximum term, in turn, defines Ξ. The result of this calculation is

$$-kT\ln\Xi = nA\Sigma[\Phi(r)-X_1(r)\mu_1-X_2(r)\mu_2], \quad 5.80$$

where

$$\Phi(1) = -6[X_1(1)\epsilon_{11}+X_2(1)\epsilon_{22}-\{\tfrac{1}{2}X_1(1)X_2(1)+\tfrac{1}{8}[X_1(1)X_2(2)+X_1(2)X_2(1)]\}\Delta\epsilon-\tfrac{1}{4}[X_1(1)\epsilon_{11}+X_2(1)\epsilon_{22})]+kT\{X_1(1)\ln[X_1(1)]+X_2(1)\ln[X_2(1)]\}\ \text{for } r=1 \quad 5.81$$

and

$$\Phi(r\neq 1) = -6\ \{X_1(r)\epsilon_{11}+X_2(r)\epsilon_{22}-\tfrac{1}{2}[X_1(r)\chi_2(r)+X_2(r)\ \chi_1(r)]\Delta\epsilon\}+kT\ \{X_1(r)\ \ln[X_1(r)]+X_2(r)\ \ln[X_2(r)]\}\quad \text{for } r=2,3\ldots\ell. \quad 5.82$$

Note that these expressions for $\Phi(r)$ are functions of the system properties $X_1(r)$, $X_2(r)$, ϵ_{11}, ϵ_{22} and ϵ_{12} only.

Now that Ξ is evaluated, the only additional information needed to evaluate the surface tension are the values of Ξ^α and Ξ^β. Using Equation 5.80, with the conditions

$$X_1(r)=X_1\ (\ell)\ ,\ X_2(r)=X_2(\ell) \quad 5.83$$

for all layers, one obtains

$$-\ kT\ln\Xi^\alpha = nA\ell[\Phi(\ell)-X_1(\ell)\mu_1-X_2(\ell)\mu_2] \quad 5.84$$

in which

$$\Phi(\ell) = -6\{X_1(\ell)\epsilon_{11}+X_2(\ell)\epsilon_{22}-[X_1(\ell)X_2(\ell)]\Delta\epsilon\}+kT\{X_1(\ell)\ln[X_1(\ell)]+X_2(\ell)\ln[X_2(\ell)]\}. \quad 5.85$$

Making use of the assumption that the density in the gas phase is effectively zero, one has

$$\Xi^\beta = 1. \quad 5.86$$

It is now possible to substitute the expressions for Ξ, Ξ^α, and Ξ^β into Equation 5.73 to obtain

$$A\gamma = An\sum_{r=1}^{\ell}[\Phi(r)-\Phi(\ell)]-n\mu_1\sum_{r=1}^{\ell}[X_1(r)-X_1(\ell)]-n\mu_2\sum_{r=1}^{\ell}[X_2(r)-X_2(\ell)] \quad 5.87$$

or

$$\gamma = n\sum_{r=1}^{\ell}[\Phi(r) - \Phi(\ell)] - \mu_1\Gamma_i - \mu_2\Gamma_2. \qquad 5.88$$

Given this result, all that is required in order to evaluate γ and the Γ_i for each layer are the values of ϵ_{11}, ϵ_{22}, and ϵ_{12}.

A typical result of such a calculation is shown in Figure 5.7. As is always the case with a nearest neighbor model, the drop-off in the surface excess concentration with distance is much too rapid, due to the short range of the interactions assumed in the model. Note, too, that because the total number of atoms per unit area in each plane is constant, any surface excess of component 1 must be balanced by a corresponding deficit in component 2.

One could apply the foregoing result equally well to the case of a pure material simply by using

$$X_1(r) = 1, \quad X_2(r) = 0 \quad r = 1,2,3,\ldots \ell. \qquad 5.89$$

The result of such a calculation is

$$\gamma = 3/2\; n\, \epsilon_{11}. \qquad 5.90$$

This is simply the number of bonds broken per unit area to form the (111) surface. Alternatively, one could consider an ideal solution, for which $\Delta\epsilon = 0$. This solution will have the same values of $X_1(r)$ and $X_2(r)$ for all layers, which leads to

$$\gamma = X_1\gamma_1 + X_2\gamma_2, \qquad 5.91$$

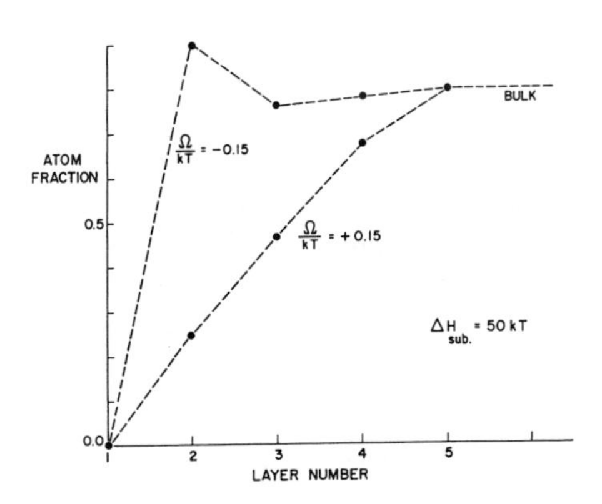

Figure 5.7 Calculated atomic fraction *vs*. depth (layer number) for one component of a substitutional binary alloy, treated as a regular solution. The component segregated to the surface is the one with the lower heat of sublimation. Positive values of Ω, the regular solution parameter, corresponds to alloys that tend to form stable solid solutions over a wide composition range.(Reprinted with permission from J.M. Blakely in: R. Vanselow, ed. *Chemistry and Physics of Solid Surfaces* II. Boca Raton, Florida: CRC Press, 1979, p. 253. Copyright CRC Press, Inc. Boca Raton, Florida.)

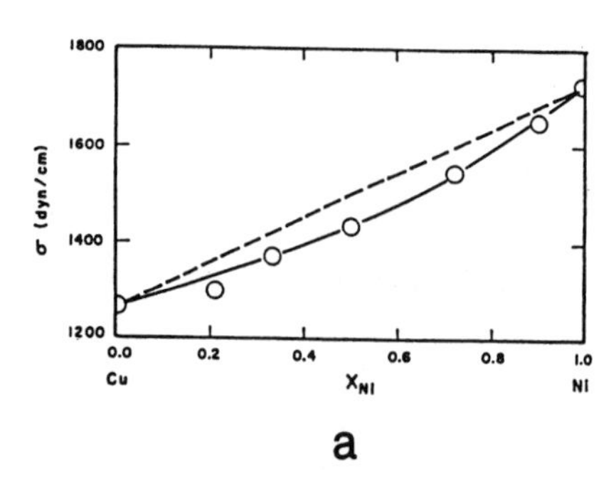

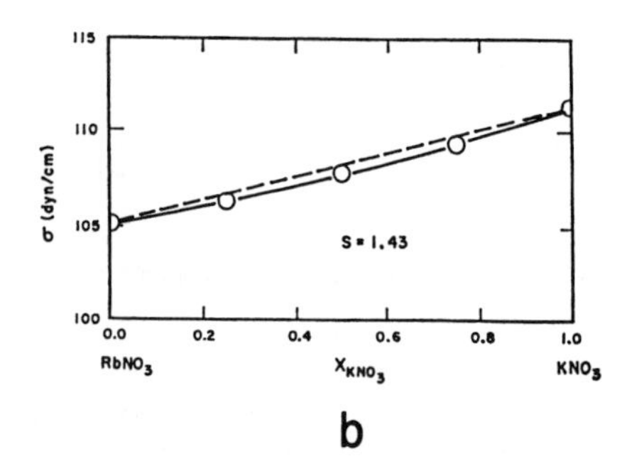

Figure 5.8 Variation of the surface tension of two-component systems as a function of composition. a. Liquid Cu-Ni alloys at 1550°C. b. Liquid mixtures of $RbNO_3$ and KNO_3 at 350°C. (Reprinted with permission from J.G. Eberhart, *J. Phys. Chem.* 1966;70:1183. Copyright 1966, American Chemical Society.)

where γ_1 and γ_2 are the surface tensions of the pure crystalline materials. Typical examples of results obtained for the variation of surface tension with composition are shown in Figure 5.8. The graphs compare the results of the theory just developed with experimental measurements of γ (open circles). These particular systems are not too far removed from ideality, and consequently the assumptions of the model do not lead to large errors. Systems in which the value of $\Delta\epsilon$ is large would be expected to show considerable deviation from the predictions of this simple model.

Problems

5.1. Discuss the Gibbs model and the Young model of the interface. Include in your discussion the necessary conditions for the two models to be consistent with one another and the physical significance of the surface of tension and the Gibbs dividing surface. Identify which parameters of the models are thermodynamic properties defining the state of the system and which are artifacts of the models.

5.2. The adsorption of sulfur on a polycrystalline nickel surface has been measured by preforming a zero-creep measurement (the Udin–Shaler–Wulff experiment) in an atmosphere of sulfur vapor at various sulfur pressures at a temperature of 1500 K. Calculate the adsorption of sulfur at $p = 400$ Torr sulfur pressure and $T = 1500$ K, using the data given below.

p (Torr)	20	50	100	200	400	800	1000
γ (dyn/cm)	1711	1637	1571	1500	1407	1295	1142

Approximately what fraction of the surface will be covered with sulfur?

5.3. The adsorption of xenon on liquid mercury has been measured using the pendant drop technique to measure surface tension in an atmosphere of xenon. The resulting data for surface tension lowering as a function of gas-phase xenon pressure at $T =$ 300 K were obtained:

p (Torr)	69	93	146	227	278
$(\gamma_0 - \gamma)$(dyn/cm)	0.8	1.10	1.75	2.75	3.3

Calculate the adsorption of xenon in atoms per square centimeter at p = 227 Torr. Roughly what fraction of the surface will be covered with xenon atoms?

5.4. A drop of water is placed on top of a clean metal plate in an inert atmosphere at 300 K. The surface energy of the clean metal is 500 erg/cm^2; the interfacial energy between the water and the metal surface is 435 erg/cm^2. Water vapor adsorbs on the metal surface according to the Langmuir isotherm, with $\chi_T = 10\ \text{Torr}^{-1}$ at 300 K.

a. What is the contact angle between the water drop and the metal plate when the drop is first deposited, before water adsorption takes place on the plate?

b. If the system comes to equilibrium with respect to adsorption, what is the final equilibrium contact angle?

5.5. Consider a system in which argon is adsorbed on carbon black to a concentration of 5×10^{14} atom/cm^2. The number of adsorption sites is 2×10^{15} sites/cm^2. If the energy of adsorption is –3000 cal/mol, the vibrational frequency of the adsorbed atoms is 10^{12} sec^{-1} for all three modes, and the temperature is 100 K, what is the equilibrium pressure in the gas phase over the adsorbed layer?

Bibliography

Adamson, AW. *Physical Chemistry of Surfaces,* 2nd ed. New York: John Wiley and Sons, 1967, Chapter 2.

Blakely, JM. *Introduction to the Properties of Crystal Surfaces.* New York: Pergamon Press, 1973, Chapter 2.

Blakely, JM, and Shelton, JC. Equilibrium adsorption and segregation. In: J.M. Blakely, ed. *Surface Physics of Metals.* New York: Academic Press, 1975:1:189–240.

Defay, R, Prigogine, I, Bellemans, A, and Everett, DH. *Surface Tension and Adsorption.* London: Longmans, Green & Co., 1966, Chapters 2 and 7.

Gibbs, JW. *The Collected Works of J. Willard Gibbs.* New York: Longmans, 1928.

Guggenheim, EA. *Thermodynamics.* Amsterdam: North Holland, 1959.

Hill, TL. *Introduction to Statistical Thermodynamics.* New York: Addison-Wesley, 1960.

Hondros, ED, and McLean, DL. Surface energies of solid metal alloys. In: *Surface Phenomena of Metals,* Monograph No. 28. London: Society of Chemical Industry, 1968.

Isett, LC, and Blakely JM. Segregation isosteres for carbon at the (100) surface of nickel. *Surface Sci.* 1976;58:397–414.

Leamy, HJ, Gilmer, GH, and Jackson, KA. Statistical thermodynamics of clean surfaces. In: JM. Blakely, ed. *Surface Physics of Metals.* New York: Academic Press, 1975;1:121–188.

CHAPTER 6

Surface Mobility

This chapter will consider for the first time kinetic processes at crystal surfaces. In particular, it will consider the rate of mass transport over the surface, or surface diffusion, beginning with a phenomenological description of surface diffusion, then considering the energetics of atom movements at a surface. The chapter will then address the measurement of surface mass transport, describing both atomic scale and macroscopic techniques.

Phenomenological Description

Look first at the consequences of surface atom mobility from a phenomenological viewpoint, to determine general relations that describe the rate of surface migration, or surface diffusion. Here the concern is with describing the overall rate of mass transport, without worrying about the transport mechanism at the atomic level.

For the case of surface diffusion, one can set up expressions that account for the effect of the presence of an interface on the overall rate of diffusive mass transport in a way similar to that which was used to account for the surface excess thermodynamic properties. Consider the total diffusive flux, J, in one direction, through a slab of thickness L, as shown in Figure 6.1. In this case, one can write, for Fick's first law

$$J = -\left(\frac{dc}{dx}\right)\int_0^L D(y)dy. \tag{6.1}$$

If the flux were uniform from zero to L, or in other words if the diffusion coefficient were constant and equal to the bulk diffusivity, D_B, one would revert to the more familiar form of Fick's first law

$$J_B = -D_B\left(\frac{dc}{dx}\right)L. \tag{6.2}$$

On this basis, the surface excess flux may be defined as

$$J_s = -½(J - J_B)\ , \tag{6.3}$$

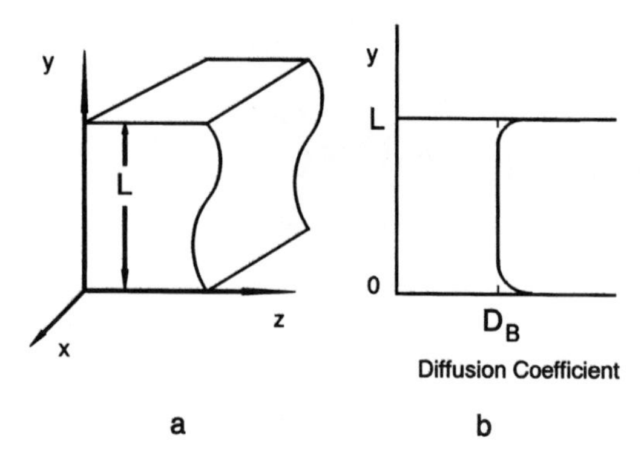

Figure 6.1 Possible variation of the atomic self-diffusivity with position across a slab of crystal.

where the factor ½ arises from the fact that the slab has two surfaces. Combining Equations 6.1 to 6.3, one may also write that

$$J_s = -\frac{1}{2}\left(\frac{dc}{dx}\right)\int_0^L \left[D(y) - D_B\right]dy \,. \qquad 6.4$$

In evaluating this equation, one usually assumes that although $D(y)$ can vary continuously with y over the thickness L, that the diffusivity can be defined by a constant value, D_s, for some fixed depth, δ (usually one atomic plane) at each surface and by the bulk value, D_B, for the balance of the slab thickness. Note that, for a slab of macroscopic thickness, δ will be very much smaller than L. Consequently, any effects associated with a value of $D_s \geq D_B$ will be observed only for cases in which D_s is very much larger than D_B.

Energetics of Atom Migration

Chapter 1 considered the atomic structure of surfaces and described the binding energy associated with atoms in various configurations in terms of a nearest neighbor bonding model. It was found that one could describe the various equilibrium configurations such as adatom, kink atom, and ledge atom, in terms of some equilibrium number of nearest neighbor bonds. Considering the path that an atom must follow in moving from one equilibrium position to another, it can be seen that there is always an intermediate position in which the number of nearest neighbor bonds is fewer than that in the equilibrium position.

As an example, consider the migration of an adatom on a BCC (100) surface, as shown in Figure 6.2. At equilibrium, this adatom has four nearest neighbors in the plane below it. In order to move from one equilibrium site to an adjacent equilibrium site, the atom must pass through a configuration such as the one shown dotted in the figure. In this position, the atom has only two nearest neighbors. The migration process will thus require an activation energy of roughly twice the nearest neighbor bond

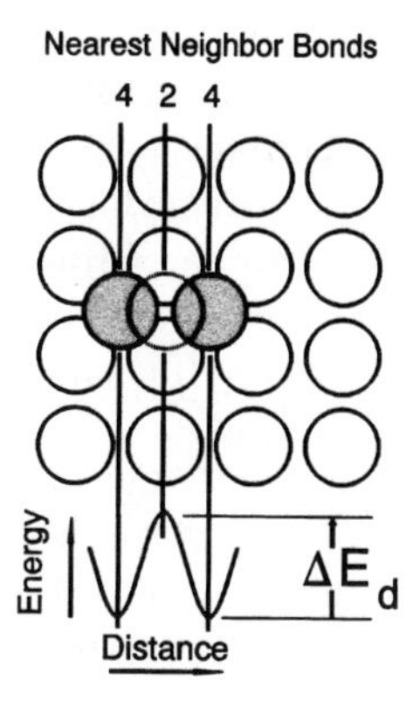

Figure 6.2 Schematic representation of the diffusion of an adatom on a BCC (100) surface, showing the system potential energy variation as diffusion takes place.

energy. The variation of system energy with adatom position, as the adatom moves from one equilibrium site to the next, is shown schematically below the model. Here ΔE_d is the activation energy for surface diffusion by adatom migration.

This example can be generalized to migration of the adatom along other paths, for example, directly over an atom in the surface, and will lead to a variety of values for ΔE_d, depending on the path. Similarly, one could go through the same kind of argument for other surface point defects, such as surface vacancies, ledge adatoms, or ledge vacancies. The general conclusion would be that an activation energy is always required for atom motion and that the energy required is always a fraction of the heat of sublimation of the crystal. The details of the variation of the binding energy of an adatom as a function of position on the surface will be considered in greater detail in Chapter 9, gas scattering.

Atomistic Mechanisms

Consider the mechanism of surface diffusion in terms of the energetic concepts discussed above in order to develop ways of calculating or measuring the surface diffusivity, D_s. Just as diffusion usually proceeds by a defect mechanism in the bulk of a crystal, it appears that in most cases defects are responsible for the observed diffusive transport at surfaces, as indicated in the previous section.

The most common potentially mobile defects are surface adatoms and surface vacancies and, conceivably, diadatoms and divacancies. The overall diffusion coefficient will involve contributions from all possible defects. That is,

$$D_s = \sum_i \left(\frac{n_i}{M}\right) D_i, \qquad 6.5$$

in which n_i is the concentration of the ith type of defect and D_i is the mobility of this type of defect.

At this point an assumption must be made concerning the concentrations of the various defects, namely, whether or not the steady state concentration is equal to the equilibrium concentration. The answer to this question will depend on the experimental situation. In the case of self-diffusion, where the moving species and the bulk lattice are the same chemical species, and where the diffusion process is being measured in some sort of mass transport process on a clean surface, the defect concentration is probably in equilibrium. For the case of heterodiffusion, where one is measuring the diffusion of one chemical species, such as an adsorbed atom, over the surface of a second chemical species, the surface defect concentrations are most likely not in equilibrium.

Fortunately, there are some cases in which the question of defect concentration can be separated from that of the inherent mobility. Looking at this situation conceptually, first, for the case of adatom mobility, assume that the adatom moves over an otherwise perfect terrace by a "random walk" process. The mean distance, $\overline{R}$, that it will travel in a time t is given by

$$\overline{R^2} = 2D_a t, \tag{6.6}$$

where D_a is the intrinsic adatom diffusivity and is given by

$$D_a = \frac{1}{4}\left(\frac{\ell^2}{\tau}\right). \tag{6.7}$$

In this equation ℓ is the adatom jump distance, a distance on the order of the interatomic spacing in the surface, and τ is the mean time between jumps.

One can develop theoretical expressions for τ using absolute reaction rate theory. This treatment leads to two different results, depending on the magnitude of E_d compared to kT. If the barrier to motion is large compared to kT, the adatoms will be localized at specific sites, and

$$\frac{1}{\tau} = 2\nu_x \exp\left(\frac{-\Delta E_d}{kT}\right), \tag{6.8}$$

in which ν_x is essentially an attempt frequency for surface diffusive jumps. If the barrier to motion is small compared to kT, adatoms will be free to move over the surface as a two-dimensional gas, and

$$\frac{1}{\tau} = \left(\frac{1}{\ell}\right)\left(\frac{kT}{2\pi m}\right)^{1/2}. \tag{6.9}$$

In this case, ℓ is the surface mean free path of the diffusing species and m is its mass. These two expressions for τ lead to

$$D_a = \frac{1}{2}\ell^2 \nu_x \exp\left(\frac{-\Delta E_d}{kT}\right) \tag{6.10}$$

for the localized case, and

$$D_a = \frac{1}{4}\ell\left(\frac{kT}{2\pi m}\right)^{1/2} \tag{6.11}$$

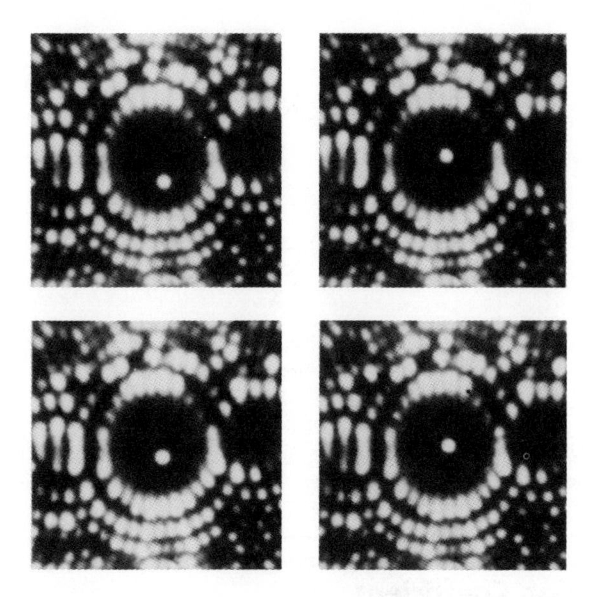

Figure 6.3 Field ion micrographs showing diffusion of rhenium atoms on a W (211) plane at T = 327 K. Field ion images were taken at 60 sec diffusion intervals. (Reprinted with permission from G. Ehrlich, *CRC Crit. Rev. Solid. State and Matls. Sci.* 1982;10:391. Copyright CRC Press, Inc. Boca Raton, FL.)

for the two-dimensional gas case. Experimental measurements of self-diffusion on crystals have in all cases shown behavior consistent with the localized adatom case.

Experimental Measurement of Adatom Diffusivity

The adatom diffusivity described above can be measured in some systems using field ion microscopy. The measurement, while simple in concept, is difficult and tedious in practice and is limited to the surfaces of those refractory metals that make suitable surfaces for field ion microscopy.

The technique involves direct measurement of the mean distance traveled in the random walk of an adatom on a small terrace of a given, low index orientation ($\overline{R^2}$ in Equation 6.6). An appropriate surface is first prepared by field evaporation. A single adatom is then deposited on the central flat terrace at the end of the tip, usually by evaporating a sample of the desired material in the system containing the field emitter tip. The evaporated material may be the same chemical species as the tip or a different species. The tip is then imaged, and the adatom position recorded. The temperature is next raised high enough to permit adatom mobility, then reduced to permit acquisition of a second image. The change in adatom position is recorded to determine the displacement. The results of such an experiment are shown in Figure 6.3, for rhenium on a W (211) surface. Values of $\overline{R^2}$ for the case of self-diffusion on rhodium are plotted against temperature in Figure 6.4.

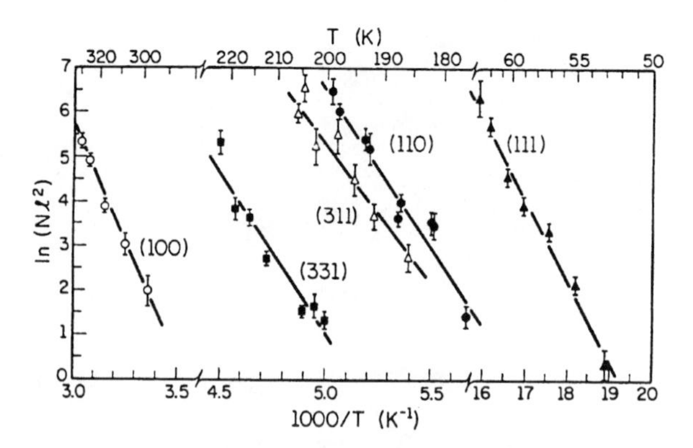

Figure 6.4 Temperature dependence of self-diffusion on rhodium single crystal planes. The diffusion interval is 3 min. N is the number of jumps per interval, ℓ is the jump distance. (Source: G. Ayrault and G. Ehrlich, *J. Chem. Phys.* 1974;60:281. Reprinted with permission.)

Experimental Measurement of the Overall Diffusion Coefficient

Consider next the question of how to measure experimentally the overall surface diffusivity. Assuming that equilibrium concentrations of all of the defects involved in the surface migration process are maintained during the diffusion process and that surface adatoms are the only mobile species, one has

$$D_s = \left(\frac{n_a}{M}\right) D_a, \tag{6.12}$$

or

$$D_s = \frac{1}{2}\ell^2 \nu_x \exp\left(\frac{-\Delta E_d}{kT}\right) \exp\left(\frac{-\Delta G_a}{kT}\right). \tag{6.13}$$

Note that the temperature dependence of D_s includes contributions from both D_a and n_a, so that the overall diffusivity will increase more rapidly with temperature than either the adatom concentration or the adatom mobility.

The most consistently successful technique for measuring D_s involves measurement of the rate of mass transport on a clean surface, preferably in ultrahigh vacuum. A number of configurations have been used to induce long range mass transport and to measure the rates of this process. All of them involve producing a gradient in chemical potential on the surface as a result of a variation in surface curvature.

In one such technique, a sinusoidal profile is formed on an otherwise flat surface, either by scratching or by etching the surface in a regular pattern. Figure 6.5 shows such a sinusoidal profile. A number of mass transport processes can lead to smoothing of this profile, including surface diffusion, bulk diffusion, and evaporation–condensation sequences. In all cases, the driving force for the mass transport is the variation in

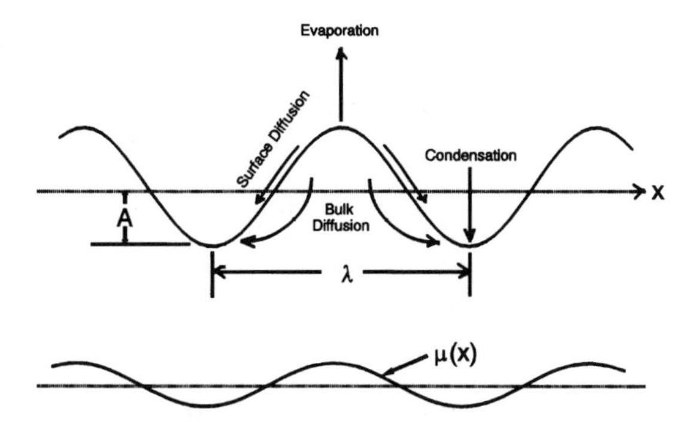

Figure 6.5 Schematic representation of a sinusoidal profile on a surface, showing the various possible mechanisms for smoothing of the profile.

chemical potential with distance over the surface due to the changing curvature over the sinusoidal profile.

Consider for the moment only the case of mass transport by surface diffusion, for by proper choice of experimental conditions, one can minimize interference from competing processes. It was determined in Chapter 4 that the chemical potential of a curved solid surface was related to the chemical potential of the flat surface by

$$(\mu - \mu^\circ) = \mathcal{V}(\gamma + \gamma'')\,K\,, \tag{6.14}$$

where K is the surface curvature, $\mathcal{V}$ the molar volume of the solid and

$$\gamma'' = \left(\frac{\partial^2 \gamma}{\partial \theta^2}\right) \tag{6.15}$$

and arises from the variation in surface tension with crystallographic orientation. The diffusive flux density, j, can be related to the chemical potential gradient at the surface by

$$j = -\left(\frac{D_s C_s}{kT}\right)\left(\frac{\partial \mu}{\partial x}\right) \tag{6.16}$$

$$j = -\left(\frac{D_s C_s}{kT}\right)\left(\frac{\partial}{\partial x}\right)\left[\mu^\circ + (\gamma + \gamma'')\mathcal{V} K\right], \tag{6.17}$$

where C_s is the number of surface atoms per unit area. For a sinusoidal profile one can write

$$K = -\left(\frac{\partial^2 y}{\partial x^2}\right) = \frac{-\partial^2}{\partial x^2}(A \sin \kappa x)$$

$$= A\kappa^2 \sin \kappa x\,, \tag{6.18}$$

where A is the amplitude of the profile,

$$\kappa = \frac{2\pi}{\lambda}, \tag{6.19}$$

and λ is the wavelength of the sinusoidal profile. Thus

$$j = -\left(\frac{D_s C_s}{kT}\right)(\gamma + \gamma'')\mathcal{V} A\kappa^3 \cos \kappa x. \tag{6.20}$$

Rather than measuring the diffusive flux directly, it is more convenient to measure the change in A with time and to determine D_s from this variation. One must thus determine dA/dt, which is related to j by

$$div\, j = -\left(\frac{1}{\mathcal{V}}\right)\left(\frac{dA}{dt}\right)\sin \kappa x, \tag{6.21}$$

where div j, the divergence of the surface flux, is given by

$$div\, j = \frac{\partial j}{\partial x} \tag{6.22}$$

for the one-dimensional case considered here. Thus,

$$-\left(\frac{1}{\mathcal{V}}\right)\left(\frac{dA}{dt}\right)\sin \kappa x = \frac{\partial}{\partial x}\left[-\left(\frac{D_s C_s}{kT}\right)(\gamma + \gamma'')\mathcal{V} A\kappa^3 \cos \kappa x\right] \tag{6.23}$$

$$= \left(\frac{D_s C_s}{kT}\right)(\gamma + \gamma'')\mathcal{V} A\kappa^4 \sin \kappa x. \tag{6.24}$$

Rearranging yields

$$\frac{dA}{dt} = -\left(\frac{D_s C_s}{kT}\right)(\gamma + \gamma'')\mathcal{V}^2 A\kappa^4$$

$$= -B\,\kappa^4 A, \tag{6.25}$$

in which

$$B \equiv \left(\frac{D_s C_s}{kT}\right)(\gamma + \gamma'')\mathcal{V}^2. \tag{6.26}$$

Integration of Equation 6.25 yields

$$A = A^\circ \exp\,(-B\kappa^4 t)\,. \tag{6.27}$$

That is, the amplitude of the profile decays exponentially with a time constant

$$\tau = \frac{1}{B\kappa^4}. \tag{6.28}$$

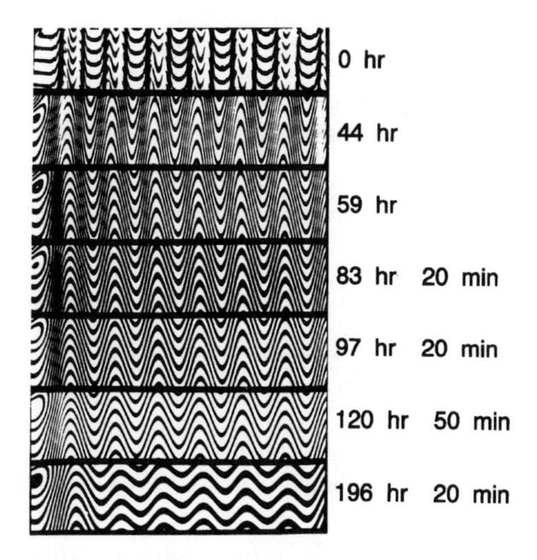

Figure 6.6 Series of interference micrographs showing the flattening of a periodic profile on a nickel surface in vacuum at 1490 K, with increasing annealing times. (Source: P.S. Maiya and J.M. Blakely, *J. Appl. Phys.* 1967;38:698. Reprinted with permission.)

Consequently, if one measures the amplitude vs. time, one can calculate τ and then use Equations 6.25 and 6.26 to calculate D_s.

Similar calculations based on the assumption that other processes, such as bulk diffusion or evaporation–condensation, control dA/dt lead to expressions for A vs. t that have a different functional dependence on the profile wavelength. The general expression, including all possible contributions, is

$$A = A^{\circ} \exp\left[-(A\,\kappa^2 + C\,\kappa^3 + B\,\kappa^4)\,t\right]. \qquad 6.29$$

In this expression, the first term in the exponential is associated with evaporation–condensation, the second with bulk diffusion, and the third with surface diffusion. Note that the shorter the wavelength (that is, the larger the value of κ) the greater the contribution of surface diffusion to the overall flux. As a practical matter, for metals, for wavelengths shorter than about 20 μm, surface diffusion is the predominant mass transport mechanism. Note too that profiles with short wavelengths decay more rapidly. Thus, if the surface profile initially formed contains higher harmonics (e.g., if it is essentially a square wave), the higher harmonics decay early in the smoothing process, leaving the desired sinusoidal profile. Figure 6.6 shows this effect for the case of the (110) face of nickel.

Surface Heterodiffusion

The diffusion of one species over the surface of another is a process that is of considerable interest, especially in the understanding of surface chemical reactivity. This process cannot be studied by the mass transport techniques discussed in the preceding section, as the required assumptions concerning equilibrium among defects can-

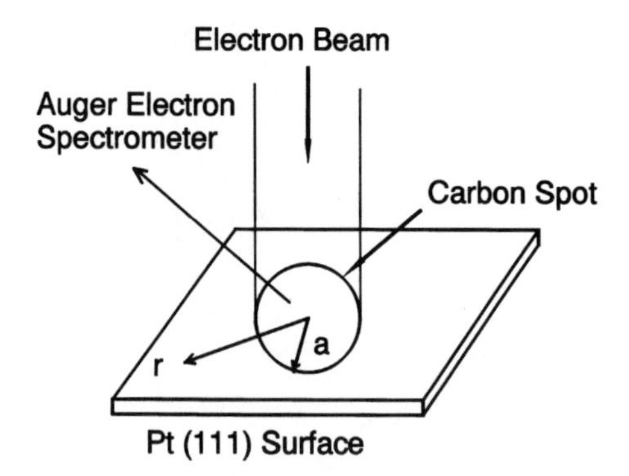

Figure 6.7 Experimental configuration for the measurement of carbon adatom diffusivity on the Pt (111) surface, using Auger electron spectroscopy.

not be made. Some information on intrinsic adatom diffusivity for hererodiffusion has been obtained using random walk measurements in the field ion microscope, as discussed in the section on experimental measurement of adatom diffusivity. Overall diffusivity measurements can also be made using more-or-less classical diffusion measurement techniques. In these studies, a concentration gradient of the diffusing species is formed on an otherwise clean surface, and the evolution of the concentration profile with increasing time at temperature is observed. Until quite recently, few such measurements have been made under conditions of known surface cleanliness. Within the past few years, however, a number of measurements have been made in ultrahigh vacuum using such surface analytical techniques as Auger electron spectroscopy and laser-induced desorption to measure the evolving concentration profile.

The use of these techniques is illustrated in a study of carbon adatom diffusion on Pt (111). Figure 6.7 illustrates the basic experimental setup. In this experiment, electron-bombardment-induced decomposition of CO molecules adsorbed on the sample surface formed the initial carbon spot. The spot formed was roughly 0.2 mm in diameter. The sample was then heated to a temperature at which diffusion was rapid, and Auger electron spectroscopy monitored the rate of migration of carbon atoms away from the spot. (This common surface analytical technique is discussed in detail in Chapter 14. For now, it is sufficient to know that this technique provides a quantitative measure of the atomic concentration of any species in the area sampled by a primary electron beam.)

The diffusion process in this case can be described phenomenologically by Fick's second law:

$$\frac{\partial^2 C}{\partial x^2}+\frac{\partial^2 C}{\partial y^2}=\left(\frac{1}{D}\right)\left(\frac{\partial C}{\partial t}\right) \tag{6.30}$$

For the case of a circular spot of initially uniform composition, the solution to Equation 6.30 is

$$C(r,t)=C_0 a\int_{u=0}^{\infty} J_1(au)\,J_0(ru)\exp\left(-Dtu^2\right)du. \tag{6.31}$$

Here J_1 and J_0 are Bessel functions, a and r are defined in Figure 6.7, and u is a dummy variable used in the integration process.

Experimentally, if one has an Auger spectrometer system with adequate lateral spatial resolution, one can measure the concentration profile and fit the resulting curve to Equation 6.31. Alternatively, one can simply measure the amount of material remaining within the boundary of the initial spot as a function of time. If one solves Equation 6.30 for this amount of remaining material, one obtains

$$Q = 2\pi a^2 C_0 \int_{u=0}^{\infty} \left[\frac{J_1^2(au)}{r} \right] \exp\left(-Dtu^2\right) du. \qquad 6.32$$

The approximate short-time solution of this equation is

$$\frac{Q}{Q_0} = 1 - \left(\frac{2}{a}\right)\left(\frac{Dt}{2\pi}\right)^{1/2}, \qquad 6.33$$

in which

$$Q_0 \equiv \pi a^2 C_0. \qquad 6.34$$

Measurements of this sort over a range of temperature permit calculation of the diffusivity as

$$D = D_0 \exp\left(\frac{-\Delta E_d}{kT}\right). \qquad 6.35$$

Figure 6.8 shows the results of this process for the case of carbon diffusion on Pt (111). Diffusion measurements using techniques of this sort are at present limited to a few

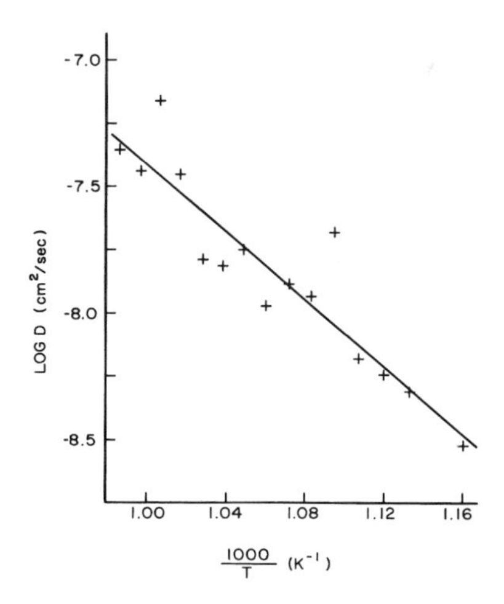

Figure 6.8 Summary of observed diffusivities for carbon adatoms on Pt (111) – log D *vs.* $^1/T$. (Source: M.T. Martin and J.B. Hudson, *J. Vac. Sci. Technol.* 1978;15:474. Reprinted with permission.)

systems. It is an area of increasing interest, as the capabilities of surface composition measurement techniques improve. Chapter 16 gives an example of the use of laser-induced desorption to produce surface concentration profiles and to measure the resulting surface diffusion process.

Problems

6.1. Calculate the adatom and surface vacancy diffusion coefficients for the (111) face of a face-centered cubic crystal, using the nearest neighbor model. Assume that the heat of sublimation of the crystal is 50000 cal/mol, that the interatomic distance is 3×10^{-8} cm, that the attempt frequency, ν, is 10^{12} $\sec^{-1}$, and that the temperature is 800 K. (Hint: Think carefully about the number of bonds that have to be broken in order to move an atom.)

6.2. The surface self-diffusion rate was measured on a metal surface using the profile smoothing technique. The following results were obtained in a measurement made at $T = 800$ K:

t (sec)	0	3	10	25	30	50	100
A (cm $\times 10^{-5}$)	10.00	9.32	7.91	5.55	4.94	3.09	0.95

The profile wavelength was 10^{-4} cm. You may assume that for the material studied that $C_s = 10^{15}$ cm^{-2}, $(\gamma + \gamma'') = 300$ dyn/cm and that the molar volume is 10 cm^3. Calculate the value of D_s.

Bibliography

Blakely, JM. *Introduction to the Properties of Crystal Surfaces*. New York: Pergamon Press, 1973;213–228.

Chen, JR, and Gomer R. Mobility of oxygen on the (110) plane of tungsten. *Surface Sci.* 1979;79:413-444.

Ehrlich, G, and Hudda, F. Atomic view of surface self diffusion: tungsten on tungsten. *J. Chem. Phys.* 1966; 44:1039.

Gjostein, NA. Surface self-diffusion. In: W.D. Robertson and N.A. Gjostein, eds. *Metal Surfaces*. Metals Park, Ohio: American Society for Metals, 1963, Chapter 4.

Gjostein, NA. Surface self diffusion in metals. In: J.J. Burke, N.L. Reed, and V. Weiss, eds. *Surfaces and Interfaces* I. Syracuse, NY: Syracuse University Press, 1967.

Martin, MT, and Hudson, JB. Surface diffusion of carbon on (111) platinum. *J. Vac. Sci. Technol.* 1978;15:474–477.

Naumovets, AG, and Vedula, YS. Surface diffusion of adsorbates. *Surface Sci. Reports* 1985;4:365–434.

PART II

Gas–Surface Interactions

CHAPTER 7

The Kinetic Theory of Gases

Prior to consideration of gas–surface interactions, this chapter and the following will present the kinetic theory of gases and the properties of molecular beams. An understanding of these two subjects is required as a background for understanding both the theoretical and experimental aspects of gas–surface interactions.

First developed by Maxwell, the kinetic theory of gases has been elaborated by many subsequent investigators. This theory explains both the bulk properties of a gas, such as the pressure, molecular impingement rate, and distribution of molecular velocities, and the transport properties, such as viscosity, thermal conductivity, and diffusion, in terms of the dynamics of motion of the individual molecules.

The theory can be developed either in terms of classical mechanics or in terms of quantum statistical mechanics. The classical description is simpler but requires that one make some questionable assumptions which can be shown to be unnecessary by the mathematically more difficult statistical mechanical approach. The classical approach will be used here, as the added rigor of the statistical treatment does not add enough to the physical understanding of gases to warrant the added complexity.

Physical Picture and Assumptions

The physical picture used in the classical approach represents the gas as a collection of a large number of hard, spherical particles, having a small size compared to their separation, in constant rapid motion in random directions. The particles collide with each other and with the walls of the container. The only result of these collisions is to change the momentum and kinetic energy of the particles involved, subject to the constraints imposed by the conservation of momentum and energy. Although the momenta and energies of individual molecules change rapidly with time, the average properties of the collection of particles are time independent if no external process takes place.

The assumptions on which this model is based may be stated formally as

1. Any finite volume of gas contains a large number of molecules.
2. Molecules exert no forces on one another except when they collide. Between collisions with each other or with the container walls they travel in straight lines.

3. Collisions of one molecule with another or with surfaces exposed to the gas are perfectly elastic. All surfaces are considered smooth.
4. In the absence of external forces, the molecules are distributed uniformly throughout the container.
5. All directions of molecular velocities are equally probable.
6. The speed of a given molecule may have any magnitude, but the number of molecules with speeds in a given range is constant with time. (That is, the speed distribution is time invariant).

Now consider the validity of these assumptions in turn:

1. The number of molecules in 1 cm^3 at atmospheric pressure and room temperature is approximately 3×10^{19} and varies directly with pressure. Consequently, except for extremely low pressures, the number of molecules per cubic centimeter will be large, and assumption 1 is valid.
2. The validity of assumption 2 depends on the definition of collision. If one defines a collision as any approach close enough that there is an interaction, then assumption 2 is valid. In practice, this is a good assumption as long as the distance between particles is very large compared to the particle diameter.
3. Assumption 3 is true only for an ideal gas. Basically, it assumes that there are no intermolecular attractive forces in the gas and no attractive forces between a gas molecule and the surface of the container. At low pressures (or equivalently, at low number densities), the effects of intermolecular attractive forces are small enough that they can be ignored in determining the properties of the gas. However, at high densities, or at low temperatures, these forces must be explicitly accounted for, as they lead to condensation in the case of gas–gas interactions and adsorption in the case of gas–surface interactions. These effects will be discussed in detail. Condensation will be discussed later in this chapter, in the section on molecular interactions. Adsorption is discussed in Chapters 10, 11, and 12.
4. Assumption 4 is generally a good approximation. It breaks down only when assumption 1 breaks down, which is at very low number densities. (This assumption does, however, break down if external forces are present, for example, for the case of an atmosphere in a gravitational field.)

5 & 6. Assumptions 5 and 6 are the least tenable of all. They are valid in practice, however, and this is the only justification for their use. There is no *a priori* justification for these assumptions, and one can show by the statistical mechanical approach that they are unnecessary and appear as a consequence of the treatment.

The Molecular Impingement Rate

The first step in the development of the kinetic theory of gases is to develop an expression for the number of gas molecules that strike a given area in the system per unit time. This is called the *molecular impingement rate* and is usually stated in units of

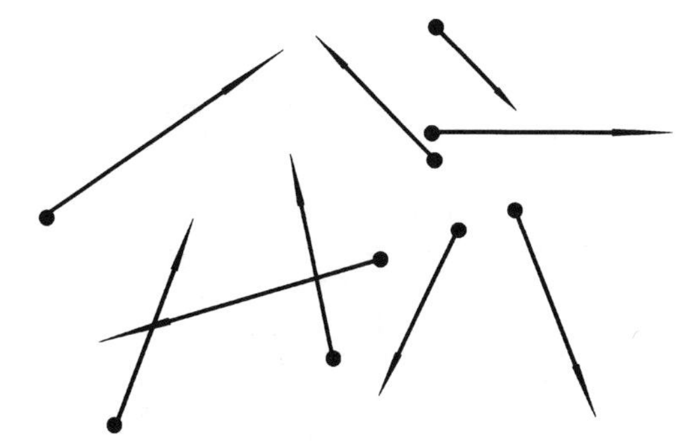

Figure 7.1 Molecular motion in a gas, showing velocity vectors attached to each molecule.

molecules/sec-cm^2. This is an important parameter in the kinetic theory, as it will be used to develop the expression for pressure and it is often the rate-controlling parameter in gas-surface interactions.

To begin this development, one must first put assumption 5, concerning the equal probability of all directions of molecular velocities, into analytical form. This is done by imagining first that there is attached to each molecule in the system, at any moment, a vector representing the direction and magnitude of its velocity, as shown in Figure 7.1. All of these vectors are transferred to a common origin, as shown in Figure 7.2, and a sphere of radius r constructed around this origin. In this construction, the velocity vectors (or their extensions for those molecules whose speeds have amplitudes less than r) intersect the surface of the sphere in as many points as there are molecules in the system. The assumption of uniform distribution of directions of velocity simply means that each unit of area on the surface of the sphere that we have constructed is intersected by the same number of vectors (or their extensions). The average number of intersections per unit area is thus

$$\frac{N}{S} = \frac{N}{4\pi r^2}, \qquad 7.1$$

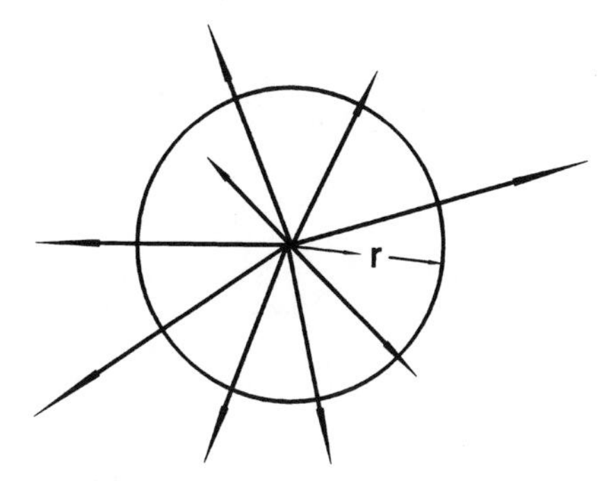

Figure 7.2 Velocity vectors transferred to a common origin.

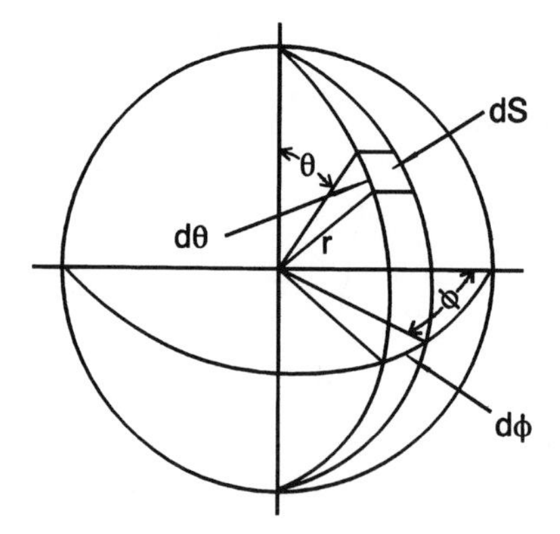

Figure 7.3 Definition of an element of surface on a sphere in terms of the polar angle θ and the azimuthal angle Φ.

where N is the total number of molecules in the system and S is the surface area of the sphere. The number of intersections on any element of the surface of the sphere, dS, is thus

$$dN = \left(\frac{N}{4\pi r^2}\right) dS, \tag{7.2}$$

independent of where dS is located on the sphere.

Carrying this development one step further and defining a specific surface element, dS, in terms of the polar coordinates θ , Φ as shown in Figure 7.3, leads to

$$dS = rd\theta \; r \sin\theta \, d\Phi,$$

or

$$dS = r^2 sin\theta \, d\theta \, d\Phi \, . \tag{7.3}$$

The number of molecules whose velocity vectors intersect this element of area, that is, the number of molecules whose velocities are directed within a polar angle from θ to $\theta + d\theta$ and azimuthal angle from Φ to $\Phi + d\Phi$, which is called $d^2N_{\theta,\Phi}$, is

$$d^2N_{\theta,\Phi} = \left(\frac{N}{4\pi r^2}\right)\left(r^2 \sin\theta \, d\theta \, d\Phi\right)$$

or

$$d^2N_{\theta,\Phi} = \left(\frac{N}{4\pi}\right)(\sin\theta \, d\theta \, d\Phi). \tag{7.4}$$

As a final step, divide both sides by the system volume, V, to get

$$d^2n_{\theta,\Phi} = \left(\frac{n}{4\pi}\right)(\sin\theta \, d\theta \, d\Phi), \tag{7.5}$$

where $n = N/V$. Equation 7.5 is the desired analytical expression of assumption 5.

The foregoing result can be used to develop the expression for molecular impingement rate. This is the rate at which molecules from the gas strike a unit area of surface exposed to the gas—either the container surface or any area constructed within the bulk of the gas. To do this, consider an arbitrary element, dA, of any surface exposed to the gas. (Note that this dA is an element of real surface, as opposed to the dS discussed previously which was an element of area on an arbitrarily constructed sphere in vector space.) Construct a line normal to the plane of dA, and a reference plane perpendicular to the normal as shown in Figure 7.4. The question now is that of how many molecules impinge on this area, dA, in the time interval, dt, traveling in a direction specified by θ, Φ, and having a specified speed, v. Such a collision of a molecule having direction defined by θ, Φ, with speed v, is called a θ, Φ, v collision (that is, a molecule with directions between θ and $\theta + d\theta$, Φ and $\Phi + d\Phi$ and speed between v and $v + dv$).

To determine the number of such collisions per unit area per unit time, construct a prism of height vdt, and base dA, inclined to the plane of dA at angles θ, Φ, again as shown in Figure 7.4. The volume of this figure is

$$dV = vdA\ dt\ cos\theta. \tag{7.6}$$

Because of the fact that this volume element was defined by direction–speed coordinates, θ, Φ, v, all of the molecules that were within this volume element at $t = 0$ will strike dA in time dt. In order to determine the contribution to the impingement rate from this volume element, the number of molecules in this element having velocities from v to $v + dv$ must be determined. One must then integrate over all possible values of θ, Φ, v to get the total impingement rate. To do this one must make one further assumption, namely, that just it was assumed that the distribution of all of the molecules throughout the volume of the system was uniform, it is now assumed that the distribution of any subgroup of molecules is uniform throughout the volume. That is, the fraction of the molecules in the prism constructed that are θ, Φ, v molecules is the same as the fraction of the molecules in the total system that are θ, Φ, v molecules.

Having made this assumption, one may now define dn_v as the number of

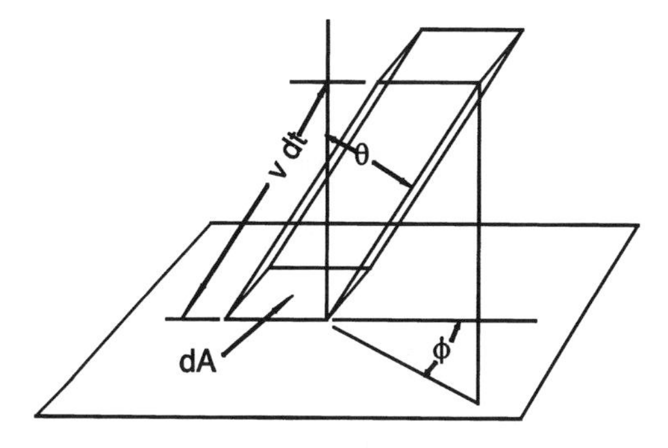

Figure 7.4 Reference element of a real surface area, dA, with prism defined by θ, Φ, v.

molecules per unit volume having speeds from v to $v + dv$, irrespective of the direction of their velocity. If the number density of molecules with directions described by θ, Φ, is

$$d^2 n_{\theta,\Phi} = \left(\frac{1}{4\pi}\right) n \sin\theta \, d\theta \, d\Phi, \tag{7.7}$$

one may similarly define

$$d^3 n_{\theta,\Phi,v} = \left(\frac{1}{4\pi}\right) dn_v \sin\theta \, d\theta \, d\Phi, \tag{7.8}$$

the number of molecules per unit volume having speeds between v and $v + dv$, traveling in a direction defined by θ, Φ. The total number of θ, Φ, v molecules in the prism constructed is simply the product of $d^3 n_{\theta,\Phi,v}$ and the volume of the prism, dV, or

$$\begin{aligned} N_{\theta,\Phi,v} &= d^3 n_{\theta,\Phi,v} dV \\ &= dA \, dt \left(\frac{v dn_v}{4\pi}\right) \sin\theta \cos\theta \, d\theta \, d\Phi. \end{aligned} \tag{7.9}$$

From this one may write the impingement rate of θ, Φ, v molecules per unit area per unit time as

$$I_{\theta,\Phi,v} = \frac{N_{\theta,\Phi,v}}{dA dt} = \left(\frac{1}{4\pi}\right) v dn_v \sin\theta \cos\theta \, d\theta \, d\Phi. \tag{7.10}$$

One may define the total impingement rate of molecules having speeds from v to v + dv, over the whole of the space above the reference area, by integrating Equation 7.10 with respect to θ from zero to $\pi/2$ and with respect to Φ from zero to 2π, yielding

$$I_v = \left(\frac{1}{4\pi}\right) v dn_v (2\pi) \left(\frac{1}{2}\right),$$

or

$$I_v = \tfrac{1}{4} v dn_v. \tag{7.11}$$

Furthermore, the total impingement rate for all molecules, irrespective of their speeds, is

$$I = \tfrac{1}{4} \int v dn_v. \tag{7.12}$$

This may finally be expressed in terms of the arithmetic mean speed of the molecules, which is defined as

$$\bar{v} = \frac{\sum_i n_i v_i}{\sum_i n_i} \tag{7.13}$$

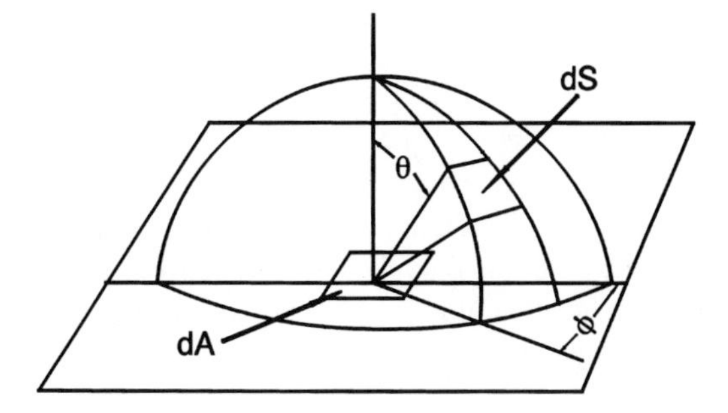

Figure 7.5 Defining geometry for calculation of impingement rate as a function of polar angle.

in which one sums over all possible values of v_i. Assuming that the velocity distribution is continuous, that dn_v represents the number density of molecules with speed v, and that the total number density of molecules is large, then one can replace the sums in Equation 7.13 by integrals and write

$$\bar{v} = \frac{\int v dn_v}{\int dn_v} = \frac{\int v dn_v}{n}. \quad 7.14$$

Thus,

$$\int v dn_v = n\bar{v}, \quad 7.15$$

and

$$I = \tfrac{1}{4}\, n\bar{v}, \quad 7.16$$

the desired result. Thus, the molecular impingement rate depends directly on the number density and the mean velocity. The next section in this chapter will show that $\bar{v}$ depends on the temperature, T, and that n is directly related to the pressure, p.

One further property related to the subject of impingment rate may be derived, namely, the impingement rate as a function of the polar angle θ. To do this, consider a reference area element located at the center of a sphere, as shown in Figure 7.5, in the plane defined by the x and y axes. Molecules impinging on dA from a direction defined by θ, Φ are those that enter the sphere drawn within the cone defined by θ to $\theta + d\theta$ and Φ to $\Phi + d\Phi$, which intersects the surface of the sphere in an area element dS. As previously shown

$$dS = r^2 \sin\theta \, d\theta \, d\Phi, \quad 7.17$$

and the solid angle subtended by the cone is

$$d\omega = \sin\theta \, d\theta \, d\Phi = \frac{dS}{r^2}. \quad 7.18$$

One may rewrite Equation 7.10 in terms of ω as

$$I_{\theta,\Phi,v} = I_{\omega,v} = \left(\frac{1}{4\pi}\right) v dn_v \cos\theta \, d\omega \,, \tag{7.19}$$

an expression giving the impingement rate of molecules having a velocity from v to $v + dv$ and coming from a direction defined by the solid angle from ω to $\omega + d\omega$. From this, one can write an expression for the impingement rate of molecules in the velocity range from $v + dv$ per unit solid angle as

$$\left(I_v\right)_\omega = \frac{I_{\omega,v}}{d\omega} = \left(\frac{1}{4\pi}\right) v dn_v \cos\theta \,. \tag{7.20}$$

The total impingement rate for all velocities per unit solid angle is thus

$$I_\omega = \left(\frac{1}{4\pi}\right) \cos\theta \int v dn_v$$

or

$$I_\omega = \left(\frac{1}{4\pi}\right) n\bar{v} \cos\theta = \tfrac{1}{4}\, n\bar{v} \frac{\cos\theta}{\pi} \,. \tag{7.21}$$

This is known as the cosine law. The greatest value of I_ω is associated with the element of solid angle located normal to the surface. I_ω decreases uniformly as the polar angle increases toward the surface. This result will be used in Chapter 8, in the discussion of effusion from orifices and molecular beam technology.

Pressure

Knowledge of the impingement rate may be used to develop an expression for the pressure exerted on the container walls arising from the collisions of the gas molecules with the walls. The pressure may be defined either as the force exerted per unit area by the gas molecules on the walls of the container or as the rate at which momentum is imparted to unit area of wall surface. Mathematically, this latter definition can be written as

$$p = \frac{\not{p}}{dAdt} \,, \tag{7.22}$$

where $\not{p}$ is the rate of momentum transfer. To develop an expression for pressure, one must thus calculate the average momentum transfer from a molecule to the wall per collision, then multiply this by the impingement rate, I, to get the pressure.

To make this calculation, begin by recalling assumption number 3, that the collisions are perfectly elastic and the walls perfectly smooth. Assume in addition that the walls are so massive relative to the molecular mass that the wall velocity is unaffected in the collision process. If this is the case, then the change in velocity experienced by a colliding molecule having an initial velocity v, coming from a direction defined by θ, Φ is, as before,

$$\Delta v = 2v \cos \theta \,. \tag{7.23}$$

This is essentially a billiard-ball collision. Because the wall is smooth, only the component of the velocity perpendicular to the wall is affected. The change in momentum per collision is then

$$\wp = m\Delta v = 2mv \cos\theta \tag{7.24}$$

for a molecule having an initial velocity v, coming from a direction defined by θ, Φ. Recall that Equation 7.9 showed that the number of such molecules is

$$N_{\theta,\Phi,v} = \left(\frac{1}{4\pi}\right) v dn_v \sin\theta \cos\theta \, d\theta \, d\Phi \, dAdt \,.$$

The total change in momentum due to collisions of this type on the area dA in the time dt is thus

$$N_{\theta,\Phi,v}\wp = \left(\frac{1}{2\pi}\right) mv^2 dn_v \sin\theta \cos^2\theta \, d\theta \, d\Phi \, dAdt \,. \tag{7.25}$$

This expression may be integrated as before over θ from zero to $\pi/2$ and over Φ from zero to 2π to get the total momentum transfer due to molecules in a given velocity range as

$$\begin{aligned} \wp_v &= \left(\frac{1}{2\pi}\right) mv^2 dn_v dAdt (2\pi)\left(\frac{1}{3}\right) \\ &= \left(\frac{1}{3}\right) mv^2 dn_v dAdt \,. \end{aligned} \tag{7.26}$$

Again, as before, one may integrate over all possible values of velocity to get the total momentum transferred to the area dA in the time dt as

$$\wp = \frac{1}{3} m \int v^2 dn_v dAdt \tag{7.27}$$

and, using the definition of pressure in Equation 7.22,

$$p = \frac{\wp}{dAdt} = \frac{1}{3} m \int v^2 dn_v \,. \tag{7.28}$$

Here again, as in the case of the impingement rate, this is an integral over velocity increments, only this time it is over increments of v^2. One may proceed as before to define the average value of v^2, the mean square velocity, as

$$\overline{v^2} = \frac{\sum_i v_i^2 n_i}{\sum_i n_i}. \qquad 7.29$$

Again, making the same assumptions as before concerning the continuity of the distributions of velocities, one may replace the summations with integrals to obtain

$$\overline{v^2} = \frac{\int v^2 dn_v}{\int dn_v} = \frac{\int v^2 dn_v}{n} \qquad 7.30$$

or

$$\int v^2 dn_v = n\overline{v^2}, \qquad 7.31$$

yielding as the expression for pressure

$$p = \tfrac{1}{3} m\left(n\overline{v^2}\right)$$

or

$$p = \tfrac{1}{3} nm\overline{v^2}. \qquad 7.32$$

Note at this point that $\bar{v}^2 \neq \overline{v^2}$. The difference between these two parameters will become more apparent in the next section of this chapter, a discussion of the distribution of velocities.

Before leaving the subject of pressure, consider pressure from another viewpoint. The assumptions made in developing the kinetic theory of gases are equivalent to those that define the ideal gas on a thermodynamic basis. For example, the assumptions of elastic collisions and no intermolecular attractive forces are equivalent to the thermodynamic criterion that

$$\left(\frac{\partial E}{\partial V}\right)_T = 0. \qquad 7.33$$

Consequently, a kinetic theory gas should have the same equation of state as an ideal gas—that is, it should obey Boyle's and Charles's laws and should have as an equation of state

$$pV = \left(\frac{N}{N_{av}}\right) RT. \qquad 7.34$$

If the equation for pressure derived above is correct, it must be identical to the pressure defined by Equation 7.34, which can be derived from purely thermodynamic considerations. That is,

$$p_{\text{kin}} = p_{\text{thermo}}$$

or

$$\frac{1}{3}nm\overline{v^2} = \left(\frac{N}{V}\right)\left(\frac{R}{N_{av}}\right)T \qquad 7.35$$

or, substituting into the right-hand side

$$n = \frac{N}{V}, \quad \frac{R}{N_{av}} = k, \qquad 7.36$$

where k is Boltzmann's constant,

$$\frac{1}{3}m\overline{v^2} = kT \qquad 7.37$$

The kinetic energy of a kinetic theory gas is thus,

$$\frac{1}{2}m\overline{v^2} = \left(\frac{3}{2}\right)kT. \qquad 7.38$$

It is apparent that this energy is a function of temperature only, as it must be if the thermodynamic criterion for an ideal gas, $(\partial E/\partial V)_T = 0$, is to be satisfied. Note in passing that one can rearrange Equation 7.37 to obtain

$$\overline{v^2} = \frac{3kT}{m}, \qquad 7.39$$

a relation that will be used the next section of this chapter, the distribution of velocities.

Finally, substitution of the identities shown in Equation 7.36 into Equation 7.34 yields

$$p = nkT. \qquad 7.40$$

This will be a very convenient form for p in much of this work, as it gives p as a function of the gas density and the temperature.

Distribution of Molecular Velocities

To this point it has been assumed that the gas molecules at equilibrium have some given distribution of velocities and that this distribution is time invariant. Now consider the form of this distribution.

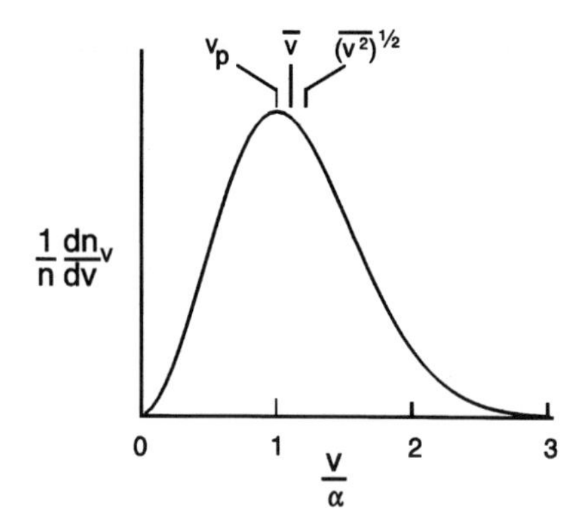

Figure 7.6 The equilibrium distribution of molecular velocities in a kinetic theory gas. v_p: $\overline{v}$: $(\overline{v^2})^{1/2}$ as 1.000 : 1.128 : 1,225.

The mathematics required in developing the distribution in detail would take more time than is justified for present purposes. The development proceeds, however, by assuming that the three orthogonal components of the velocity of a given molecule are independent of one another, then uses a variational process involving the use of an undetermined multiplier, α. This involves essentially varying the velocity components of the various molecules, subject to the constraint of constant total system energy, in order to determine the equilibrium distribution. The result of this process gives the relation for the fraction of the molecules in the system in a given velocity range in terms of the velocity and the multiplier α. The resulting relation is

$$\frac{dN_v}{N} = \frac{dn_v}{n} = \left(\frac{4}{\pi^{1/2}\alpha^3}\right)\exp\left(\frac{-v^2}{\alpha^2}\right)v^2 dv. \qquad 7.41$$

Figure 7.6 is a graph of this relation. From the graph one may define three characteristic velocities:

1. The most probable velocity, v_p—the velocity associated with the maximum in the distribution. That is, the velocity for which

$$\left(\frac{1}{n}\right)\left(\frac{dn_v}{dv}\right) = 0 \qquad 7.42$$

This can be found by differentiation to be

$$v_p = \alpha \,. \qquad 7.43$$

2. The mean velocity, $\overline{v}$. Using the definition of $\overline{v}$,

$$\overline{v} = \frac{\int v dn_v}{\int dn_v}, \qquad 7.44$$

and plugging in dn_v from Equation 7.41 it can be determined that

$$\bar{v} = \left(\frac{2}{\pi^{1/2}}\right)\alpha \qquad 7.45$$

3. The mean square velocity $\overline{v^2}$. Again, using the definition

$$\overline{v^2} = \frac{\int v^2 dn_v}{\int dn_v} \qquad 7.46$$

one determines that

$$\overline{v^2} = \frac{3}{2}\alpha^2. \qquad 7.47$$

Recall that it has already been determined, by comparing kinetic theory results to thermodynamics, that

$$\overline{v^2} = \frac{3kT}{m}; \qquad 7.48$$

thus one may write that

$$\bar{v} = \left(\frac{8kT}{\pi m}\right)^{1/2} \qquad 7.49$$

and

$$v_p = \left(\frac{2kT}{m}\right)^{1/2}. \qquad 7.50$$

A comparison of the numerical values of these three characteristic velocities yields:

$$v_p : \bar{v} : (\overline{v^2})^{1/2} = 1.000 : 1.128 : 1.225\,. \qquad 7.51$$

Note finally that now that $\bar{v}$ has been evaluated in terms of observables, one may write as the final expression for the molecular impingement rate that

$$I = \tfrac{1}{4}\, n\bar{v}$$

$$I = \frac{1}{4}\left(\frac{p}{kT}\right)\left(\frac{8kT}{\pi m}\right)^{1/2}$$

or

$$I = \frac{p}{(2\pi mkT)^{1/2}}. \qquad 7.52$$

This expression will be used throughout the book.

Finite Molecular Size Effects

Up to this point, the gas molecules have been treated as point masses. Allowing for the fact that the molecules, in reality, are of finite size, leads to three new concepts, namely, excluded volume, collision frequency, and mean free path.

Excluded volume arises from the fact that because the molecules take up space, not all of the volume of the container is available to any given molecule at any given time. One may account for this effect by assuming first that the molecules are spheres of radius ρ. Consequently, each molecule excludes the centers of all of the other molecules in the system from a sphere of radius 2ρ, or a volume of $8V_m$, where

$$V_m = \left(\frac{4}{3}\right)\pi\rho^3. \tag{7.53}$$

If one calculates the average volume available to a given molecule on this basis, by effectively adding molecules one at a time to an empty volume and calculating the volume available to each incremental molecule, the result is

$$V_A = V - 4NV_m \, . \tag{7.54}$$

The equation of state for the gas, including the effect of this term is

$$p(V - b) = NkT \, , \tag{7.55}$$

where

$$b = 4NV_m = \left(\frac{16}{3}\right)\pi N\rho^3 \tag{7.56}$$

and is a constant for any given molecular species.

Consider next the concept of *collision frequency*. Because the molecules are of finite size, they will collide if they get within a certain distance of one another. Empirically, one would expect this collision frequency, Z, to be proportional to the number of molecules present, how close they are to one another (the product of these two terms is given by n), the mean molecular velocity, $\overline{v}$, and the cross-sectional area that one molecules "sees" as it approaches another. That is,

$$Z = \sigma n\overline{v} \, , \tag{7.57}$$

where σ is the collision cross-section. One can develop this relation rigorously, but the intuitive argument given above will suffice here.

Finally, consider the so-called *mean free path*, λ. This is defined as the average distance a molecule travels between collisions. One can get at this parameter simply by dividing the average distance traveled per unit time by the average number of collisions per unit time, or

$$\lambda \propto \frac{\bar{v}dt}{Zdt} \propto \frac{\bar{v}}{\sigma n\bar{v}} = \frac{1}{\sigma n}. \tag{7.58}$$

A more rigorous treatment would yield

$$\lambda = \frac{1}{\sqrt{2}n\sigma}. \tag{7.59}$$

Note that the mean free path turns out to be independent of temperature and inversely proportional to pressure, through its dependence on n.

Molecular Interactions

To this point, it has been assumed that there were no forces acting between molecules in the gas. This is clearly an unrealistic assumption but one which gives results in agreement with observation at high temperatures and low densities. Now consider these forces and their effect on the behavior of real gases.

The forces in question are the permanent or induced dipole interactions that are always present when molecules approach one another. These are the forces that lead to condensation at low enough temperatures. The net effect of these forces in the case of a gas is to reduce the measured pressure below that which would be calculated using the kinetic theory equations developed above. The magnitude of this effect depends on the type of forces involved in any given case. In general, permanent dipole forces will be stronger than induced dipole, or van der Waals, forces, and the forces will be stronger for heavy molecules than for light molecules. The net effect on the equation of state of allowing for these forces is to modify the pressure dependence, leading to

$$\left(p + \frac{a}{V^2}\right)(V - b) = NkT, \tag{7.60}$$

the so-called *van der Waals equation of state.*

Figure 7.7 is a plot of the van der Waals equation of state, showing the relation of p to V for a range of temperatures. At high temperatures, the curves are essentially the

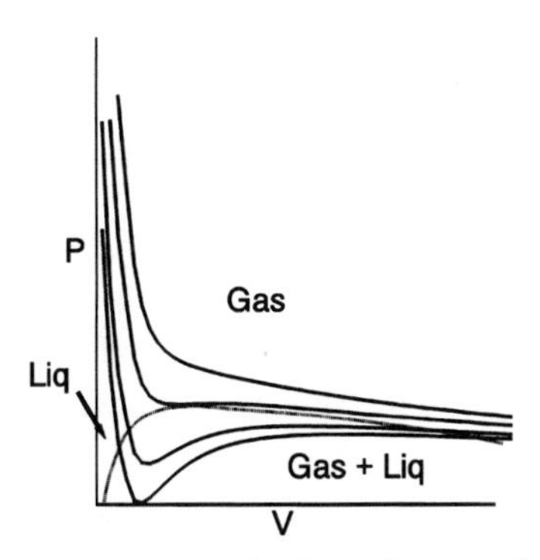

Figure 7.7 Relation between pressure and volume for a gas that obeys the van der Waals equation of state. Isotherms are plotted for a range of temperatures. Below T_c, a gas and a condensed phase may coexist at equilibrium. Two phases will be present inside the dotted area.

hyperbolas predicted by the kinetic theory. As the temperature is reduced, there are increasing deviations in the direction of lower pressure for a given volume. Eventually, a point is reached where the curve becomes double valued in V for a given p. At this point, called the critical temperature, T_c, condensation occurs, and a two-phase mixture of gas and condensed phases is formed. With further decreases in temperature, the equilibrium pressure over the condensed phase is further reduced.

Polyatomic Molecules

As a final topic under the behavior of gases, consider the complications that arise when the details of molecular structure are taken into account. In particular, the concern is the way in which the average total energy per molecule depends on the number and kind of atoms that make up the molecule. For the case of translational kinetic energy, the average energy per molecule is

$$E_{tr} = \frac{1}{2} m\overline{v^2} = \left(\frac{3}{2}\right) kT. \qquad 7.61$$

This leads to a value for the heat capacity of $(3/2)N_{Av}k$ per mole.

For the case of polyatomic molecules, one must consider in addition contributions to the total energy from the so-called *internal degrees of freedom,* namely, rotational and vibrational energies.

Rotational motion can generally be treated classically, as the motion of a rigid rotor. The result of such a treatment is a contribution to the total average energy per molecule of kT for the case of diatomic or linear polyatomic molecules and of $(3/2)kT$ per molecule for nonlinear polyatomic molecules. This leads to a total average energy per molecule of $(5/2)kT$ (linear) or $3kT$ (nonlinear) and to heat capacities of $(5/2)N_{Av}k$ and $3N_{Av}k$ per mole, respectively. These considerations will be important in the discussion of molecular beam formation.

Because the separation between adjacent vibrational energy levels is generally large compared to kT, vibrational energies must be treated quantum mechanically. The net result of such a treatment is that the fraction of molecules to be found in a given excited vibrational state, i, will be

$$\frac{n_i}{n} = (const)\exp\left(\frac{-h\nu}{kT}\right). \qquad 7.62$$

Because of the large energy differences from state to state, vibrational excitation is rare, except at high temperatures. The contribution of vibrational energy to the heat capacity is, in general, correspondingly small.

Summary

By way of summary, Table 7.1 shows the values expected for n, I, Z, λ, the intermolecular spacing, and the time required to form a monolayer of gas on a surface for

Table 7.1

Pressure (P), Number Density (n), Impingement Rate (I), Intermolecular Spacing, Collision Frequency (Z), Mean Free Path (λ) and Monolayer Time for a Gas Having m = 28 AMU at T = 273 K.

P *(Torr)*	P *(Pascal)*	n *(molec/cm^3)*	I *(molec/cm^{2}sec)*	*Intermolecular Spacing (cm)*	Z *sec^{-1}*	λ *(cm)*	*Monolayer Time (sec)*
760	10^5	2.7×10^{19}	3×10^{23}	3.3×10^{-7}	7×10^9	6×10^{-6}	3.3×10^{-9}
1	1.3×10^2	3.5×10^{16}	4×10^{20}	3×10^{-6}	9×10^6	4.5×10^{-3}	2.5×10^{-6}
10^{-3}	1.3×10^{-1}	3.5×10^{13}	4×10^{17}	3×10^{-5}	9×10^3	4.5	2.5×10^{-3}
10^{-6}	1.3×10^{-4}	3.5×10^{10}	4×10^{14}	3×10^{-4}	9	4.5×10^3 (150 ft)	2.5
10^{-9}	1.3×10^{-7}	3.5×10^7	4×10^{11}	3×10^{-3}	9×10^{-3}	4.5×10^6 (28 mi)	2.5×10^3 (42 min)
10^{-2}	1.3×10^{-10}	3.5×10^4	4×10^8	3×10^{-2}	9×10^{-6}	4.5×10^9 (28000 mi)	2.5×10^6 (700 hr)
10^{-15}	1.3×10^{-13}	35	4×10^5	3×10^{-1}	9×10^{-9}	4.5×10^{12} (2.8×10^7 mi)	2.5×10^9 (79 yr)

the case of a gas with $m = 28$ AMU at $T = 273$ K. This table warrants extensive examination, as it contains the information necessary to get a feeling for the physical situation in systems at various pressure levels.

Problems

7.1. Derive an expression equivalent to the kinetic theory expression for pressure ($p = 1/3nm\overline{v^2}$) for a two-dimensional gas, that is, one whose molecules can move only in a plane. Note that instead of force/area in this case it is force/length of the boundary enclosing the available area.

7.2. The pressure in a vacuum system is 10^{-9} Torr, the external pressure is 1 atm, and the temperature is 300 K. There is a pinhole in the wall of the system having an area of 10^{-10} cm^2. Assuming that every molecule that strikes the hole passes through it:

a. How many molecules leak in in 1 hour?

b. If the system volume is 2 liters, how great a pressure rise will result? (Assume no pumping in the system.)

7.3. A spherical bulb 10 cm in diameter is pumped continuously to a high vacuum. In the bulb is a small vessel, closed except for a 0.2 mm diameter hole, located at the center of the larger bulb. The small vessel contains mercury at 100°C, where its vapor pressure is 0.28 Torr.

a. Calculate $\overline{v}$ for the mercury in the vessel.

b. Calculate the rate of efflux of mercury through the hole in grams per hour.

c. How long a time is required for 10^{-6} g of mercury to be deposited on 1 cm^2 of bulb surface in a direction making a 45° angle with the normal to the hole? Assume that all mercury atoms that strike the bulb surface stick to it.

7.4. At 300 K, what fraction of the molecules in argon gas have a kinetic energy greater than 1.0 eV/atom?

7.5. The density of N_2 at 273 K and 3000 atm is 0.835 g/cm^3. Calculate the average distance between the centers of the molecules. How does this compare with the molecular diameter calculated from the van der Waals equation of state, for which $b =$ 39.1 cm^3/mole ?

Bibliography

Clarke, JF, and McChesney, M. *The Dynamics of Real Gases*. London: Butterworths, 1964.

Dushman, S, and Lafferty JM. *Scientific Foundations of Vacuum Technique*. 2nd ed. New York: John Wiley and Sons, 1962.

Jeans, JH. *Introduction to the Kinetic Theory of Gases*. London: Cambridge, 1940.

Kennard, EH. *Kinetic Theory of Gases*. New York: McGraw Hill, 1938.

Knudsen, M. *The Kinetic Theory of Gases*. New York: McGraw Hill, 1927.

Loeb, LB. *Kinetic Theory of Gases*. London: Methuen, 1950.

CHAPTER **8**

Molecular Beam Formation

Chapter 7 developed equations that describe the properties of a gas in equilibrium—the so-called *bulk properties*. These equations apply to real gases over a wide range of temperature and pressure, provided that the gas is not subjected to any external forces.

In much of the work that has been done in the study of gas–surface interactions, the experimental techniques used involve exposing the surface of interest to a gas described by the kinetic theory equations. This presents some problems, insofar as determining the details of the interaction process at the molecular level is concerned, because, as Chapter 7 showed, this equilibrium kinetic theory gas contains molecules having a wide range of velocities, impinging on the surface from a wide range of angles. This makes detailed calculation or measurement of the interaction process, in terms of gas molecule–surface atom interactions, difficult.

A practical way around this process involves the use of what are called *molecular beams*. Basically, a molecular beam is a collection of gas molecules, all moving in essentially the same direction. In some cases it is possible to prepare beams that also are made up of molecules all of which have essentially the same translational kinetic energy and in some cases also all having essentially the same rotational and vibrational energies. Using molecular beam techniques, carried to the extreme capabilities of the current state of the art, it is possible to study the interaction of a molecule having closely specified impingent angle and translational, vibrational, and rotational energies (i.e., a "state selected" molecule), with a clean, well-ordered surface, and to determine the translational, vibrational, and rotational energies of the molecules leaving the surface, also on a state-by-state basis.

Effusion Beams

A wide range of techniques can be used to generate molecular beams. The simplest, both in principle and in practice, is the so-called *effusion cell,* or *Knudsen cell.* The operation of such a source is shown schematically in Figure 8.1. A closed cell, or oven, contains the material to be used to form the beam. This is connected to the vacuum system in which the interaction study is to be carried out through a small hole or narrow slit. The surface being studied is located with a line-of-sight to the oven apperture. From kinetic theory, the flow through the aperture is simply the molecular

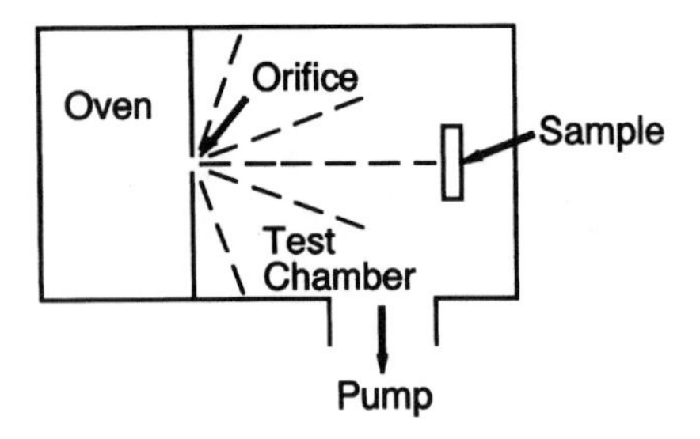

Figure 8.1 Schematic picture of a simple Knudsen cell effusion beam system.

impingement rate on the area of the orifice. Assuming that the pressure on the vacuum system side is negligible, this impingement rate is

$$I_{so} = \tfrac{1}{4}\, n\bar{v}\,, \tag{8.1}$$

or, in terms of observables

$$I_{so} = \frac{1}{4}\left(\frac{p_{so}}{kT_{so}}\right)\left(\frac{8kT_{so}}{\pi m}\right)^{1/2} \tag{8.2}$$

The total flux through the hole will thus be

$$Q_{so} = I_{so}A_{so} = \frac{p_{so}\pi r_{so}^2}{(2\pi mkT_{so})^{1/2}} \tag{8.3}$$

where a sharp-edged circular orifice of radius r_{so} has been assumed.

Sources of this sort are usually operated at low enough pressures, or with small enough orifices, that the mean free path in the cell is long compared to the size of the orifice. This condition can be expressed quantitatively in terms of a parameter called the Knudsen number, defined as

$$Kn \equiv \lambda / d\,, \tag{8.4}$$

where d is the diameter of the orifice, and λ, the mean free path, is given by

$$\lambda = \frac{1}{\sqrt{2}n\sigma}, \tag{8.5}$$

in which σ is the collision cross section, defined in Chapter 7.

The spatial distribution of molecules effusing from the orifice of a Knudsen cell will be the same as the distribution of the molecules striking the orifice area, normally a cosine distribution. That is,

$$I_{\omega} = \frac{1}{4} n\bar{v}\left(\frac{\cos\theta}{\pi}\right). \tag{8.6}$$

Because of the geometry of the system, the intensity of the beam will drop off as the square of the distance from the orifice. The intensity at a surface located at a distance L from the orifice will thus be

$$I_{su} = I_{so}A_{so}\left(\frac{\cos\theta}{\pi}\right)\left(\frac{1}{L^2}\right), \qquad 8.7$$

or

$$I_{su} = \left[\frac{p_{so}}{(2\pi mkT_{so})^{1/2}}\right](\cos\theta)\left(\frac{r_{so}}{L}\right)^2. \qquad 8.8$$

Note that because molecules having a high velocity have a greater probability of striking the orifice than molecules having a low velocity, the velocity distribution in the beam will differ from that calculated in Chapter 7 for a static gas. The appropriate distribution for the beam effusing from a Kundsen cell is

$$\frac{dn_v}{n} = 2\left(\frac{v^3}{\alpha^4}\right)\exp\left(\frac{-v^2}{\alpha^2}\right)dv, \qquad 8.9$$

where α, again, is $(2kT/m)^{1/2}$.

Integration over the distribution yields

$$E_{tr} = 2kT \qquad 8.10$$

as the mean translational energy of the molecules in the beam. Note that the intensity is a maximum in the direction normal to the orifice and decreases with increasing θ, the angle between the direction of interest and the normal to the plane containing the orifice. This dependence can be represented as a polar plot of intensity as shown in Figure 8.2.

The shape of this polar plot demonstrates two major problems with this type of

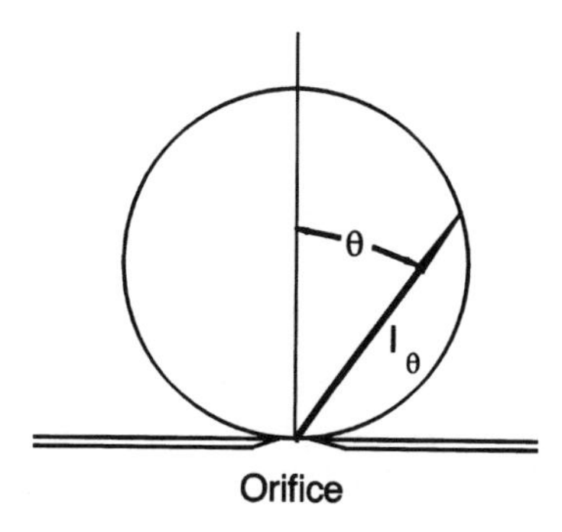

Figure 8.2 Polar plot of the effusion intensity as a function of polar angle, θ, for a thin, sharp-edged orifice.

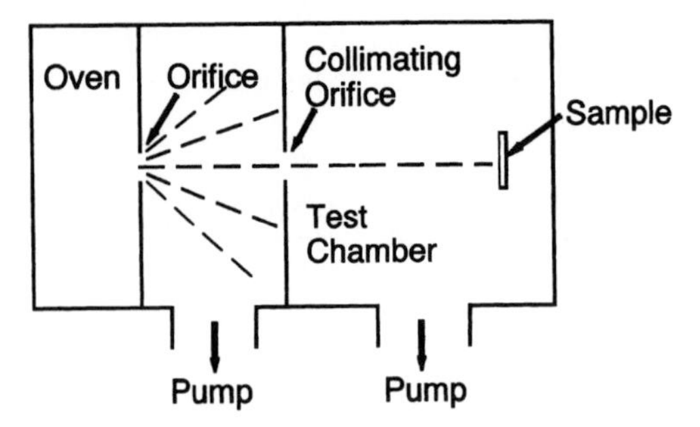

Figure 8.3 Schematic picture of Knudsen cell effusion beam system with a collimating orifice and differential pumping.

source and, to a greater or lesser extent, with all molecular beam sources. If the source is far away from the sample, the beam intensity will be low, and most of the material effusing through the orifice will be wasted. Improvements to the Knudsen cell technique have been mostly directed toward minimizing these limitations.

The simplest solution to the problem of introducing unwanted material into the system, in the form of material not directed toward the sample, is to use a concept known as *collimation.* In its simplest form, collimation involves putting a barrier between the source and the sample, with a small hole in it, as shown in Figure 8.3. This *collimator,* as it is called, intercepts all of the flow from the source except for that traveling directly toward the sample and, thus, sharply defines the direction of the beam. If the beam is made up of a readily condensible material, as is the case with the metal vapor beams used in many molecular beam epitaxy systems, for example, this may be all that is necessary. If the beam material is a gas at ordinary temperatures, however, it must be removed from the system to avoid a contribution to the flux to the surface from molecules that have collided with the collimator and rebounded. This is done by closing off the volume between the oven and the collimator completely and pumping it with a separate pump. This technique is called differential pumping, and it is used very often in surface research systems to improve the ratio of signal to background.

Multicapillary Sources

The simplest step that can be taken to improve the fraction of the material leaving the source that exits in the direction of the sample is to modify the shape of the orifice. The cosine distribution discussed in the preceding section will be observed only for an orifice whose thickness is small compared to its diameter. Any departure from this condition both decreases the total flux through the orifice and changes the spatial distribution to one which favors the straight-ahead direction even more than does the cosine distribution. Figure 8.4 shows a simple case for an orifice which is a short tube, rather than a sharp-edged orifice. The total flow through such a tube, Q_t, for the case of large Knudsen number (long mean free path, small orifice) is given by

$$Q_t = KQ_{orf}, \tag{8.11}$$

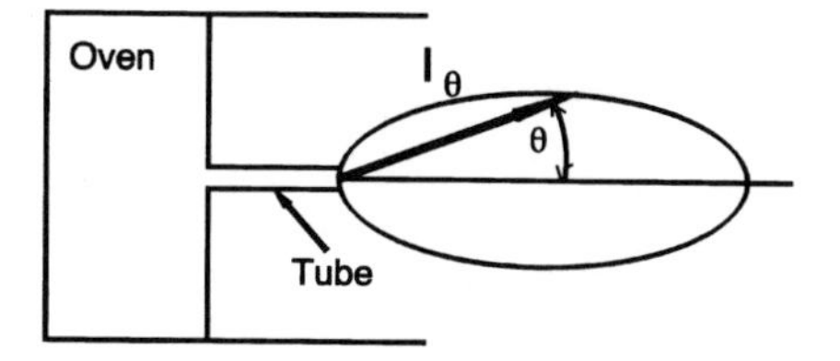

Figure 8.4 Knudsen cell system with a single tube orifice.

where K, the so-called Clausing factor, or transmission probability for the tube, is a complicated function of the tube geometry, but is approximately

$$K \approx \left(\frac{8}{3}\right)\left(\frac{r}{\ell}\right) \qquad 8.12$$

for a tube of length ℓ and radius r. This factor arises from the fact that, in the case of a tube, a molecule that enters the tube at an angle may return to the source, rather than passing on to the sample.

In addition to the reduction in total flow, one also finds that the spatial distribution of material exiting the tube is more sharply peaked than for the case of a sharp edged orifice. This is shown in Figure 8.5a for the case of a tube having $r = 0.019$ cm, $\ell = 0.48$ cm. The relative improvement in straight-ahead intensity can be stated as a "peaking factor," which is the ratio of the straight-ahead intensity obtained with the tube to that expected from a thin orifice having the same total gas flow. For the case of a single tube, the maximum value observed for the peaking factor is about 10, at low pressure, and decreases with increasing pressure.

In practice, an array of many capillaries, each with a very small diameter, is used

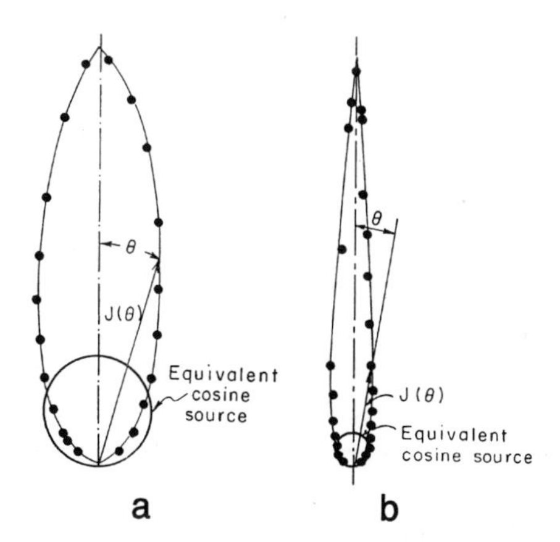

Figure 8.5 Polar plots of the angular distribution of gas effusing from single capillary (a) and multicapillary (b) sources. (Source: R.H. Jones, D.R. Olander and V.R. Kruger, *J. Appl. Phys.* 1969;40:4641. Reprinted with permission.)

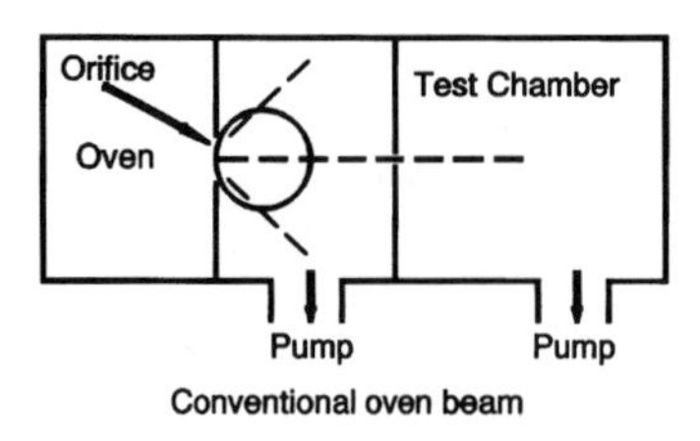

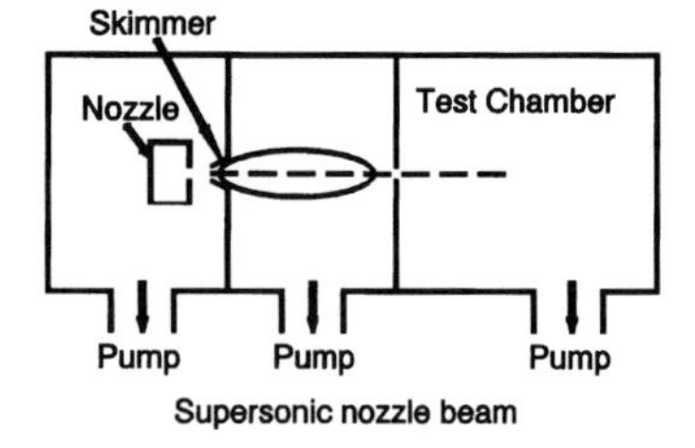

Figure 8.6 Comparison of system geometries for Knudsen cell (conventional oven beam) and free jet expansion (supersonic nozzle beam) source systems.

to make up the multicapillary source. These so-called *multicapillary arrays* are made by fusing together many small glass tubes and drawing them down to a smaller size. Several of the resulting arrays are then joined, and the drawing operation is repeated. The final result of this process is an array of roughly 10^6 of these tubes, each of which is about 1 μm in diameter and 1mm long, covering a total area of roughly 1 mm in diameter The peaking factor for such an array can be as high as 20.

The major limitation of this type of source is that in order to obtain a high peaking factor, the pressure behind the array must be kept relatively low, so that the flow in the tube occurs at a large Knudsen number. As a practical matter, this means an operating pressure in the vicinity of 10 Torr for a typical array. Sources of this type are used commonly as a means of dosing gases directly onto a surface in ultrahigh vacuum systems. However, the velocity distribution and mean energy per particle in sources of this type are the same as obtained from a simple orifice at large Knudsen numbers. This source thus improves the efficiency of gas utilization but does not reduce the wide spread in particle velocities that is characteristic of effusion sources.

Free Jet Expansion Sources

The source conditions in free jet expansion sources differ significantly from the effusion sources discussed in the previous section, as does the mechanical configuration of the source. In this case, the source gas passes through the orifice, or nozzle, at high pressure, with a very small Knudsen number, so that the gas in the nozzle is in continuum flow. As the gas passes into the much lower pressure of the vacuum system, it expands, essentially reversibly and adiabatically. As a result, the temperature of the gas drops sharply, to a few degrees Kelvin in practical systems. The expanded flow is then sampled by a *skimmer*—a hollow cone with a small hole in the end. Figure 8.6 shows the configuration of this source, along with the classical Knudsen cell configuration for comparison.

In order to understand the behavior of such a source, it is helpful to look at a model of the flow emerging from the nozzle. Such a model is shown in Figure 8.7. The central region of the flow is essentially a region in which gas is undergoing a reversible adiabatic expansion. This region is bounded on the sides and at its end by shock waves, which are set up when the density of the gas in the flow drops to a value comparable with the density of the gas in the vacuum chamber. All of the useful and interesting properties of this source are related to the flow in the isentropic core, inside the region

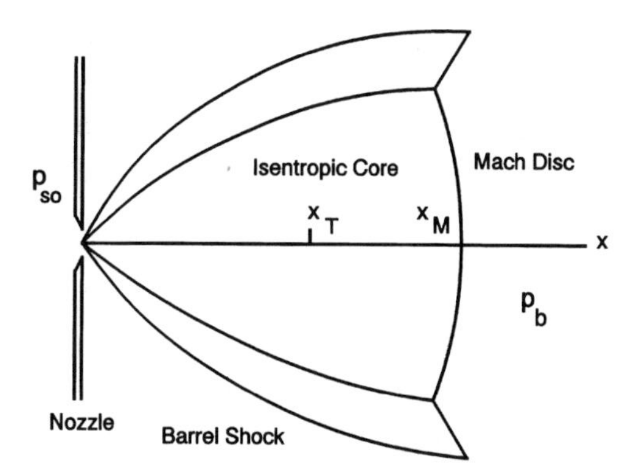

Figure 8.7 Model of the flow from a supersonic nozzle into a low pressure region. In the isentropic core, the flux is essentially a reversible adabatic expansion.

bounded by the shock waves. The only account that must be taken of the shock waves is to ensure that the skimmer is located closer to the nozzle than the *Mach disc* shock wave at the end of the isentropic region. Empirically, it has been found that this condition will be met if the nozzle-to-skimmer distance is less than x_M, which is given by

$$x_M \approx 0.67\left(\frac{p_{so}}{p_b}\right)^{1/2}, \qquad 8.13$$

where p_{so} is the pressure behind the nozzle and p_b is the pressure in the vacuum chamber.

In the isentropic core, the behavior of the gas can be described in terms of a parameter called the Mach number, M, defined by

$$M \equiv \frac{u}{c}, \qquad 8.14$$

where u is the mean molecular velocity along the direction of flow and c is the speed of sound in the gas and is in turn given by

$$c = \left(\frac{\gamma RT}{m}\right)^{1/2}. \qquad 8.15$$

Here γ is the specific heat ratio, C_p/C_v, which will depend on the type of molecule in the flow.

The variation of the Mach number with x, the distance from the nozzle, is given by

$$M \approx 3.20\left(\frac{x}{d}\right)^{(\gamma-1)}, \qquad 8.16$$

where d is the nozzle diameter, out to the point in the expansion where the density becomes so low that there are essentially no more molecular collisions. This point is reached when

$$M = M_T \approx 1.17(Kn_o)^{\left[\frac{1-\gamma}{\gamma}\right]} \tag{8.17}$$

or, for a monatomic gas

$$M_T \approx 1.17\,(Kn_o)^{-0.4}. \tag{8.18}$$

The other properties in the flow such as the temperature, T, the speed of sound, c, the pressure, p, and the density, n, are related to M by

$$\frac{T_{so}}{T} = 1 + \left(\frac{\gamma - 1}{2}\right)M^2, \tag{8.19}$$

$$\frac{T_{so}}{T} = \left(\frac{c_{so}}{c}\right)^2 = \left(\frac{p_{so}}{p}\right)^{\left[\frac{\gamma-1}{\gamma}\right]} = \left(\frac{n_{so}}{n}\right)^{(\gamma-1)}. \tag{8.20}$$

This last set of equations follows directly from the condition that, for an isentropic expansion of an ideal gas,

$$pV^{\gamma} = \text{const}, \tag{8.21}$$

and from

$$n = \frac{N}{V}. \tag{8.22}$$

The importance of these relations, as far as the properties of the beam is concerned, is that the higher the Mach number, the lower the temperature, and thus the more nearly monoenergetic the resulting molecular beam. This is of great importance in developing beams to be used for diffraction studies and for probing surface vibrational structure.

Consider next the beam intensity to be expected from a free jet expansion source. This can be expressed in terms of the properties of the beam at the skimmer entrance, the skimmer size, and the skimmer to sample distance as

$$I_{FJ} = n_{sk}\,\pi r_{sk}^2 u_T \left(\frac{1}{\pi x^2}\right)\left(\frac{x_{sk}}{x_T}\right)^2 \left(\frac{1}{2}\gamma M_T^2 + \frac{3}{2}\right), \tag{8.23}$$

where the subscript T refers to conditions at the point where the terminal Mach number is reached and the subscript sk refers to the skimmer location. The number density at the skimmer is related to that behind the nozzle by the expression, developed from Equations 8.16, 8.19, and 8.20,

$$\frac{n_{so}}{n_{sk}} = \left[1+\left(\frac{\gamma-1}{2}\right)M_T^2\right]^{\left[\frac{1}{\gamma-1}\right]}\left(\frac{x_{sk}}{x_T}\right)^2. \quad 8.24$$

The flow velocity at this point is

$$u_T = M_T c_T = M_T c_{so}\frac{1}{\left[1+\frac{(\gamma-1)}{2}M_T^2\right]^{1/2}}. \quad 8.25$$

The impingement rate based on the number density behind the nozzle is thus

$$I_{FJ} = n_{so}\left(\pi r_{sk}{}^2\right)\left(\frac{1}{\pi x^2}\right)M_T c_{so}\left\{\frac{1}{\left[1+\left(\frac{\gamma-1}{2}\right)M_T^2\right]^{\left(\frac{1}{\gamma-1}+\frac{1}{2}\right)}}\right\}\left(\frac{1}{2}\gamma M_T^2+\frac{3}{2}\right) \quad 8.26$$

or

$$I_{FJ} = n_{so}\left(\frac{r_{sk}}{x}\right)^2 M_T c_{so}\frac{\frac{1}{2}\gamma M_T^2+\frac{3}{2}}{\left[1+\left(\frac{\gamma-1}{2}\right)M_T^2\right]^{\left(\frac{1}{(\gamma-1)}+\frac{1}{2}\right)}}. \quad 8.27$$

As an indication of the relative beam intensity obtainable with a free jet expansion source, compared to an effusion source, compare the expression for I_{FJ}, based on n_s, with the expression for the effusion source

$$I_{EF} = n_{so}\left(\pi r_{so}^2\right)\frac{1}{4}\left(\frac{8kT}{\pi m}\right)^{1/2}\left(\frac{1}{\pi x^2}\right), \quad 8.28$$

where the intensities have been normalized by assuming that n is the same at the skimmer and at the orifice of the effusion source. The resulting ratio is

$$\frac{I_{FJ}}{I_{EF}} = \left(\frac{\pi C_p}{k}\right)^{1/2}\gamma\left(\frac{x_{sk}M_T}{x_T}\right)^2. \quad 8.29$$

In practical nozzle systems, one can obtain M_T as high as 25. This would lead to a ratio of intensities of roughly 2000.

One may also determine the distribution of molecular velocities and the average energy per molecule in a free jet expansion beam. The distribution of velocities can most simply be thought of as a Maxwellian distribution about the mean flow velocity determined from the Mach number. That is,

$$\frac{dn_v}{n} = \left(\frac{1}{\pi}\right)^{3/2}\left(\frac{1}{\alpha}\right)^3 \exp\left[\frac{-(v-u_T)^2}{\alpha^2}\right] d^3v, \tag{8.30}$$

where

$$u_T = M_T c_T = M_T\left(\frac{\gamma R T_T}{m}\right)^{1/2} \tag{8.31}$$

and

$$T_T \approx T_{so}\left(\frac{2}{\gamma-1}\right) M_T^{-2}. \tag{8.32}$$

Figure 8.8 shows this velocity distribution. As an example, for $M_T = 20$, $T_{so} = 300$ K, $\gamma = 5/3$, one would have a terminal temperature

$$T_T = 2.25\ K. \tag{8.33}$$

The mean flow velocity can be related to the mean velocity in the unexpanded gas by substituting Equation 8.32 into Equation 8.31 to yield

$$u_T \approx \left(\frac{2}{\gamma-1}\right)^{1/2} c_{so}, \tag{8.34}$$

or, for $\gamma = 5/3$,

$$u_T \approx \sqrt{3}\, c_{so}. \tag{8.35}$$

Thus, the terminal temperature can be very low, yielding a correspondingly narrow range of molecular velocities about the average velocity u_T. This mean velocity will depend on γ but will always be on the order of twice the mean velocity in the gas upstream of the nozzle.

The average energy per molecule in such a beam is given by

$$E = \left(\frac{\gamma}{\gamma-1}\right) kT, \tag{8.36}$$

which for $\gamma = 5/3$ yields

$$E = \left(\frac{5}{2}\right) kT. \tag{8.37}$$

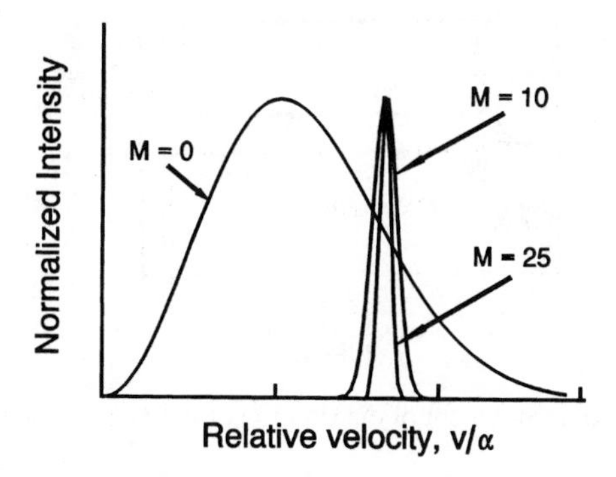

Figure 8.8 Comparison of velocity distributions for a Knudsen cell source and free jet expansion sources having terminal Mach numbers of 10 and 25.

This may be compared with the previously-determined value of $E = 3/2kT$ for the static gas and $2kT$ for the flux from a Knudsen cell.

Finally, consider what happens in the expansion of a beam containing two different molecular species. In this case, if small amounts of a heavy molecule are seeded into a large amount of a low molecular weight gas, the heavier molecules will be carried along in the expansion and will reach essentially the same value of u_T as the lighter molecules. Since the kinetic energy of a molecule is proportional to $m(u_T)^2$, the energy of the heavier molecules will be increased above that of the lighter molecules by the factor M_H/M_L. This provides a means of forming molecular beams having very high energy per molecule. For example, with a very low concentration of xenon seeded into helium we would have

$$\frac{E_{Xe}}{E_{He}} \approx \frac{M_{Xe}}{M_{He}} \approx \frac{120}{4} = 30. \qquad 8.38$$

It is possible to reach energies as high as 30 eV per molecule using this technique.

Problems

8.1. What fraction of the molecules in a molecular beam of N_2 formed by effusion of N_2 gas initially at 300 K from an orifice at a large Knudsen number will have kinetic energies greater than 8 kcal/mol?

8.2. In a free jet expansion of 1% N_2 molecules seeded into a helium carrier gas, what must the temperature of the gas be, before expansion, in order for the average kinetic energy per N_2 molecule, after the expansion, to be 8 kcal/mol?

8.3. Figure 8.9 shows a classical experimental system for measuring the distribution of molecular velocities in a gas. The vapor effusing from the source in the horizontal direction is condensed on a cooled collector surface at a distance r_2 from the defining orifice, a horizontal slit in the cooled collimator plate. In this case, the source material is zinc, the oven temperature is 700 K, and the pressure inside the oven is

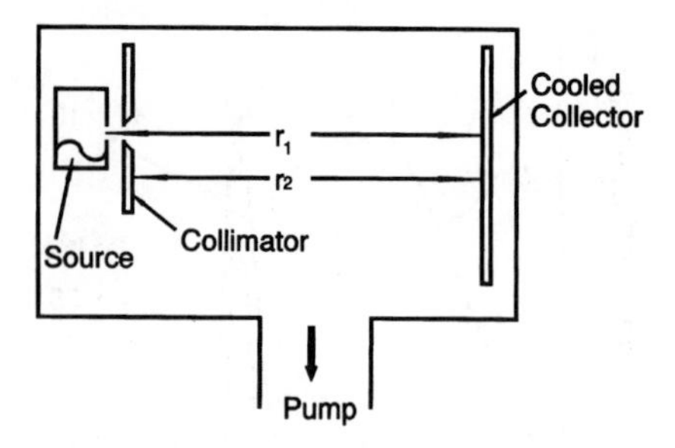

Figure 8.9 System for measurement of distribution of molecular velocities.

2×10^{-3} Torr. The distances $r_1 = 102$ cm and $r_2 = 100$ cm. The collimator slit is 0.1 mm wide and 1.00 cm long. The acceleration due to gravity is 980 cm/sec^2.

a. What is the most probable velocity in the oven?

b. What is the most probable velocity in the beam?

c. How far does a zinc atom having the most probable velocity in the beam drop between the collimator plate and the collector?

d. How long must the system operate in order to deposit a total of 10^{-6} g of zinc on the collector plate?

8.4. A typical free jet expansion molecular beam source is operated with neon gas on the upstream side of the nozzle at a total pressure of 3 atm and a temperature of 300 K. The nozzle is a 0.05 cm diameter hole in a 0.001 cm thick plate, and the skimmer hole diameter is 0.20 cm. Assume that the collision cross section for neon atoms is 21×10^{-16} cm^2.

a. What will be the total neon flux from the nozzle (molecules per second)?

b. What will be the terminal Mach number, M_T, for the flow from the nozzle?

c. What will be the beam impingement rate on a surface located 30 cm downstream from the skimmer?

Bibliography

Anderson, JB, Andres, RP, Fenn, JB, and Maise, G. Studies of low density supersonic jets. In: J.H. de Leeuw, ed. *Rarefied Gas Dynamics*. New York: Academic Press, 1965;106–127.

Anderson, JB, Andres, RP, and Fenn, JB. Supersonic nozzle beams. In: J. Ross, ed. *Molecular Beams*. New York: Interscience, 1966;275–318.

Campargue, R. High intensity supersonic molecular beam apparatus. In: J.H. de Leeuw, ed. *Rarefied Gas Dynamics*. New York: Academic Press, 1965;279–298.

Fenn, JB, and Deckers, J. Molecular beams from nozzle sources. In: JA. Lauerman, ed. *Rarefied Gas Dynamics*. New York: Academic Press, 1963;497–515.

Kusch, P. A study of the free evaporation of alkali halide crystals by velocity selected molecular beams. In: E. Rutner, P. Goldfinger, J.P. Hirth, eds. *Condensation and Evaporation of Solids*. New York: Gordon and Breach, 1964;87–98.

CHAPTER 9

Gas Scattering

This chapter begins discussion of the various interactions that can take place when molecules from a gas phase interact with the surface of a condensed phase. This chapter will consider those processes that do not result in the molecule being trapped at the surface for a significant period of time. Subsequent chapters will consider the stronger interaction processes that lead to adsorption, surface chemical reaction, and crystal growth.

Scattering Modes

If one considers the impact of an atom from a gas phase on a crystal surface, one can classify the possible outcomes of this impact in terms of one of three processes:

1. Elastic scattering, in which there is a single encounter between the gas atom or molecule and the surface, and in which there is no excitation of the internal modes of the surface.
2. Direct inelastic scattering, in which again there is a single encounter between the gas atom or molecule and the surface, but in this case the encounter involves energy interchange between the gas phase species and the surface.
3. Trapping, in which the incident atom or molecule is held in a potential well at the surface for a long time compared to the surface vibration rate.

These three cases are shown schematically in Figure 9.1. The current chapter will consider in detail only the first two possibilities.

The Interaction Potential

When an atom or molecule approaches a surface, it will experience an energetic interaction with the surface, much as two atoms or molecules in a gas phase experience an attractive interaction when they approach each other, as described in Chapter 7. In the case of a gas–surface interaction, however, the incoming particle will interact with a relatively large number of surface and near-surface atoms, as shown schematically in Figure 9.2. The total interaction energy in this case can be described in terms of the

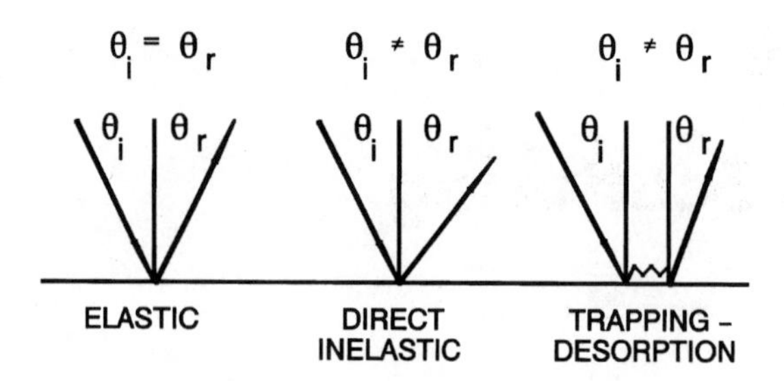

Figure 9.1 Possible outcomes of the collision of a gas atom with a surface. Elastic: low temperature; weak interaction; light molecule; energetically smooth, regular surface. Direct inelastic: higher temperature; weak interaction; heavier molecule; energetically rough, irregular surface. Trapping-desorption: low temperature, strong interaction, heavier molecule.

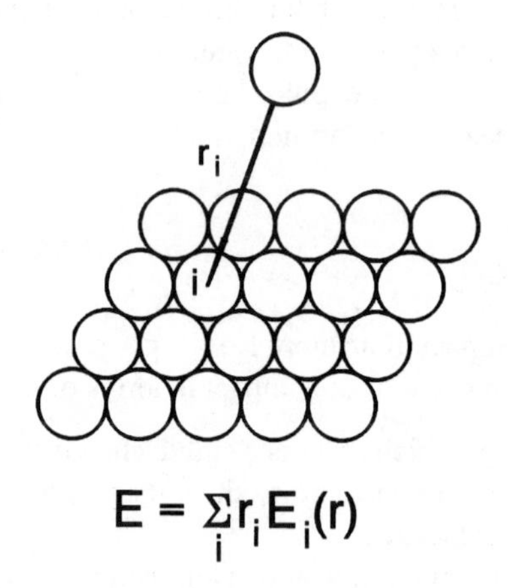

$$E = \sum_i r_i E_i(r)$$

Figure 9.2 Interaction of an impinging gas molecule with the atoms of a surface.

sum of all of the pairwise interactions between the incoming particle and each of the surface atoms involved. That is,

$$E = \sum_i r_i E_i(r), \tag{9.1}$$

where the sum is taken over all atoms that interact significantly with the incoming particle. This number will depend on the details of the interaction in any specific system.

There are a number of ways of looking at this interaction schematically. The simplest is to plot the interaction energy as a function of the perpendicular distance between the particle and the surface, as the particle approaches the surface along a path perpendicular to the surface. Figure 9.3 shows such a plot. The interaction potential will, in general, contain an attractive term, which tends to reduce the total system energy, and a repulsive term, which tends to increase the total system energy. Taking

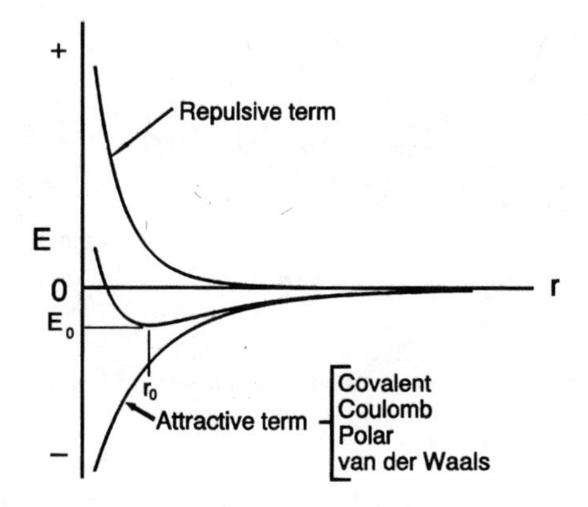

Figure 9.3 One-dimensional potential energy diagram for an atom approaching a solid surface along a path perpendicular to the surface.

the algebraic sum of these two contributions leads to the total interaction energy curve shown, which generally shows a minimum in system energy, E_0, at some equilibrium distance, r_0.

The repulsive part of the potential is due primarily to the overlap of filled electron orbitals in the surface with those of the incoming particle. Recall, from the calculations presented in Chapter 2, that the electron charge density rises rapidly as one approaches the surface from the gas phase.

The attractive term will depend on the nature of the interaction between the surface and the particle, which will differ significantly from system to system. Table 9.1 summarizes the four general types of interaction that will be considered in terms of the relative strength of the interaction and its range of effectiveness.

A number of factors complicate the simple picture shown in Figure 9.3. The details of the curve shown there will depend on where the incoming particle strikes the surface. In terms of Equation 9.1, for a given r, the various r_i will depend on where the particle strikes the surface, relative to the equilibrium surface atom positions. This point of impact relative to the surface mesh is called the *impact parameter*. Several different impact sites for various surface meshes of an FCC crystal are shown schemati-

Table 9.1

Interaction	*Range*	*Strength*
Covalent	Very short, $1/r^6$	Very strong (E_o> 100 kcal/mol)
Coulomb (ionic)	Long, $1/r$	Strong
Polar		Weak
van der Waals	Long, $1/r^3$	Very weak (E_o < 5 kcal/mol)

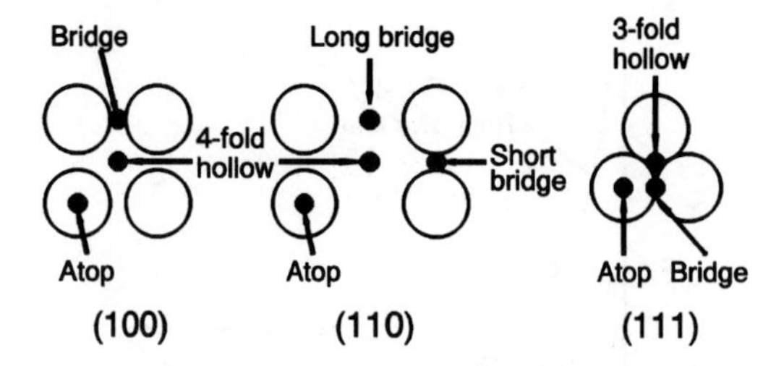

Figure 9.4 Possible impact sites for an atom approaching a solid surface, for various FCC surface geometries.

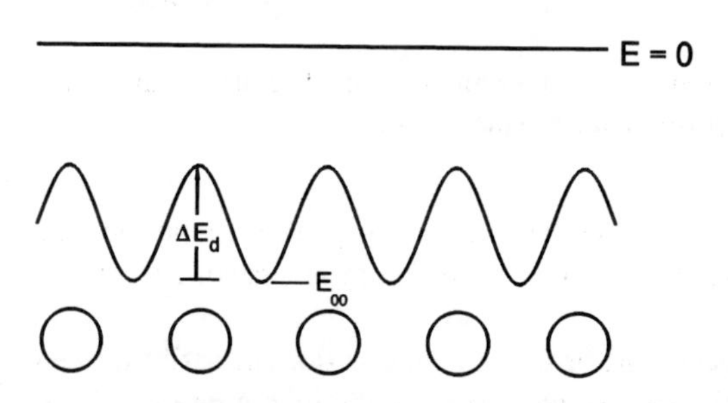

Figure 9.5 Variation of one-dimensional potential well depth with position along a row of surface atoms.

cally in Figure 9.4. It is clear from the figure that the various points of impact give rise to different values of r_i, and consequently different values of E_0.

The variation of E_0 with distance along a line of surface atoms can be shown schematically as in Figure 9.5. Such a plot shows a regular variation in E_0 with position. The points of maximum E_0, labeled E_{00}, represent sites of minimum system energy and, thus, stable positions on the surface. The hills between these sites, characterized by the energy ΔE_d, represent the energy barrier that must be surmounted in order for surface diffusion to take place, as explained in Chapter 6.

A similar, albeit more complicated, plot of energy vs. impact parameter may be developed in three dimensions, to show the variation in E_0 with position on the whole two-dimensional surface. Such a plot is shown in Figure 9.6, a so-called *potential energy surface*. Surfaces of this type are of great importance in understanding adsorption, surface reactions and surface diffusion, that is, any case in which trapping occurs at the surface.

For cases of elastic and direct inelastic scattering, where no trapping occurs, it is generally found that effects associated with the attractive part of the potential are weak or negligible. In this case, it is possible to explain the scattering behavior simply in terms of the repulsive part of the potential, or as it is often called, the *surface corrugation.* This surface corrugation can be shown schematically in a manner similar to the potential energy surface. In this case, in two dimensions, one generally plots the surface electron charge density as a function of distance along a line across the surface, with contours representing equal charge densities, as shown in Figure 9.7.

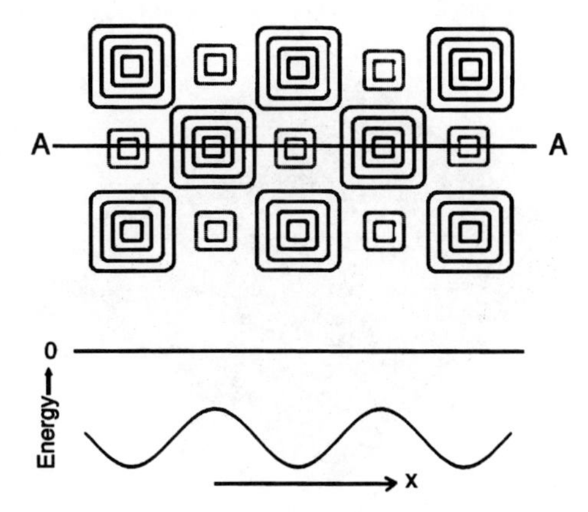

Figure 9.6 A two-dimensional potential energy surface, showing the variation in potential well depth with impact parameter for an atom approaching a solid surface. Dotted contours indicate regions of strongest attractive interaction, solid contours indicate regions of weakest interaction.

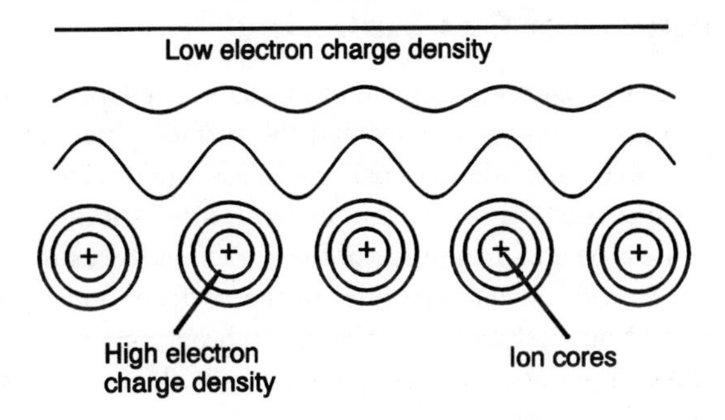

Figure 9.7 Electron charge density contours near the surface of a typical metal.

Recall from Chapter 2 that in the treatment using the Jellium model, the charge density varied only with distance perpendicular to the surface. More sophisticated treatments, which take account of the charge density variation about each of the ion cores, lead to a variation in charge density along the surface, as shown in the figure. Note that the contours drawn are relatively rough close to the ion cores and become much flatter for increasing distance from the surface.

It is also possible to represent the surface corrugation in a three-dimensional plot, by plotting the height of a given charge density contour above the surface, as a function of position over the surface. Such a surface corrugation plot for the (110) face of nickel, determined by helium atom scattering, is shown in Figure 9.8. Note that the degree of roughness of such a plot will depend on which charge density contour is being plotted. Contours far from the surface (low charge density) will be flatter than contours close to the surface (high charge density).

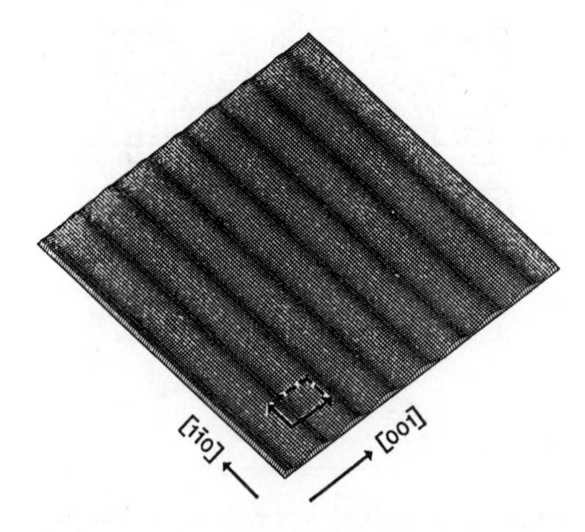

Figure 9.8 Three-dimensional plot of a given charge density contour over the surface of a metal—the Ni (110) surface. (Source: T. Engel and K.H. Rieder, *Surface Sci.* 1981;109:140. Reprinted with permission.)

Measurement of Gas Scattering

The details of any scattering event at the surface will depend on a number of variables pertaining to the incoming species and the surface potential. For the incoming particle, these variables include the particle mass and velocity, the direction of approach to the surface, and the impact parameter. For the surface potential, the important variables are the type and magnitude of the attractive forces involved and the variation of these forces with distance parallel to the surface; and the magnitude of the repulsive part of the potential and its variation both perpendicular and parallel to the surface. The relative importance of these parameters will vary, depending on the particle–surface combination involved.

Consider now the measurements that must be made in order to characterize these parameters.

For the case of elastic scattering, one must characterize the flux to the surface in terms of impingement rate, polar and azimuthal angles relative to the surface mesh, and velocity distribution. These same parameters must also be characterized for the flux leaving the surface.

For the case of direct inelastic scattering, one must characterize all of the parameters listed for the case of elastic scattering and, in addition, the amount of energy transferred or rearranged in the collision process.

For the case of trapping, one must know in addition the trapping probability as a function of the parameters characterizing the incoming flux, the surface lifetime of the trapped particle, the equilibrium position of the trapped particle on the surface, the details of any molecular rearrangements associated with the trapping or desorption process, the parameters characterizing the desorbing flux, including the energy distribution in internal modes of the particle, and the vibrational structure of the surface.

Numerous experimental arrangements have been developed to study the scatter-

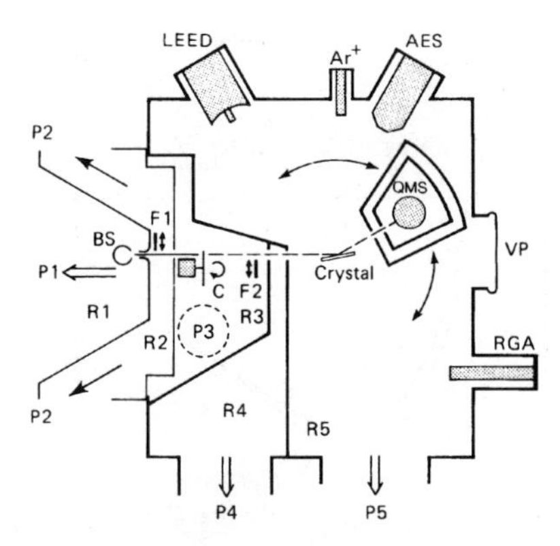

Figure 9.9 Schematic diagram of a typical molecular beam system used to study gas–surface scattering. *BS* is the beam source, *C* is a chopper assembly, *F*'s are beam flags, *QMS* is a quadrupole mass spectrometer detector, *VP* is a viewport. *R*'s and *P*'s indicate differentially pumped regions of the system and their associated pumping systems. (Source: C.T. Rettner, L.A. DeLouise and D.J. Auerbach, *J. Chem. Phys.* 1986;85:1131. Reprinted with permission.)

ing process in greater or lesser detail. Generally, the apparatus used is a molecular beam system similar to that shown in Figure 9.9. The impinging gas flux is generated in an arrangement of vacuum chambers that generally employs a nozzle source (effusion or free jet) and a series of skimmers or orifices separated by differential vacuum pumping stages. One of these stages may contain a chopper wheel for time-modulating the molecular beam. The surface under study is contained in an ultrahigh vacuum chamber, mounted to permit sample motion, heating, and cooling. One or more detectors for the scattered particles is required. These are generally mass spectrometers, which produce an output current proportional to the flux of scattered species of the mass of interest entering the detector. Ideally, the detector should be rotatable in the plane of the molecular beam and the sample normal, but much information can be obtained even from a fixed detector. The sample chamber often also contains other equipment for sample cleaning or for the detection of adsorbed particles. In its most versatile configuration, such a system permits measurement of all of the parameters discussed above.

Elastic Scattering

Consider now measurements of elastic scattering, in which the direction of motion of the incoming particle is changed in the scattering event, but its total energy and the energy distribution of any internal modes are not. Three basic phenomena are observed in scattering events of this type, namely, specular scattering, rainbow scattering, and diffraction. The type of scattering observed in any case will depend on the mass and velocity of the incoming particle and on the surface corrugation.

In all cases to be discussed, the attractive interaction between the particle and the surface is weak. For example, consider the potential well deduced for the interaction of

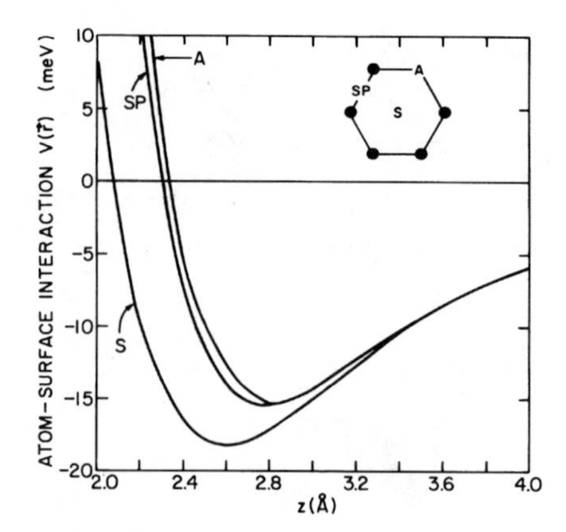

Figure 9.10 Potential energy diagram for a helium atom on the basal plane of graphite. Curves *A, SP,* and *S* give the potential energy as a function of distance for perpendicular approaches to a carbon atom center (*A*), the bridge position between two carbon atoms (*SP*) and the center of the 6-membered carbon ring (*S*). (Source: W. Carlos and M.W. Cole, *Surface Sci.* 1980;91:339. Reprinted with permission.)

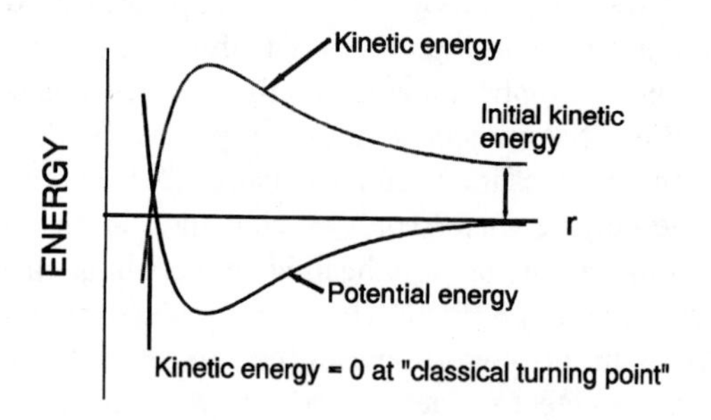

Figure 9.11 Potential energy (solid curve) and kinetic energy (dotted curve) for an atom approaching a solid surface.

helium with the graphite (0001) face, as shown in Figure 9.10. The potential well depth in this case is 0.014 eV. By comparison, the kinetic energy per atom in a helium beam produced in a free jet expansion from 300 K is 0.063 eV—a much larger value.

Because the attractive well depth is so small, elastic scattering can be explained quite well by ignoring the attractive part of the potential well and looking at scattering due to the interaction between the particle and the repulsive part of the potential, as suggested earlier in this chapter.

If one considers the interaction of an incoming particle with the surface in terms of the relation of the particle kinetic energy to the system potential energy, one gets behavior similar to that shown in Figure 9.11. The particle approaches the surface with some initial kinetic energy. As the particle approaches closely enough to sense the surface potential, it is first accelerated by the attractive part of the potential, increasing its

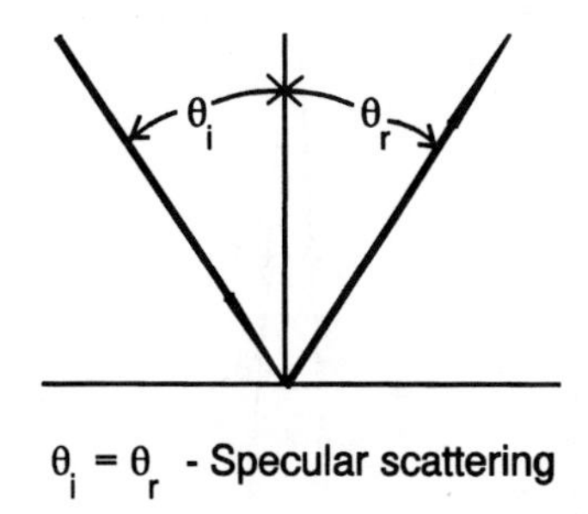

Figure 9.12 Specular atom scattering from a flat corrugation.

kinetic energy, then decelerated by the repulsive part of the potential, eventually reaching zero kinetic energy. At this point, the particle is reflected and returns to the gas phase. The point at which this reflection event takes place is called the *classical turning point*.

Note that as the incident initial kinetic energy increases, the particle penetrates farther into the repulsive region of the potential before being reflected and thus "sees" a stronger corrugation. The results of this behavior will be evident in the next section. Scattering in this regime is often treated using what is called a *classical corrugated hard wall potential*.

Classical Scattering

Scattering from the surface corrugation can be treated either classically or quantum mechanically. Consider first the classical approach, which can explain both specular and rainbow scattering.

If a particle is scattered at very low incident energy, so that the classical turning point will be far from the surface, in the region in which the surface corrugation is essentially flat, or is scattered from a surface having an inherently small corrugation amplitude, the scattering will be specular, as shown in Figure 9.12. The plot of scattered intensity vs. angle for this case is sharply peaked at $\theta_r = \theta_i$, as shown in Figure 9.13.

If a particle penetrates close enough to the surface to see a significant corrugation amplitude, significantly different behavior is observed. In this case, classically, the particle behaves as if it were scattered in the local specular direction, as shown in Figure 9.14. This leads to what is called *rainbow scattering*. The maximum possible scattering angle relative to the vertical in such a situation occurs for scattering from the impact parameter associated with the inflection point in the surface corrugation, as shown in Figure 9.15. The trend to large maximum angles for steep corrugations is obvious. If one analyzes all of the possible classical trajectories in such a case, one obtains a plot of scattered intensity vs. angle such as that shown in Figure 9.16. The sharp maxima occur at the maximum scattering angles, or the *rainbow angles* as they are sometimes called. An experimental example of this behavior is shown in Figure 9.17, for the case of neon scattered from the LiF (001) surface.

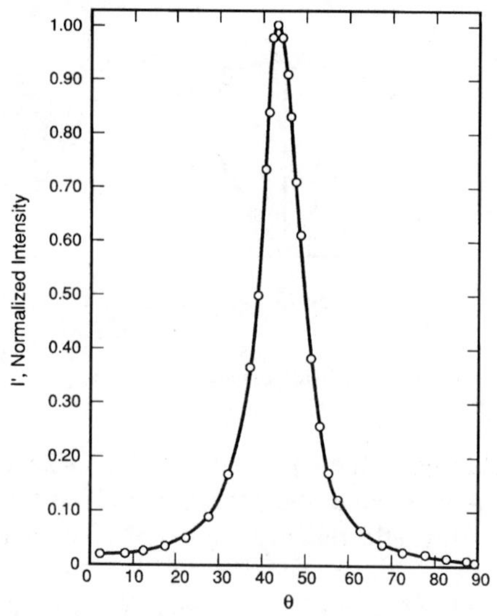

Figure 9.13 Plot of scattered intensity as a function of scattering angle for specular scattering: Helium scattering from W (110). (Source: W.H. Weinberg and R.P. Merrill, *J. Chem. Phys.* 1972;56:2881. Reprinted with permission.)

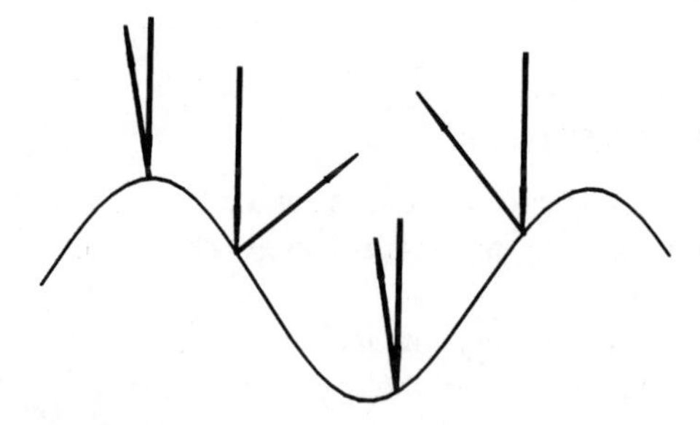

Figure 9.14 Scattering of atoms from a rough surface corrugation. Scattering occurs at the local specular angle, giving rise to "rainbow scattering."

Quantum Scattering

The observation of diffractive scattering cannot be explained on the basis of classical scattering. To understand this behavior, one must have recourse to a quantum mechanical description of the scattering process. In this treatment one must realize that atoms can be treated as having a wavelength, just as was the case for electrons treated in Chapter 1. Again, the wavelength is related to the particle momentum by the DeBroglie equation

$$\lambda = \frac{h}{mv} = \left(\frac{h^2}{2mE}\right)^{1/2}. \qquad 9.2$$

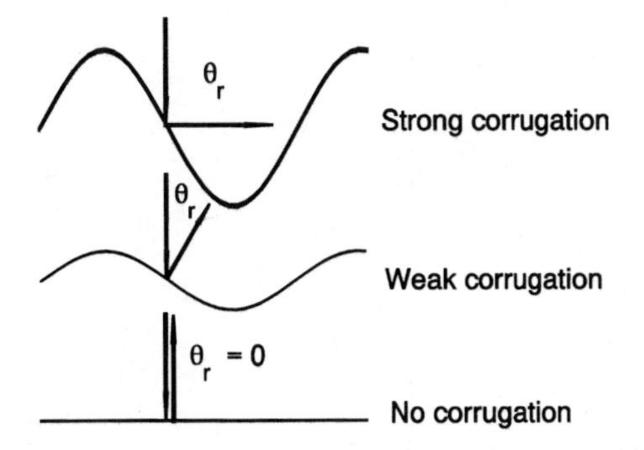

Figure 9.15 Scattering trajectory plots for the impingement of a gas atom on surfaces of differing corrugation amplitude. The maximum deviation from specular scattering is associated with the inflection point of the corrugation.

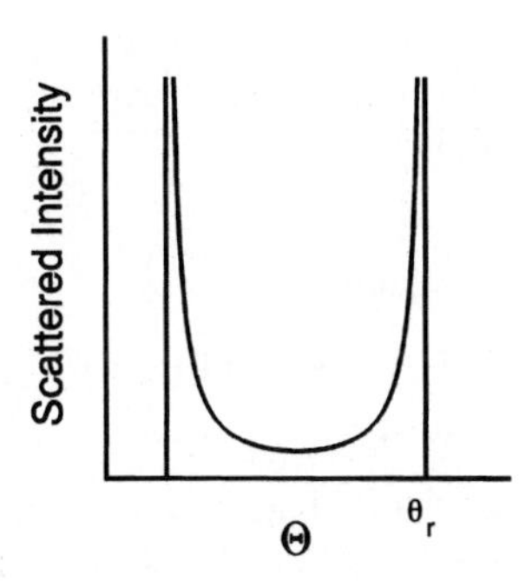

Figure 9.16 Calculated scattered intensity distribution for scattering from a surface with a large corrugation amplitude. Note the singularities at the rainbow angles.

Alternatively, one can describe the impinging atom in terms of its momentum vector, $\mathbf{k}_0$, again as in the case of electrons, as

$$\mathbf{k}_0 = \frac{2\pi}{\lambda} = 2\pi\left(\frac{2mE}{h^2}\right)^{1/2}. \qquad 9.3$$

Recall that for the case of electrons

$$\lambda \approx \left(\frac{150}{V}\right)^{1/2} (\text{Å}), \qquad 9.4$$

where V was the electron energy in electron volts. For the case of atoms, the wavelength will depend on both m and E (or, alternatively, the velocity, v). For the case of a monatomic gas expanded from 300 K in a free jet expansion,

$$E = \left(\frac{5}{2}\right)kT = 0.063\,eV. \qquad 9.5$$

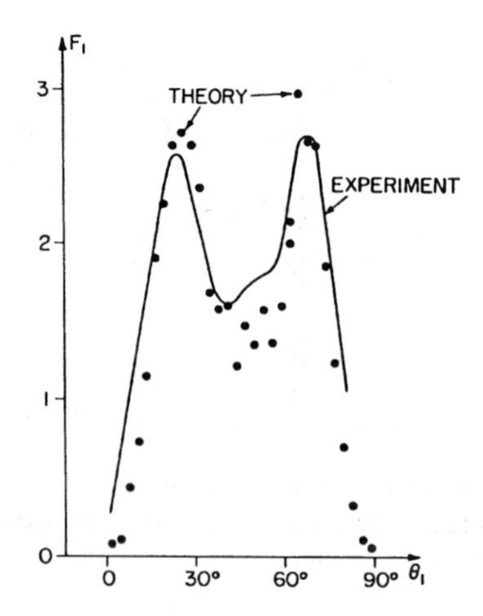

Figure 9.17 Experimental observation of rainbow scattering for neon scattered from LiF (001). Dotted curve: theory; solid curve, observation. (Source: F.O. Goodman, *Surface Sci.* 1971;26:327. Reprinted with permission.)

For the case of helium, $m = 4/N_{Av}$, yielding

$$\lambda = 0.56 \text{ Å}, \quad \mathbf{k}_0 = 11.22 \text{ Å}^{-1}. \qquad 9.6$$

Again, as in the case of the 150 eV electrons, the wavelength is comparable with the interatomic distance on a surface. Note that, at a given temperature, the wavelength will decrease with increasing particle mass and that, for a given atom, the wavelength will decrease with increasing temperature.

The principal consequence of the wave nature of atoms is that they will scatter from the surface corrugation just as electrons scatter from the ion cores. Again, a surface mesh can be defined based on the periodicity of the corrugation, and from this one can construct a surface reciprocal lattice. Again, as in the case of electrons, the Ewald sphere construction determines what diffracted beams will be seen in atom scattering and the scattering angles at which they will appear. The only significant difference in this case is that, for reasons of system geometry, most atom scattering studies scan the detector in a plane defined by the incoming atomic beam direction and the surface normal. The net result of this is that the Ewald sphere is reduced essentially to a circle of radius $R = \mathbf{k}_0$ and will intersect only those reciprocal lattice rods that lie in the scattering plane. These rods will be defined by the azimuthal orientation of the crystal surface relative to the scattering plane.

As an example, consider the (110) surface of an FCC material, as shown in Figure 9.18. The surface corrugation along the [10] direction is shown schematically. If an atomic beam is incident on this surface at a polar angle θ and at the azimuthal angle, Φ, corresponding to this [10] direction, the Ewald sphere construction will be as shown in Figure 9.19. Diffracted beams will be observed wherever the circle intersects the reciprocal lattice rods and at the scattering angle, θ, defined by the vector $\mathbf{k}'$. Note that only those diffracted beams associated with back scattering from the surface will be

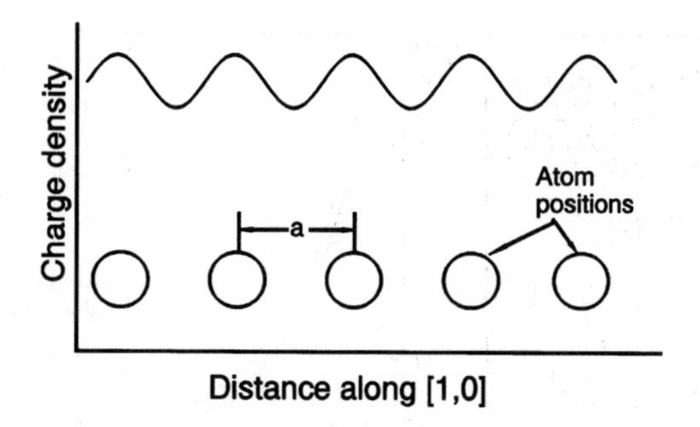

Figure 9.18 Surface corrugation along the [10] direction of the (100) surface of an FCC crystal.

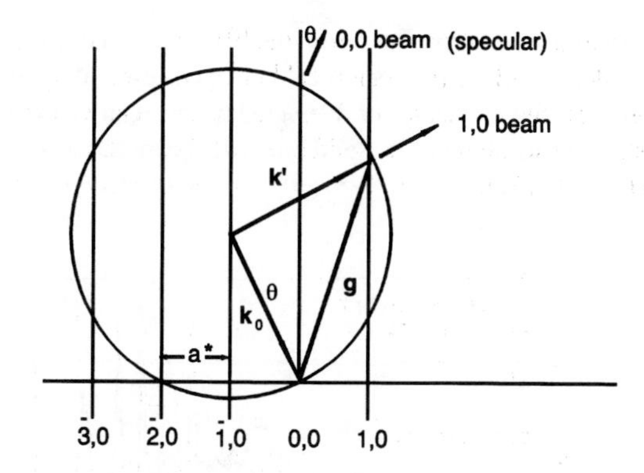

Figure 9.19 Ewald sphere construction for atom scattering along the [10] direction of the (110) face of an FCC crystal.

observed, and that the diffracted beams will move toward the specular beam as $\mathbf{k}_0$ is increased (i.e., as λ is decreased, or the particle energy increased). The intensity of the scattered beams will generally increase as the amplitude of the corrugation increases. Consequently, intense diffractive scattering will be observed on surfaces having inherently strong corrugations or as the particle energy is increased so that it penetrates closer to the surface, and thus into a region of stronger corrugation. That is, stronger diffraction effects are observed as E is increased and λ correspondingly decreased. This effect is seen in the results of a study of helium diffraction from Ni (110), shown in Figure 9.20. It is through analysis of diffraction patterns such as this, obtained as a function of particle energy, θ and Φ, that the surface corrugation plots like the one shown in Figure 9.8 are obtained.

Diffractive scattering can also be used to obtain information on surface reconstructions or the formation of ordered adsorbed layers, as was the case for LEED. As examples of these sorts of studies, Figure 9.21 shows data for helium scattering from the Si (111) - 7 × 7 structure and Figure 9.22 for an ordered layer of hydrogen adsorbed on Ni (110).

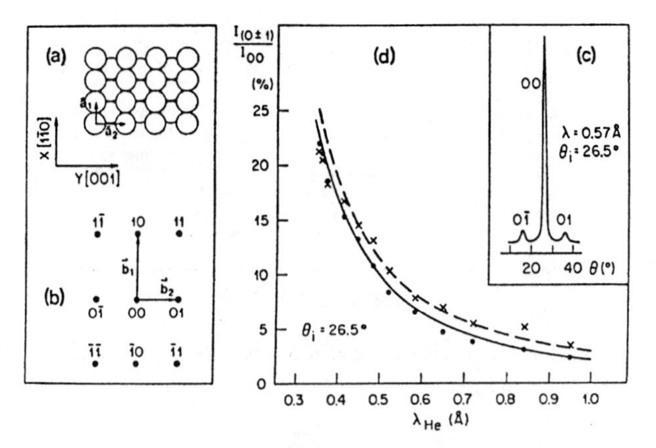

Figure 9.20 Helium atom diffraction from the [01] azimuth of the Ni (110) surface, showing hard sphere model of surface (a), reciprocal lattice corresponding to this surface (b), helium scattering spectrum (c), and intensities of diffracted beams relative to the specular beams as a function of the particle wavelength (d). Solid line: (01) beam, Dashed line: (0$\bar{1}$) beam. (Source: T. Engel and K.H. Rieder, *Surface Sci.* 1981;109:140. Reprinted with permission.)

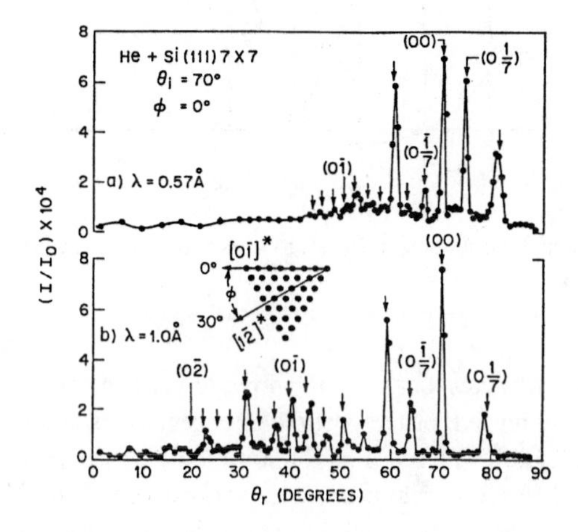

Figure 9.21 Diffractive scattering of helium from the Si (111)-7×7 structure. The beam was incident at a polar angle of 70°, along the [01] azimuth of the crystal; (a) λ = 0.57Å, (b) λ = 1.0Å. (Source: M.J. Cardillo and G.E. Becker, *Phys. Rev. Lett.* 1979;42:508. Reprinted with permission.)

Finally, note the parallelism of the quantum and classical approaches to elastic scattering. In the case of smooth surfaces (very weak corrugation), the diffraction process will give rise to a very large specular peak, with the intensity of the higher order peaks approaching zero as the intensity of the corrugation approaches zero. This is essentially the behavior observed classically for an ideal flat surface. For more strongly corrugated surfaces, a parallel exists between the observed intensities of the various

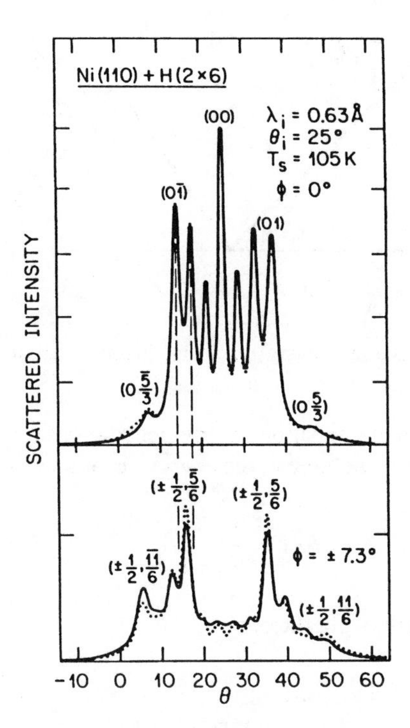

Figure 9.22 Diffractive helium scattering from a hydrogen-covered Ni (110) surface. Curves are shown both for in plane ($\Phi = 0$, top) and out of plane ($\Phi = 7.3°$, bottom) scattering. (Source: K.H. Rieder and T. Engel, *Phys. Rev. Lett.* 1980;45:824. Reprinted with permission.)

higher order reflections and the intensity as a function of scattering angle predicted for classical rainbow scattering. This trend is shown in Figure 9.22, for helium scattering from Ni (110).

Direct Inelastic Scattering

Next consider processes in which there is still only a single collision with the surface but one that involves finite energy exchange with the surface. One can look at this process either in terms of energy transfer or of momentum transfer. Unfortunately, at this point, the theoretical treatments that successfully describe energy transfer do not provide information on momentum transfer, so that two different treatments of the scattering process will have to be developed.

Consider first energy transfer. One can look at this process in terms of the processes that take place when a gas in thermodynamic equilibrium at a temperature T_i impinges on a surface at $T_s \neq T_i$ as shown in Figure 9.23. In general, if there were a completely elastic collision, with no energy exchange between the gas molecules and the surface, then the gas leaving the surface after striking it would have $T_r = T_i$. At the opposite extreme, if these gas molecules came to complete thermal equilibrium with the surface before returning to the gas phase, the result would be $T_r = T_s$. The degree of

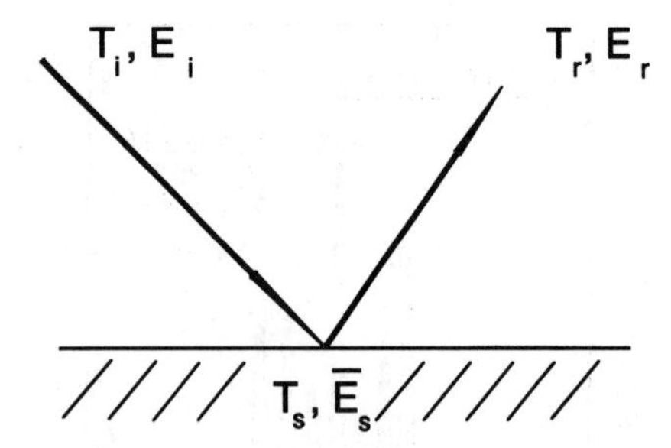

Figure 9.23 Schematic picture of energy exchange in direct inelastic scattering.

temperature or energy accommodation in any specific case can be characterized in terms of an accommodation coefficient, the energy accommodation coefficient,

$$EAC = \frac{E_r - E_i}{E_s - E_i}, \tag{9.7}$$

or the thermal accommodation coefficient

$$TAC = \frac{T_r - T_i}{T_s - T_i}, \tag{9.8}$$

Note the limits on the value of the *EAC* and *TAC*: if $T_r = T_s$, implying complete accommodation, the accommodation coefficients will be equal to unity; if $T_r = T_i$, implying no energy exchange, the accommodation coefficients will equal zero. Intermediate cases will yield values between zero and one.

The Hard Sphere Model

Consider next theoretical approaches that can be used to characterize the expected value of the thermal accommodation coefficient. The simplest way to treat the problem is to assume that the impinging gas atom interacts with a single atom in the surface, and that the interaction follows the classical laws of energy interchange in a collision between hard spheres. This situation is shown in Figure 9.24. Carrying out this calculation, assuming that the surface atom is initially at rest, leads to

$$TAC = \frac{2\mu}{(1+\mu)^2}, \tag{9.9}$$

where μ is the ratio of the masses of the gas atom, m, and the surface atom, M. Note that when m is very small compared to M, the accommodation coefficient will be small. For increasing values of μ, the accommodation coefficient will go through a maximum at $\mu = 1$, $TAC = 0.5$. This model gives reasonable agreement with experiment for cases in which the incident particle energy is large compared to the depth of the attractive

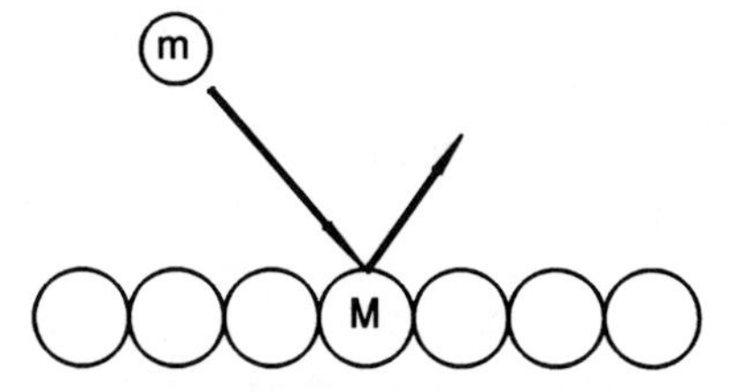

Figure 9.24 The hard sphere model of atom scattering.

potential well for the surface interaction, that is, at high temperatures or in systems in which the attractive part of the potential is weak.

The "Bedspring" Model

For cases in which the incident energy is not large compared to the depth of the potential well, a more sophisticated treatment is required. Several such treatments have been developed, all of which treat the interaction of the gas atom with a lattice having orthogonal symmetry and bound together by nearest neighbor interactions only. Treatments of this sort have been developed for one-, two- and three-dimensional lattices. The model in one such three-dimensional case is shown schematically in Figure 9.25. The surface atomic bonds are represented as springs, which either stretch or contract when the impact of the gas atom takes place. (Such a model is often referred to as a *bedspring* model.)

This model has a number of limitations. It is based on a classical mechanical model of the atom–atom interaction. It assumes that the surface is initially at rest, or equivalently, that the atom–atom interaction and the thermal modes of the lattice are only loosely coupled. As mentioned, a cubic lattice with near-neighbor bonding forces is assumed, and the collision is assumed to take place along the line of centers of the colliding atoms, with only two atoms participating in the collision. (This limitation means that no information on the direction of the scattering process can be obtained.)

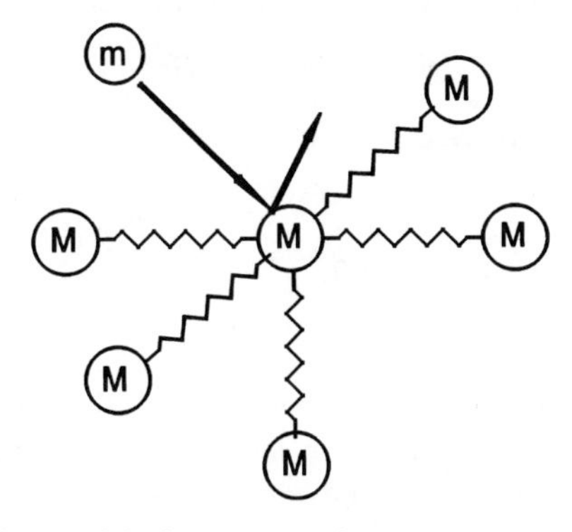

Figure 9.25 The bedspring model of atom scattering.

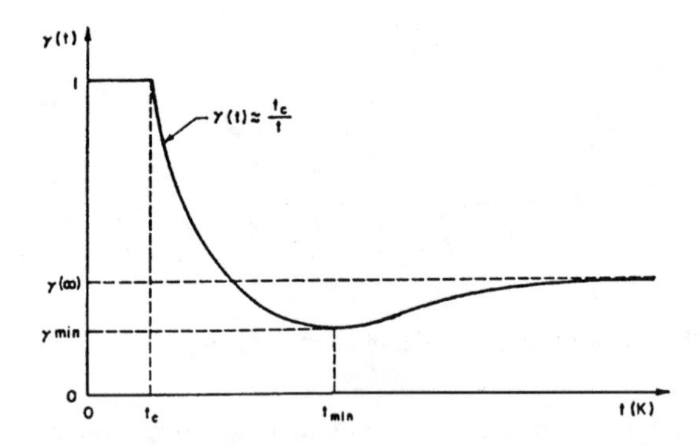

Figure 9.26 Qualitative behavior of the EAC as a function of incoming particle energy as predicted by the bedspring model. γ is the EAC. (Source: F.O. Goodman and H.Y. Wachman, *J. Chem. Phys.* 1967;46:2376. Reprinted with permission.)

Finally, a model must be assumed for the form of the attractive part of the potential between the gas atom and surface atom.

In spite of the foregoing limitations, theories of this sort can give reasonable agreement with experiment for relatively simple systems, for example, the rare gases on clean metal surfaces. The general predictions of the theory are summarized in Figure 9.26. At extremely low temperatures, such that kT is small compared to the assumed potential well depth, the atom is trapped, and complete accommodation results. As the incident energy increases and approaches the well depth, the accommodation coefficient decreases and goes through a minimum at $E_i \approx E_0$. At larger values of incident energy, the hard sphere limit of Equation 9.9 is approached. The expected hard sphere limit is observed for rare gases on many clean metals. The minimum in accommodation coefficient has been observed for helium and neon. The expected increase in accommodation coefficient with increasing gas atom mass has also been observed.

Note that the results of this treatment depend only on the temperature of the gas. The temperature of the lattice is implicitly assumed to be 0 K. This treatment has been extended to develop a so-called *reduced model,* intended to fit a wide range of gas–surface combinations. The result of this treatment is a single equation of the form

$$(B\mu)^{-1/2}(EAC) = H(\rho), \qquad 9.10$$

in which μ is the reduced mass defined above and

$$B = \frac{8a^2 D}{m_s\left(\dfrac{k\theta_D}{2\hbar}\right)}, \qquad \rho = \frac{kT_g}{D}, \qquad 9.11$$

in which D and a are parameters describing the attractive part of the potential, and θ_D, the Debye temperature, is a parameter related to the lattice vibrational frequency, which will be discussed in more detail later in this chapter. This treatment permits correlation of the behavior of a range of systems, as shown in Figure 9.27.

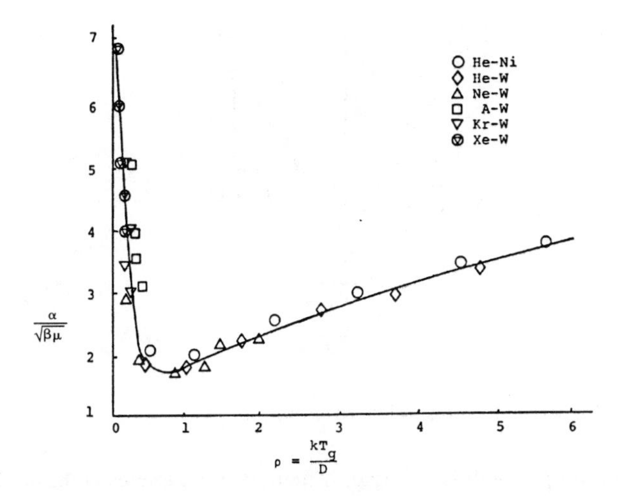

Figure 9.27 Comparison of experimental measurements of the EAC for the rare gases on tungsten with the prediction of Equation 9.10. (Source: A. McFall, in: S.S. Fisher, ed. *Rarefied Gas Dynamics*. New York: American Institute of Aeronautics and Astronautics, 1981. Reprinted with permission.)

The "Hard Cube" Model

Momentum accommodation, on the other hand, must be characterized in terms of the extent to which an incoming atom remembers the direction from which it came. Consequently, a molecular beam experiment is required in order to characterize momentum accommodation. The results of such a measurement of scattering probability as a function of scattering angle can be presented in a polar plot, as shown in Figure 9.28. This plot shows the intensity of the scattered atom flux as a function of the detector angle relative to the surface normal, for the case where the incoming gas atoms all came from a narrow range of directions. In the case of complete momentum accommo-

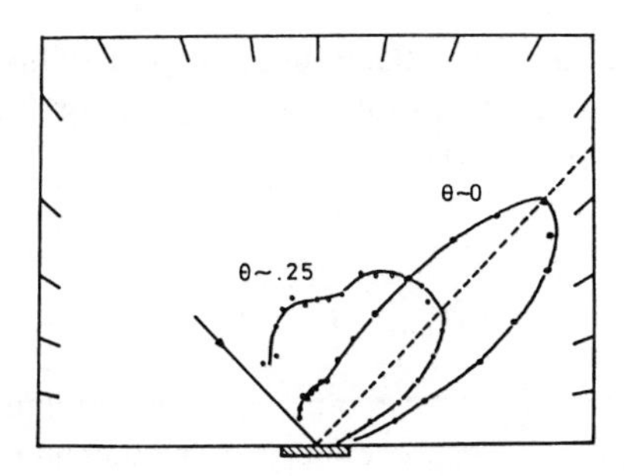

Figure 9.28 Angular scattering distribution in the scattering plane for O_2 scattering from Pd (111) at oxygen coverages of 0 and 0.25 monolayers, showing lobular scattering due to incomplete momentum accommodation. (Source: T. Engel, *J. Chem. Phys.* 1978;69:373. Reprinted with permission.)

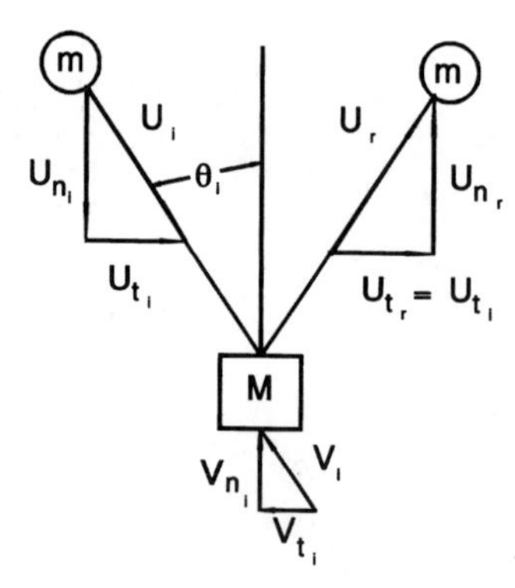

Figure 9.29 The hard cube model of atom scattering.

dation, the resulting plot will be a circle. If there is incomplete accommodation, a lobe will be observed, pointed more-or-less in the specular scattering direction. Note that for a given incident beam intensity, the area inside the curve must be constant, independent of the degree of accommodation. (Actually, the polar plot is three dimensional, as it is possible for scattering to occur out of the plane of the beam and detector. Strictly speaking the volume inside the three-dimensional curve is what must be conserved.)

One model that has been developed to explain observed momentum accommodation behavior, called the *hard cube model,* is illustrated in Figure 9.29. The atoms in the surface are assumed to be hard cubes, moving with a Maxwellian distribution of velocities, V_i. The atoms from the gas are assumed to impinge with velocity U_i and to leave with a velocity U_r. It is assumed that the component of U_i parallel to the surface, U_t, is not changed in the collision, and that the collision between the gas atom of mass m and the surface atom of mass M follows the hard sphere law for momentum exchange. That gives, for the normal component of the velocity of the scattered particle,

$$U_{n_r} = \left(\frac{1-\mu}{1+\mu}\right)U_{n_i} + \left(\frac{1}{2\mu}\right)V_{n_i}. \tag{9.12}$$

Using this model one can show that the ratio of the probability of gas molecules colliding with surface atoms having V_i positive (rising) to the probability of colliding with a surface atom having V_i negative (falling) is

$$\frac{P_{\text{rise}}}{P_{\text{fall}}} = \frac{U_{n_i} + V_{n_i}}{U_{n_i} - V_{n_i}}. \tag{9.13}$$

That is, as U_{n_i} increases, the above ratio approaches unity; as V_{n_i} increases and approaches U_{n_i}, the above ratio becomes very large compared to unity.

By applying the above ideas, one can determine the probability that a molecule approaching a surface at an angle θ_i with a velocity U_i will leave with a given θ_r, U_r as a function of θ_r, U_r. The result of this treatment is that a lobe-type scattering pattern is predicted, as shown in Figure 9.30. Moreover, as U_{n_i} increases relative to V_{n_i}, the value

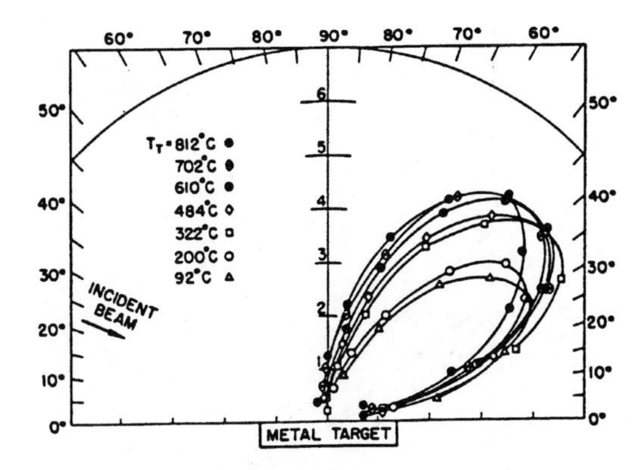

Figure 9.30 Comparison of the hard cube model with observed angular scattering intensities for the scattering of argon at 22°C from platinum for a range of surface temperatures. Arrow indicates the incident beam direction. (Source: J.J. Hinchen and W.M. Foley, in J.H. DeLeeuw, ed. *Rarefied Gas Dynamics,* Vol 2. Academic Press, 1966, p. 505. Reprinted with permission.)

of θ_r associated with the maximum scattered intensity increases for a given θ_i. That is, the lobe lies closer to the surface than the specular angle, and vice versa. Such results have been observed in practice in a few systems. Further improvements in the theory are required in order to describe all of the cases studied.

Quantum Inelastic Scattering

In order to understand the mechanism of energy transfer involved in direct inelastic scattering, it is necessary once again to look at the scattering process in quantum mechanical terms. In this case, we will look at the transfer of quantized amounts of energy to or from the impinging gas atom. The quantization of the energy transferred arises from the fact that the energy stored in the solid is due to lattice vibrations and that only certain vibrational frequencies are quantum mechanically allowed.

This discussion will be couched in terms of quanta of vibrational energy known as *phonons,* whose energy is defined by

$$\varepsilon = h\nu = \frac{hv}{\lambda}, \qquad 9.14$$

or in terms of the phonon wave vector (momentum):

$$\mathbf{k} = \frac{2\pi}{\lambda} = \frac{2\pi\nu}{v}, \qquad 9{:}15$$

where ν is the phonon frequency and v is its velocity of propagation in the crystal. Note that v will depend on the direction of propagation of the wave in the crystal, so that there will not be a one-to-one correspondence between frequency and wave vector in

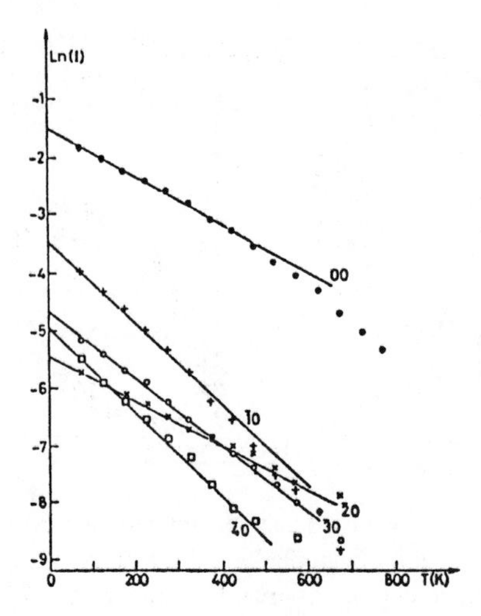

Figure 9.31 Variation of the specular peak and diffracted beam intensities with temperature for the scattering of helium from Cu (110). The decrease is due to the surface Debye–Waller effect. (Source: J. Lapujoulade, Y. LeCruer, M. Lefort, Y. Lejay, and E. Maurel, *Surface Sci.* 1982;118:103. Reprinted with permission.)

this case. (That is, there will be a "dispersion curve" for any acoustic wave relating v and **k** over a wide range of crystallographic orientations.)

The simplest way to observe the effect of lattice vibrations on gas scattering is to look at the intensity of the specularly scattered elastic peak or diffraction peaks as a function of surface temperature. The general observation, as shown in Figure 9.31, is that the peak intensity decreases exponentially with increasing surface temperature. The reason for this decrease is similar to the so-called Debye–Waller effect in x-ray or electron scattering. As the temperature of the solid increases, the vibrational amplitude of the atoms increases, and the probability of finding the atom exactly on its lattice point, so that it can contribute to diffractive scattering, decreases. This leads to scattered intensity vs. temperature behavior of the form

$$I = I_0 \exp(-2W), \qquad 9.16$$

where the physical significance of W depends on the type of scattering being considered. For the case of x-rays,

$$W = \tfrac{1}{2}\langle (u \cdot \Delta\mathbf{k})^2 \rangle_T, \qquad 9.17$$

where u is the displacement of the atom from the lattice site and $\Delta\mathbf{k}$ is the change in x-ray wave vector associated with the scattering event. For a classical harmonic oscillator, $\langle u^2 \rangle$ increases linearly with temperature, leading to the behavior predicted by Equation 9.16.

For the case of atom scattering, the situation is more complicated. The two principal complications are what is known as the Armand effect, arising from the fact that, because of finite atomic size, the incoming atom may interact with more than one surface atom in the course of the collision, and the Beebe effect, involving refraction of the incident particle due to the attractive part of the surface potential. Taking account of these complications leads to

$$\langle u^2 \rangle = \frac{3\hbar^2 T_s}{m_s k \theta_D^2} \tag{9.18}$$

and

$$(\Delta \mathbf{k})^2 = \frac{4m_g{}^2 v^2 \cos^2\theta_i}{\hbar^2} + \frac{8m_g D}{\hbar^2}. \tag{9.19}$$

In these relations, T_s is surface temperature; θ_D the surface Debye temperature, a fictitious temperature related to atomic vibration frequencies in the solid; m_s and m_g the surface and gas atom masses; θ_i the polar angle of impingement; and D the potential well depth. The second term on the right of Equation 9.19 is the correction for the Beebe effect. Inserting the above relations in the expression for specular peak intensity leads to

$$I = I_o \exp\left[-24\mu\left(\rho^2+1\right)\frac{DT_s}{k\theta_D^2}\right], \tag{9.20}$$

where

$$\rho = \left(\frac{E_i}{D}\right)^{1/2} \cos\theta_i. \tag{9.21}$$

This relation gives a reasonably good description of experiment for many cases of elastic scattering, if D is used as an adjustable parameter. However, the value of D deduced in this way does not always agree with values determined by other experimental techniques. This is still an area of active study.

Trapping

For completeness, the phenomenon of *trapping* should be mentioned at this time. Trapping implies that the incoming particle remains at the surface long enough that it comes to thermal equilibrium with the lattice. In terms of scattering behavior, this means that except for cases of activated adsorption, which will be discussed in Chapter 12, the *EAC* and *TAC* will be unity and the angular distribution of scattered atoms or molecules will show the $\cos\theta$ distribution predicted by the kinetic theory of gases.

Problems

9.1. Determine the diffracted beams that would be seen and the angles relative to the surface normal at which they would be seen for the scattering of a helium beam having an energy of 50 meV incident at an angle of 30° from the surface normal along the [1, 0] azimuth of a (100) surface of a BCC crystal. Assume that a_0 for the crystal is 0.25 nm.

9.2. What is the difference in time of flight over a 1 m path length for a helium atom that has struck a surface and excited a phonon having an energy of 12 meV, relative to the time of flight for an atom in the same beam elastically scattered from the same surface? Assume that the incident beam energy is 63 meV.

9.3. Helium gas at 500 K strikes a clean molybdenum surface maintained at 300 K. What will be the temperature of the helium gas after the interaction, assuming that the thermal accomodation coefficient is given by the hard sphere collision model?

9.4. A platinum single crystal is cut to expose a surface that is misoriented from the (111) plane by 3.11° along a [110] direction. A helium scattering experiment is carried out on this surface, using an atomic helium beam having an energy of 15 meV. The helium beam is incident along the [110] azimuth at a polar angle of 45°. For a helium beam of this energy, the surface corrugation of the Pt (111) face is essentially flat. What diffracted beams will appear in the scattered helium flux? At what polar angles will the specular, (1,0), and $(\bar{1},0)$ beams appear?

Bibliography

Barker, JA, and Auerbach, DJ. Gas-surface interactions and dynamics; thermal energy atomic and molecular beam studies. *Surface Science Reports* 1985;4:Nos.1/2.

Brusdeylins, G, Doak, RB, and Toennies, JP. High resolution inelastic scattering of He atoms from LiF (100). In: S.S. Fisher, ed. *Rarefied Gas Dynamics,* 74:1:166–177.

Goodman, FO. Theoretical aspects of atom–surface diffraction, inelastic scattering and accomodation coefficients. In: S.S. Fisher, ed. *Rarefied Gas Dynamics,* 74;1:3–49.

Hinchen, JJ, and Foley, WM. Scattering of molecular beams by metallic surfaces. In: J.H. de Leeuw, ed. *Rarefied Gas Dynamics*. New York: Academic Press, 1966;2:505.

Hutchison, J, Celli, V, Hill, NR, and Haller, M. Inelastic processes in low energy atom-surface scattering. In: S.S. Fisher, ed. *Rarefied Gas Dynamics,* 74:1;129–141.

Thomas, LB. Accomodation of molecules on controlled surfaces—experimental developments at the University of Missouri 1940-1980. In S.S. Fisher, ed. *Rarefied Gas Dynamics.* 74:1;83–108.

CHAPTER **10**

Adsorption—The Kinetic View

A thermodynamic description of multicomponent capillary systems was developed in Chapter 5, in which the concept of adsorption was presented in terms of a surface excess quantity of one or more of the components in the system. It was shown on a strictly thermodynamic basis that the adsorption, Γ, would not in general be zero. In particular, for the case of a two-component system containing only flat interfaces, in which one phase was a gas, one could arbitrarily set the adsorption of one component equal to zero and deduce that

$$\Gamma_2 = -\left(\frac{1}{RT}\right)\left(\frac{\partial \gamma}{\partial \ln p_2}\right)_T. \qquad 10.1$$

This relation would apply, for example, to the case of a crystalline solid in contact with a gas phase composed of some other chemical species. The current chapter will consider the process of adsorption at the molecular level, using the picture, discussed in Chapter 9, of the change in system potential energy that occurs when a molecule from a gas phase approaches a surface. It will be seen that the thermodynamic and kinetic treatments lead to equivalent results.

The Kinetic Picture

Consider again the process of adsorption, this time from a kinetic, rather than a thermodynamic viewpoint. This approach is mostly readily visualized by taking as a model the adsorption process at the surface of a solid exposed to an adsorbable gas. If one looks at such a system at the molecular level, one sees that there will be a continual process of the impingement of gas molecules from the gas phase onto the surface and a corresponding flux of molecules from the surface to maintain the net overall rate of accumulation at the surface equal to zero at equilibrium. Figure 10.1 shows this process schematically.

The rate of molecular impingement, I (molecules per square centimeter-second), as developed in the treatment of the kinetic theory of gases in Chapter 7, is

$$I = \frac{p}{(2\pi mkT)^{1/2}}. \qquad 10.2$$

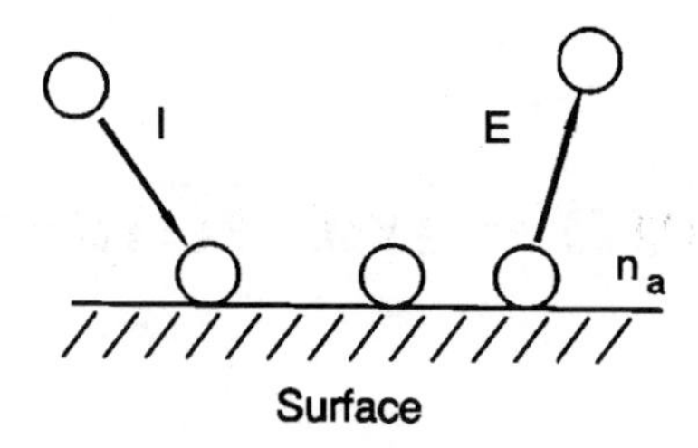

Figure 10.1 Dynamic equilibrium at a gas–solid interface. At equilibrium, $I = E = n_a/\tau_a$.

To get the impingment rate in the units given above, the pressure, p, molecular mass, m, temperature, T, and Boltzmann's constant, k, must all be in consistent units. This is an expression that is true irrespective of the nature of the gas or the surface involved. Typical values for I as a function of pressure for $T = 300$ K, $m = 28$ AMU are given in Table 10.1.

As discussed in Chapter 9, when a gas molecule approaches a surface, it will in all cases experience an initial attractive force. As the distance from the surface decreases, the magnitude of this attractive force at first increases, then passes through a maximum, and at even closer distances of approach becomes a repulsive force. Figure 10.1 shows a typical plot of potential energy vs. distance for such a system. Here the potential energy of the system when the molecule is far from the surface is arbitrarily chosen as being equal to zero. Thus, a negative value of E signifies attraction and a positive value repulsion. The shape of the curve, with the potential well shown, indicates that a potential minimum exists near the surface, having a depth ΔE_a. The magnitude of ΔE_a depends on the nature of the interaction between the gas and the surface. Observed values cover the range from 100 cal/mol for systems of light molecules where only van der Waals forces are involved to 150 kcal/mol for systems in which strong chemical bonds are formed between the gas molecule and the surface. The position of the minimum, r_0, is generally within a few Angstrom units of the surface.

The shape of the well shown in Figure 10.2, which is appropriate for the case where no barrier exists to adsorption, indicates that as a molecule approaches the sur-

Table 10.1 Molecular Impingement Rate on a Surface for Various Pressures

Pressure (Torr)	*Impingement Rate (molecules/cm²-sec)*
760	3×10^{23}
1	4×10^{20}
10^{-3}	4×10^{17}
10^{-6}	4×10^{14}
10^{-9}	4×10^{11}
10^{-12}	4×10^{8}

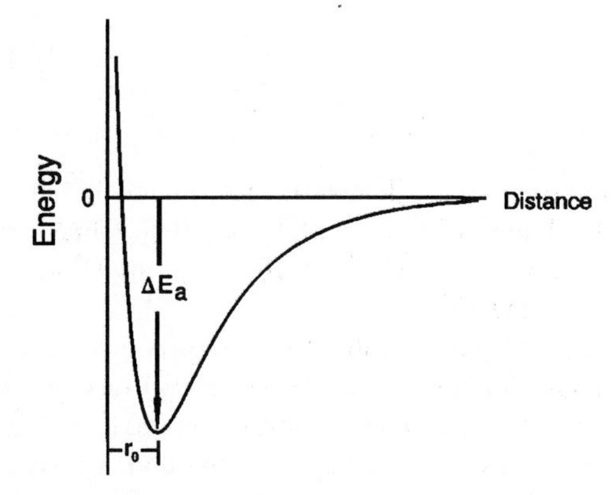

Figure 10.2 System potential energy as a function of the distance between an approaching gas molecule and the surface.

face, it will reach a position of minimum potential energy and may be trapped in this position. If such trapping takes place, then the probability of escape per unit time is given by absolute reaction rate theory as

$$\upsilon_{des} = \upsilon_o \exp\left(\frac{-\Delta G^*_{des}}{kT}\right), \tag{10.3}$$

in which ν_{des} is the desorption frequency, sec^{-1}, ν_0 an *attempt frequency,* on the order of the atomic vibrational frequency of the crystal lattice (about 10^{13} sec^{-1}), and ΔG^*_{des} is the free energy of activation for the escape process. We can approximate this last term by

$$\begin{aligned} \exp\left(\frac{-\Delta G^*_{des}}{kT}\right) &= \left(\frac{f^*}{f}\right)\exp\left(\frac{-\Delta H_{des}}{kT}\right) \\ &\approx \left(\frac{f^*}{f}\right)\exp\left(\frac{\Delta E_a}{kT}\right), \end{aligned} \tag{10.4}$$

in which f and f^* are the molecular partition functions of the system in the equilibrium and activated states, respectively. Thus,

$$\nu_{des} = \nu_o\left(\frac{f^*}{f}\right)\exp\left(\frac{-\Delta H_{des}}{kT}\right). \tag{10.5}$$

Alternatively, rather than looking at the desorption frequency, it is often more useful to use the reciprocal of this term, which is known as the *mean stay time for adsorption* or *mean surface lifetime*. This is given by

$$\tau_a = \frac{1}{\nu_{des}} = \tau_o \exp\left(\frac{\Delta H_{des}}{kT}\right), \tag{10.6}$$

where

$$\tau_0 = \frac{1}{\nu_0}\left(\frac{f}{f^*}\right). \quad 10.7$$

Note that this implies that the deeper the potential well, that is, the greater the magnitude of ΔH_{des}, the longer will be the mean stay time. The values of τ_0 measured experimentally range from about 10^{-16} to 10^{-9} sec, implying that the ratio f/f^* ranges between about 10^{-3} and 10^4.

The existence of a finite value for τ_a means that a given molecule will spend a proportionally larger fraction of its time near the surface than anywhere else in the system. Thus, the time-averaged concentration is going to be higher near the surface than in bulk of the gas phase. This is just the condition we have described previously as adsorption.

The value of τ_a in any situation is a strong function of both ΔH_{des} and T. The value of ΔH_{des} is in turn a function of the type of attractive forces present in the system. Systems in which the only attractive forces are of the van der Waals, or dispersion, type show values in the range from 100 cal/mol to 5,000 cal/mol. Customarily, adsorption processes involving forces of this magnitude are called *physical adsorption*, or *physisorption processes*. Systems in which hydrogen bonding, covalent chemical bonding, or metallic bonding can take place will show values of ΔH_{des} ranging from 5 kcal/mol to as high as 150 kcal/mol. Adsorption processes involving forces of this type are referred to as *chemisorption processes*. The effect of this wide range of observed ΔH_{des} values on the stay time is summarized in Table 10.2, based on the assumption that T = 300 K and $\tau_0 = 10^{-13}$ sec. As can be seen from the table, the values of τ_a cover a range from times essentially equal to τ_0, at the low end, to inconceivably long times at the high end. Note, however, that there is a fairly wide range of ΔH_{des} values for which τ_a is within a few orders of magnitude of 1 sec, and that this range can be greatly extended

Table 10.2 Mean Stay Time for Adsorbed Molecules at 300 K for Various Values of the Adsorption Energy, Assuming $\tau_o = 10^{-13}$ sec

ΔH_{des}	*Typical Cases*	τ_a *(sec)*
100 cal/mol	Helium	1.2×10^{-13}
1.5 kcal/mol	H_2 physisorbed	1.3×10^{-12}
3.5-4 kcal/mol	Ar, CO, N_2, CO_2 (physisorbed)	1×10^{-10}
10–15 kcal/mol	Weak chemisorption, Organics physisorbed	3×10^{-6}, to 2×10^{-2}
20 kcal/mol	H_2 chemisorbed	100
25 kcal/mol		6×10^5 (1 week)
30 kcal/mol	CO chemisorbed on Ni	4×10^9 (> 100 yr)
40 kcal/mol		1×10^{17} (≈ age of the earth)
150 kcal/mol	O chemisorbed on W	10^{1100} (≈ 10^{1090} centuries)

by changing the temperature. For example, for a temperature of 600 K instead of 300 K, the value of τ_a associated with ΔH_{des} = 40,000 cal/mol drops from 10^{17} sec, which is approximately the age of the earth, to 1 sec, a readily conceivable and experimentally measurable value.

The Fundamental Adsorption Equation

When the system is in adsorptive equilibrium, the net rate of accumulation of gas molecules on the surface is zero. This is a dynamic equilibrium, in which the impingement rate, I, is equal to the desorption rate, E. This E may be expressed as

$$E = \left(\frac{n_a}{\tau_a}\right) \qquad \text{molec / cm}^2\text{-sec}, \tag{10.8}$$

for the case of a first-order desorption process, in which n_a is the surface population of adsorbed molecules per square centimeter. The basic equation for equilibrium is thus

$$E = I \tag{10.9}$$

or

$$\left(\frac{n_a}{\tau_a}\right) = \frac{p}{(2\pi mkT)^{1/2}} \tag{10.10}$$

or

$$n_a = I\tau_a = \left[\frac{p}{(2\pi mkT)^{1/2}}\right]\tau_o \exp\left(\frac{\Delta H_{des}}{kT}\right). \tag{10.11}$$

This is a fundamental kinetic equation for adsorptive equilibrium. It must always be remembered in applying this equation that τ_a may be a function of n_a. That is, the value of ΔH_{des} may depend on surface coverage. Also, this equation may be applied only in cases in which equilibrium can be attained at the surface and in which the desorption process is first order.

The Henry's Law Model

In order to use the fundamental equation developed above to describe the adsorptive behavior in any real system, it is necessary to make some assumptions about the nature of the adsorption process, especially insofar as the question of the effect that the presence of one adsorbed molecule has on the behavior of another molecule. The simplest assumption one can make is that all atoms adsorb independently of one another. This implies that ΔH_{des}, and consequently τ_a, are independent of n_a. If one assumes further that all positions on the surface are equivalent, independent of where an incoming gas molecule strikes, then one has defined the conditions for what is called Henry's law behavior for the adsorption process.

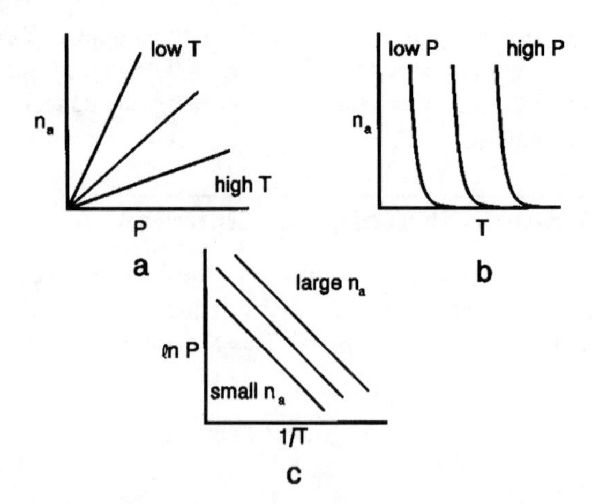

Figure 10.3 Expected equilibrium adsorption behavior in a system following Henry's law: adsorption isotherms (a), adsorption isobars (b), and adsorption isosteres (c).

Subject to the assumptions made above, the fundamental equation reduces to

$$n_a = C\left(\frac{p}{T^{1/2}}\right)\exp\left(\frac{\Delta H_{des}}{kT}\right), \qquad 10.12$$

in which

$$C = \frac{\tau_o}{(2\pi mk)^{1/2}}, \qquad 10.13$$

and will be a constant for any given gas–surface combination.

Look at the form of this equation, varying one parameter at a time. For the case of a variation in p at constant T, the equation reduces to

$$n_a = C_1 p. \qquad 10.14$$

This case is shown graphically in Figure 10.3a, in which n_a is plotted vs. p for a variety of values of T. A plot of this sort, showing behavior at constant temperature, is called an *adsorption isotherm.*

Alternatively, one can look at the variation of n_a with T at constant p. For this case, shown in Figure 10.3b, the fundamental equation reduces to

$$n_a = C_2\left(\frac{1}{T^{1/2}}\right)\exp\left(\frac{\Delta H_{des}}{kT}\right). \qquad 10.15$$

It can be seen from the figure that the exponential term in T dominates, with the value of n_a rising rapidly over a fairly narrow temperature range. A plot of this sort is known as an *adsorption isobar*.

Finally, consider the relation between p and T for a constant amount adsorbed. For this case

$$p = C_3\left(\frac{1}{T^{1/2}}\right)\exp\left(\frac{-\Delta H_{des}}{kT}\right) \tag{10.16}$$

or

$$\ell n\, p = \frac{-\Delta H_{des}}{kT} + C_4, \tag{10.17}$$

where the influence of the $T^{1/2}$ term relative to the exponential term has been neglected. This relation is plotted in Figure 10.3c as ln p vs. $1/T$ for a range of values of n_a. Note that all of these plots are straight lines having a slope of $-\Delta H_{des}/k$. A plot of this sort is known as an *adsorption isostere*.

The Langmuir Model

Consider now a slightly more complicated model of the adsorption process, namely, the Langmuir model. In this model, which was treated in Chapter 5 using the methods of statistical thermodynamics, one assumes that adatoms are held at specific sites on the surface, that the value of ΔH_{des} is the same for all sites, that there is no motion of atoms over the surface, and that the only effect that the presence of one adatom has on any other atom is that atoms striking a surface site that is already occupied do not adsorb. This last assumption is the one which makes the Langmuir model significantly different from Henry's law. It says, in effect, that τ_a has a finite value for atoms striking unoccupied sites and zero value for atoms striking occupied sites. This implies that the total adlayer coverage can never exceed the total number of sites, and coverage is limited to a single monatomic layer.

In order to apply the fundamental equation to this model, one must multiply the total impingement rate, I, by the factor $[1 - (n_a/n_0)]$, where n_0 is the number of adsorption sites per cm^2, to account for the fact that atoms striking occupied sites are not adsorbed. The resulting equation is

$$n_a = I\left[1 - \left(\frac{n_a}{n_o}\right)\right]\tau_a \tag{10.18}$$

or, solving for n_a,

$$n_a + \left(\frac{I\tau_a n_a}{n_o}\right) = I\tau_a \tag{10.19}$$

$$n_a\left[1 + \left(\frac{I\tau_a}{n_o}\right)\right] = I\tau_a \tag{10.20}$$

$$\left(\frac{n_a}{n_o}\right)\left[1+\left(\frac{I\tau_a}{n_o}\right)\right]=\frac{I\tau_a}{n_o} \qquad 10.21$$

or

$$\theta \equiv \left(\frac{n_a}{n_o}\right)=\frac{\dfrac{I\tau_a}{n_o}}{\left[1+\left(\dfrac{I\tau_a}{n_o}\right)\right]}. \qquad 10.22$$

Or, substituting $I = p/(2\pi mkT)^{1/2}$, and defining

$$\chi=\left(\frac{\tau_a}{n_o}\right)\left(\frac{1}{(2\pi mkT)^{1/2}}\right), \qquad 10.23$$

one has

$$\frac{I\tau_a}{n_o}=\chi p, \qquad 10.24$$

and thus

$$\theta=\frac{\chi p}{1+\chi p}, \qquad 10.25$$

which is the Langmuir adsorption isotherm equation and, as expected, has the same form as the equation developed previously using statistical thermodynamics (Equation 5.58). The form of this solution is shown in Figure 10.4. Note that at low pressures, where $\chi p << 1$, the equation reduces to the Henry's law relation

$$n_a = n_0\chi p. \qquad 10.26$$

At high pressures, where $\chi p >> 1$, the equation reduces to

$$\theta=\frac{\chi p}{\chi p}=1, \qquad 10.27$$

and all the surface sites are filled.

The Langmuir model provides a good description of the adsorption process in many systems in which strong chemisorption occurs. It does not in general provide a very good description of the process in systems involving relatively weak adsorptive forces (physical adsorption), as it neglects lateral interactions among adatoms, surface mobility, surface heterogeneity, and the possibility of adlayers thicker than one monolayer. Other techniques and models are required to handle these complications.

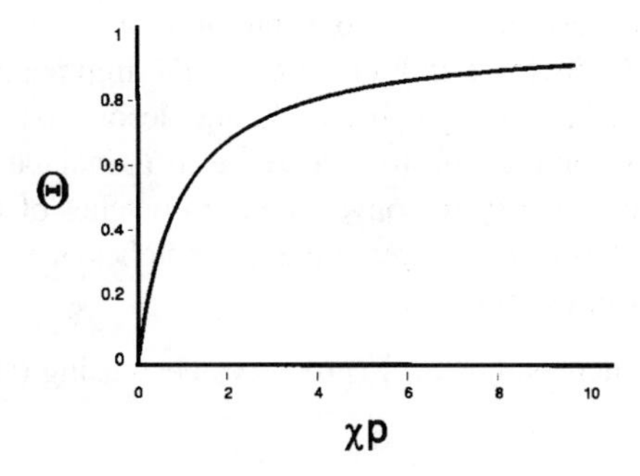

Figure 10.4 The adsorption isotherm according to the Langmuir model.

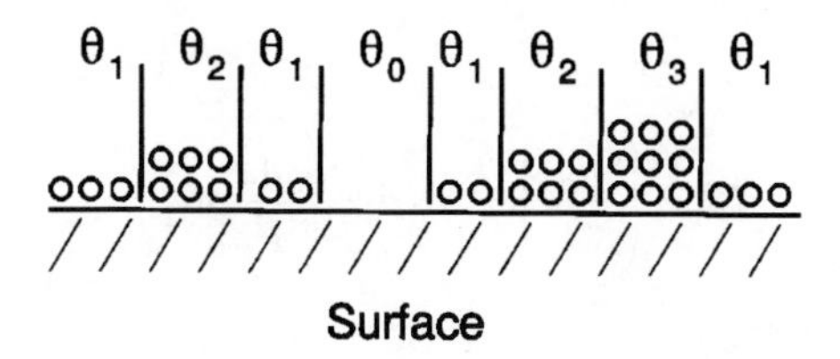

Figure 10.5 The model used for the development of the B.E.T. treatment of adsorption. The surface is broken up, conceptually, into patches covered with 1, 2, 3, . . . *i* adsorbed layers.

The B.E.T. Model

One of the principle defects of the Langmuir model is that it explicitly forbids adlayers having a value of $\theta > 1$, whereas adlayers many molecular layers in thickness are commonly observed in systems in which physical adsorption takes place. The case of multilayer adsorption has been treated explicitly by Braunauer, Emmett, and Teller [1938], who formulated the so-called *B.E.T. model* of the adsorption process.

Basically, this model assumes that when adsorptive equilibrium has been attained, the total surface area can be broken down into patches, each of which is covered with zero, one, two, or more monolayers of adsorbed material, as shown in Figure 10.5. Over a period of time, the fraction of the surface having a given coverage will remain constant. That is, if θ_i is that fraction of the surface covered by i monolayers, at equilibrium, $d\theta_i/dt = 0$ for all i, and the total amount of material adsorbed is

$$n_a = n_0\theta_1 + 2n_0\theta_2 + 3n_0\theta_3 + \cdots \qquad 10.28$$

or

$$n_a = n_o\sum_i i\theta_i. \qquad 10.29$$

Considering the dynamic nature of the adsorptive equilibrium, it is evident that this implies that the net rate of processes contributing to an increase of a given θ_i must be the same as the net rate of those processes contributing to a decrease in that same θ_i. Processes contributing to the increase in a given θ_i are the impingement of molecules on patches of $\theta_{(i-1)}$ to convert them to patches of θ_i, and desorption of molecules from patches of $\theta_{(i+1)}$. Processes contributing to a decrease in θ_i include the adsorption of molecules on patches of coverage θ_i, to convert them to patches of $\theta_{(i+1)}$, and desorption of molecules from patches of θ_i to convert them to patches of $\theta_{(i-1)}$. Thus, for each θ_i one may write an expression stating that

$$\Sigma(\text{processes increasing}\,\theta_i) = \Sigma(\text{processes decreasing }\theta_i). \tag{10.30}$$

Thus, for θ_0:

$$\left(\frac{n_o}{\tau_{a_1}}\right)\theta_1 = I\theta_o, \tag{10.31}$$

for θ_1:

$$I\theta_o + \left(\frac{n_o}{\tau_{a_2}}\right)\theta_2 = I\theta_1 + \left(\frac{n_o}{\tau_{a_1}}\right)\theta_1, \tag{10.32}$$

for θ_2:

$$I\theta_1 + \left(\frac{n_o}{\tau_{a_3}}\right)\theta_3 = I\theta_2 + \left(\frac{n_o}{\tau_{a_2}}\right)\theta_2 \tag{10.33}$$

for θ_i:

$$I\theta_{(i-1)} + \left(\frac{n_o}{\tau_{a_{(i+1)}}}\right)\theta_{(i+1)} = \theta_i + \left(\frac{n_o}{\tau_{a_i}}\right)\theta_i. \tag{10.34}$$

Subtracting the expression for the change in θ_0 from that for the change in θ_1, leaves

$$I\theta_1 = \left(\frac{n_o}{\tau_{a_2}}\right)\theta_2. \tag{10.35}$$

By a similar process for the other θ_i, subtracting the expression for changes in $\theta_{(i-1)}$ from the expression for changes in θ_i, leaves a set of simultaneous equations of the form

$$I\theta_i = \left[\frac{n_o}{\tau_{a_{(i+1)}}}\right]\theta_{(i+1)}. \tag{10.36}$$

When this set of equations is solved, subject to the assumption that $\tau_{a_2} = \tau_{a_3} = \tau_{a_i}$, one obtains an expression for n_a of the form

$$n_a = \frac{k_B n_o x}{(1+x)(1-x+k_B x)}, \tag{10.37}$$

in which

$$k_B = \tau_{a_1}, \quad x = \frac{I\tau_{a_2}}{n_o}. \tag{10.38}$$

Substituting $I = p/(2\pi mkT)^{1/2}$ in the expression for x, and making the further substitution that $\beta = 1/(2\pi mkT)^{1/2}$, one obtains

$$x = \frac{\beta\tau_{a_2}p}{n_o}. \tag{10.39}$$

Note that x is dimensionless, as $\beta\tau_{a2}/n_0$ has units of $(\text{pressure})^{-1}$. It is thus convenient to define another parameter q, having units of pressure, as

$$q \equiv \frac{n_o}{\beta\tau_{a_2}}, \tag{10.40}$$

to give

$$x = \frac{p}{q} \tag{10.41}$$

The final expression for adlayer coverage is thus

$$\theta = \frac{n_a}{n_o} = \frac{\left(\frac{k_B p}{q}\right)}{\left(1 - \frac{p}{q}\right)\left(1 - \frac{p}{q} + \frac{k_B p}{q}\right)}. \tag{10.42}$$

Equation 10.42 has the property that as $p \to q$, the denominator in the expression approaches zero. Thus, as $p \to q$, $\theta \to \infty$.

It is observed in many systems in which physical adsorption occurs that as p approaches p_0 (the equilibrium vapor pressure of the bulk condensed phase of the adsorbed species at the temperature of the experiment), the amount adsorbed rises sharply and appears headed for infinity. It has thus been customary to associate q with p_0 and, consequently, to associate τ_{a2} with ΔH_v, the enthalpy of vaporization of the condensed adsorbate. No fundamental significance is to be attached to this choice—it is merely a convenient choice based on empirical observation. The form of the final expression for the adsorption isotherm in the B.E.T. model is thus as shown in Figure 10.6.

Problems

10.1. Carbon monoxide adsorbs nondissociatively on Pd (111). The adsorption behavior follows the Langmuir isotherm.

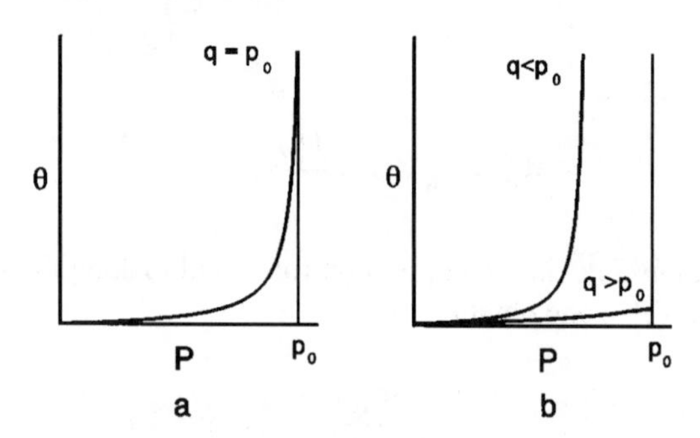

Figure 10.6 Adsorption isotherms for systems according to the B.E.T. model: (a) $q = p_0$; (b) $q < p_0$, $q > p_0$.

The heat of desorption is 20 kcal/mol. At a carbon monoxide pressure of 10^{-7} Torr, at what temperature would the equilibrium coverage be equal to one half monolayer?

10.2. If a surface is initially covered with 1×10^{14} molec/cm^2 of a monatomic gas having a heat of desorption of 30 kcal/mol, how long will it take for the coverage to decrease to 1×10^{13} molec/cm^2 at a temperature of 600 K? Assume $\tau_0 = 10^{-13}$ sec, and that readsorption may be neglected.

10.3. Copper adsorbs on and desorbs from tungsten as individual atoms. The desorption energy is 60 kcal/mol, and the activation energy for surface diffusion is 20 kcal/mol. Calculate the mean distance that an adatom migrates over the surface while it is adsorbed, at a temperature of 800 K. The jump distance for adatom diffusion may be taken as 0.3 nm.

10.4. Consider a system in which a gas having a molecular weight of 30 adsorbs on a surface having 10^{15} adsorption sites per square centimeter with a heat of desorption of 5000 cal/mol and $\tau_0 = 10^{-13}$ sec. For a temperature of 200 K:

a. Plot the isotherm, θ vs. p, for the pressure range from 0 to 250 Torr, assuming that the system obeys Henry's law.

b. Plot the isotherm, over the same pressure range, assuming that the system follows the Langmuir isotherm.

10.5. Using the B.E.T. model, determine the equilibrium adlayer coverage for carbon dioxide molecularly physisorbed on graphite at $T = 77$ K, $p = 100$ Torr. The measured heat of adsorption in this system, at low coverage, is –4000 cal/mol, and the area occupied by a carbon dioxide molecule may be assumed to be 0.15 nm^2.

Bibliography

Brunauer, S, and Copeland, LE. Physical adsorption of gases and vapors on solids. In: Symposium on properties of surfaces. Philadelphia: ASTM, 1963;59–79.

Brunauer, S, Emmett, PH, and Teller, E. *J. Am. Chem. Soc.* 1938; 60:309.

De Boer, JH. *The Dynamical Character of Adsorption.* Oxford: Clarendon Press, 1953.

Glasstone, S, Laidler, KJ, and Eyring, H. *The Theory of Rate Processes.* New York: McGraw-Hill, 1941.

Ross, S, and Olivier, JP. *On Physical Adsorption.* New York: Interscience Publishers, 1964.

CHAPTER **11**

Physical Adsorption

This chapter will consider in more detail adsorption in those systems in which the attractive forces between the adatoms and the surface are relatively nonspecific van der Waals or dispersion forces. Systems of this sort are of interest primarily because the interactions among the adsorbed species, the so-called *lateral interactions*, are comparable in strength to the interactions with the substrate atoms, which give rise to adsorption. In addition, the variation in ΔE_a with position on the surface is in many cases small compared to the value of ΔE_a. This results in a highly mobile adsorbed phase and the concept of a two-dimensional gas.

Surface Potential Variation

Chapter 6 introduced the question of variation of system potential energy as a function of distance parallel to the surface in the discussion of surface mobility. Consider this concept again, in more detail. The maximum depth of the attractive potential well that gives rise to adsorption as a function of distance along a real surface, exhibits behavior similar to that shown schematically in Figure 11.1. There will be a regular variation in the well depth, ΔE_a, with distance, reflecting the periodic nature of the surface. In addition, on any real surface, there will be sites where the minima will be either greater or less than the value for the ideal surface. These sites represent interruptions in the ideal surface due to inhomogeneities such as ledges, kinks, sites of dislocation emergence, or impurity atoms. This chapter considers the consequence of these variations in surface potential, first in terms of the regular variation associated with ideal surfaces, and second in terms of the disturbances associated with surface heterogeneity.

The detailed kinetic behavior of any adsorbed species depends on the amplitude of the variation of the surface potential relative to kT. Three possible situations exist, as shown in Figure 11.2. The case shown in Figure 11.2a, where the potential variation is small compared to kT is called *mobile adsorption*. The case of Figure 11.2b, where the variation is larger than kT but not so large that surface diffusion is prohibited, is called *localized adsorption*. If the barrier is so large compared to kT that the probability of surface diffusion is essentially zero, as shown in Figure 11.2c, *immobile adsorption* results. Experimental studies of adsorption in specific systems indicate that all three types of adsorption are observed in practice.

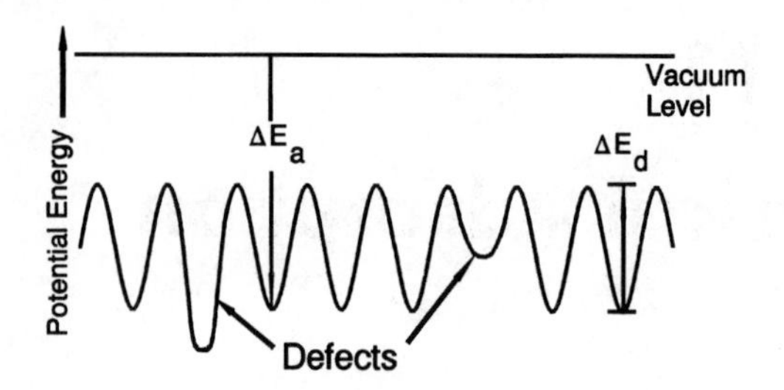

Figure 11.1 Potential energy of an adatom at a real surface as a function of position along a row of surface atoms.

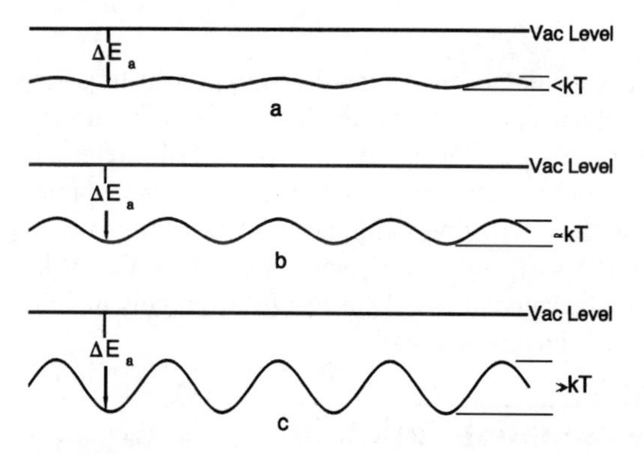

Figure 11.2 Potential energy of an adatom at a surface for three general classes of adsorption: Mobile adsorption (a), localized adsorption (b), and immobile adsorption (c).

Two-Dimensional Gases

Look now in detail at the case of mobile adsorption, where the barrier to lateral motion is small compared to kT. This will lead to the concept of lateral interactions among the adsorbed species.

If the barrier to lateral motion is small compared to kT, atoms are free to move across the surface as a two-dimensional gas. One may begin by assuming that adatoms interact only with the edges of the two-dimensional surface and make no allowance for the fact that each adatom occupies a finite amount of surface area. (That is, assume that the whole surface is available to each adatom). Further, assume that adatoms collide elastically on the surface with no energetic interaction, and that this two-dimensional gas can be described by the two-dimensional analog of the three-dimensional ideal gas law.

Consider the application of these assumptions to a clean surface bounded by some kind of barrier—for example the surface of a liquid in a vessel. There will be a force, γ, exerted on the barrier because of the surface tension of the liquid. When adsorption takes place on this surface, γ will be reduced, as described by the Gibbs adsorption isotherm. The tension on the barrier is consequently reduced as γ is reduced. One can equally well look on this reduction in tension as a pressure exerted on the barrier by the two-dimensional gas.

One can develop this relation mathematically by beginning with the Gibbs adsorption isotherm.

$$d\gamma = \sum_i \Gamma_i d\mu_i. \qquad 11.1$$

For the case of adsorption from a one-component gas phase onto a surface, using

$$d\mu_i^s = d\mu_i^g = RT\, d\, \ell n\, p_i, \qquad 11.2$$

one can determine that

$$d\gamma = -\Gamma RT\, d\, \ell n\, p. \qquad 11.3$$

However, one may substitute

$$\Gamma R = n_a k, \qquad 11.4$$

where n_a is the number of adatoms per unit area. If, now, a two-dimensional pressure, Π, is defined as

$$d\Pi = -d\gamma, \qquad 11.5$$

This leads to

$$d\Pi = n_a kT\, d\, \ell n\, p. \qquad 11.6$$

Integration of this expression yields

$$\Pi = kT \int_o^p n_a\, d\ell n\, p = kT \int_o^p n_a \frac{dp}{p}, \qquad 11.7$$

in which n_a may be a function of p.

Consider the case of Henry's law adsorption, as discussed in Chapter 10. In this case

$$n_a = \left[\frac{\tau_o}{2\pi \mathrm{mkT}^{1/2}}\right] \exp\left(\frac{\Delta H_{des}}{kT}\right) p = \kappa p \qquad 11.8$$

or

$$p = \frac{n_a}{\kappa} \qquad 11.9$$

or

$$dp = \frac{dn_a}{\kappa}. \qquad 11.10$$

Substituting these results into Equation 11.7 yields

$$\Pi = kT\int_o^{n_a} n_a\left(\frac{\kappa}{n_a}\right)\left(\frac{dn_a}{\kappa}\right) = kT\int_o^{n_a} dna \qquad 11.11$$

or

$$\Pi = n_a kT \qquad 11.12$$

or

$$\Pi\sigma = kT\,, \qquad 11.13$$

where $\sigma \equiv 1/n_a$ is the surface area per adatom. This expression has the same form as the three-dimensional ideal gas law,

$$p(V/N_{av}) = kT. \qquad 11.14$$

Those familiar with the results of the statistical mechanical treatment of the ideal lattice gas will also note the similarity between Equation 11.13 and the expression for the two-dimensional analog of pressure, Φ, that arises in that treatment. The general result in that case is

$$\Phi = -kT\ln\left(1-\frac{N}{M}\right), \qquad 11.15$$

in which N is equivalent to n_a in the present treatment, M is equivalent to n_o, the number of adsorption sites per unit area, and $\Phi = (\partial F/\partial M)_{N,T}$. At low coverage this reduces to

$$\Phi = \left(\frac{N}{M}\right)kT = \left(\frac{n_a}{n_o}\right)kT, \qquad 11.16$$

or

$$\Phi = \frac{\Pi}{n_o}. \qquad 11.17$$

The behavior of the ideal two-dimensional gas is thus seen to be exactly analogous to that of the ideal three-dimensional gas.

Alternatively, one may make more realistic assumptions concerning the behavior of the two-dimensional gas. Assume now that each molecule occupies some finite area on the surface, b_2, and that there is an interaction energy between any two adsorbed molecules, leading to a lateral interaction force, a_2. These are essentially the same assumptions that are made in developing the van der Waals equation of state for a three-dimensional gas and lead in this case to the two-dimensional equation of state

$$\left[\Pi + \left(\frac{a_2}{\sigma^2}\right)\right](\sigma - b_2) = kT. \qquad 11.18$$

Making the additional assumption that b_2 is $1/n_o$, and using

$$\theta = \frac{b_2}{\sigma} = \frac{n_a}{n_o}, \tag{11.19}$$

Equation 11.18 may be rearranged to yield

$$\left[\Pi + \frac{a_2}{\left(\frac{b_2^2}{\theta^2}\right)}\right]\left[\frac{b_2}{\theta} - b_2\right] = \mathrm{kT} \tag{11.20}$$

$$\left[\Pi + \frac{a_2}{\left(\frac{b_2^2}{\theta^2}\right)}\right] b_2(1-\theta) = \theta \mathrm{kT} \tag{11.21}$$

$$\Pi = \left(\frac{kT}{b_2}\right)\left[\frac{\theta}{1-\theta}\right] - \frac{\theta^2 a_2}{b_2^2} \tag{11.22}$$

$$\Pi = \left(\frac{kT}{b_2}\right)\left\{\left[\frac{\theta}{1-\theta}\right] - \left(\frac{a_2\theta^2}{kTb_2}\right)\right\}. \tag{11.23}$$

Using this relation and the previously stated relation between Π and p (Equation 11.6), one can determine that

$$p = K\left[\frac{\theta}{1-\theta}\right]\exp\left\{\left[\frac{\theta}{1-\theta}\right] - \left(\frac{2a_2\theta}{kTb_2}\right)\right\}, \tag{11.24}$$

where K is a constant related to the adsorption energy, ΔE_a. This relation, known as the Hill–DeBoer equation, is the adsorption isotherm equation for a two-dimensional gas with lateral interactions on a homogeneous surface. The effect of the lateral interaction energy appears in the last term in the brackets on the right-hand side of Equation 11.24.

The form of the Hill–DeBoer Equation is shown in Figure 11.3. At high temperature and low coverage, where the effects of excluded area and lateral interactions will be minimal, the exponential term approaches unity, $(1 - \theta)$ approaches unity, and the isotherm equation reduces to

$$p = K\theta, \tag{11.25}$$

which is the Henry's law isotherm. At lower temperatures, or for larger values of a_2, and at higher coverages, increasing deviations from Henry's law are observed. Below a

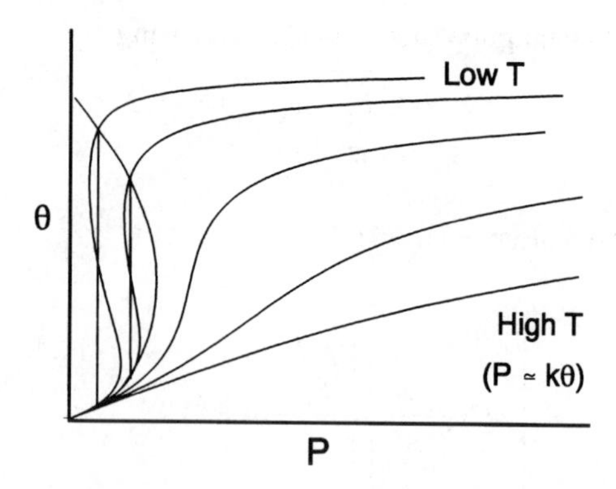

Figure 11.3 Adsorption isotherms according to the Hill–DeBoer equation, showing two-dimensional condensation.

certain temperature, T_{C2}, the two-dimensional critical temperature, the curve described by Equation 11.24 becomes double valued, and a first-order phase change is observed, just as in the case of a three-dimensional van der Waals gas. Within the region bounded by the dotted curve, two phases exist at equilibrium: a two-dimensional gas and a two-dimensional solid or liquid. At higher pressure only the condensed phase will be present.

Physical Adsorption Measurements

Behavior of the sort described in the preceding section for physical adsorption on energetically homogeneous surfaces has been observed in two general types of systems, namely, insoluble molecules adsorbed on liquid surfaces, which are uniform energetically because of their high mobility, and gases adsorbed on highly perfect crystal surfaces.

In the case of molecules adsorbed on a liquid surface, a technique has been developed which permits direct measurement of the force-area relation described by Equation 11.18. The apparatus typically used for such measurements, known as a film balance, is shown in Figure 11.4. It consists of a trough, generally coated with wax and filled with water. At one end is a barrier which prevents passage of adsorbed material from one side to the other. This barrier is attached to a force measuring system, usually an optical lever system, to permit measurement of small displacements of the barrier.

If no adsorbed material is present, the surface tensile forces will be the same on both sides of the barrier. Adsorption of insoluble organic molecules on one side of the barrier will reduce the surface tension, as indicated by the Gibbs adsorption isotherm. The resulting change in surface tensile force is measured by the system and is related to the two-dimensional pressure, Π, by Equation 11.5. If the number of molecules spread on the surface and the surface area are known, all of the information required to construct the Π-σ relation is at hand. Generally, the area per molecule, σ, is changed by using a second barrier to reduce the area available to the adsorbed molecules.

Experiments of this sort, using molecules such as long-chain fatty acids, show the behavior predicted by Equation 11.18, including the observation of a two-phase

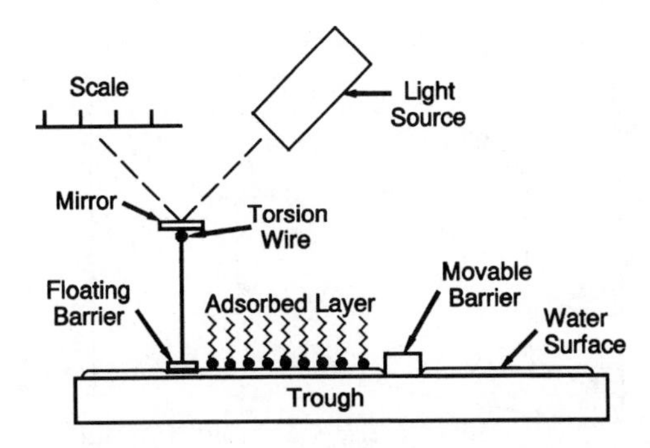

Figure 11.4 Film balance for the direct measurement of two-dimensional Π-σ behavior.

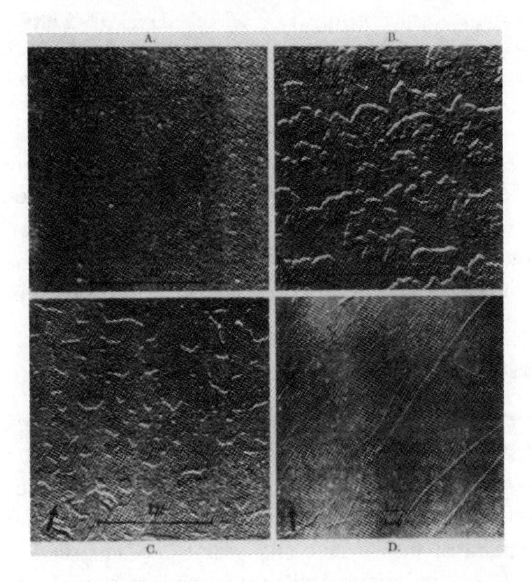

Figure 11.5 Electron micrographs of monolayer films of n-hexatriacontanoic acid on water, showing two-dimensional condensation: (a) no film, (b) about 1/2 monolayer coverage; (c) about 90% coverage; (d) "collapsed" film above monolayer coverage. (Reprinted with permission from H.E. Ries, Jr. and W.A. Kimball, *J. Phys. Chem.* 1955;59:94. Copyright 1955, American Chemical Society.)

region and eventual total condensation. An example of this behavior is shown in Figure 11.5. Here, replica electron micrographs of the adsorbate covered surface show an essentially featureless image at low values of Π, a two-phase region at intermediate Π values, and finally mechanical failure of the film at the highest pressures studied.

In the case of the physical adsorption of gases on highly uniform solid surfaces, the appropriate isotherm relation is the Hill–DeBoer isotherm, Equation 11.24. In this case, the amount of material adsorbed on a surface of known surface area is measured, either gravimetrically or volumetrically, as the pressure of the adsorbate in the gas phase is increased. The resulting isotherms show the sharp, stepwise increases in cov-

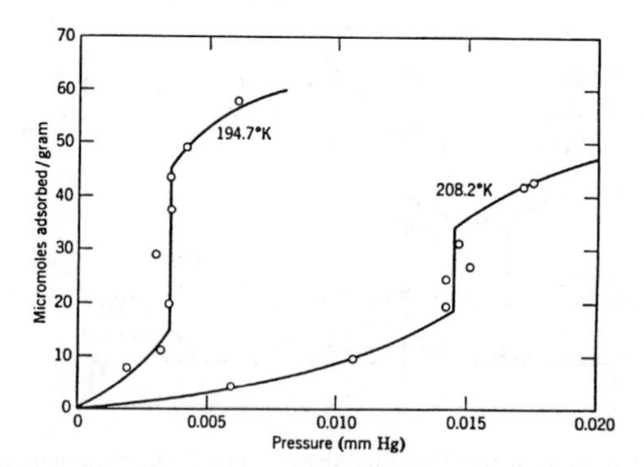

Figure 11.6 Adsorption isotherms for $CFCl_3$ adsorbed on P33 (2700°C), a very energetically uniform graphite. Note the decrease in the height of the discontinuity in coverage as the critical temperature is approached. (Source: W.D. Machin and S. Ross, *Proc. Roy. Soc. (London)* 1962;265A:455. Reprinted with permission.)

erage at well-defined pressures predicted by Equation 11.24. An example of this behavior is shown in Figure 11.6. for the case of $CFCl_3$ adsorbed on a very energetically uniform graphite surface. Note that the extent of the discontinuous rise in coverage with pressure decreases in extent with increasing pressure and would disappear at T_{C2}.

Surface Heterogeneity

There is an effect associated with the fact that the adsorption potential well depth, on any real surface, will not be the same for all adsorption sites. This effect is known as *surface heterogeneity*. It arises from the fact that the interaction energy between an adsorbed atom or molecule and a solid surface will depend on the details of the atomic arrangement at the surface site where the adsorbed species is held. For example, a ledge site on the surface will offer more potential near neighbors for an adsorbed atom, and the disruption in the order of the surface associated with the site of emergence of a dislocation will lead to modified interatomic distances between the adatom and the surface atoms.

The net effect of these heterogeneities on the adsorptive process is that rather than having a uniform value of ΔE_a for all potential adsorption sites, some sites will have larger values of ΔE_a (tighter binding) and some will have smaller values of ΔE_a (weaker binding). One can understand the effect of this surface heterogeneity on the equilibrium adlayer coverage in any situation by considering the kinetic picture of adsorption in a system that has two different types of adsorption sites, call them A and B, in which A sites have larger values of ΔE_a than B sites. At any given temperature, the mean stay time will thus be longer for A sites than for B sites, and on a long time average the occupancy of A sites will be greater than B sites. Another way of looking at it is to say that as the temperature is reduced, A sites fill first and B sites fill at a lower temperature. Similarly, if the pressure is increased at constant temperature, A sites fill at a lower pressure than B sites.

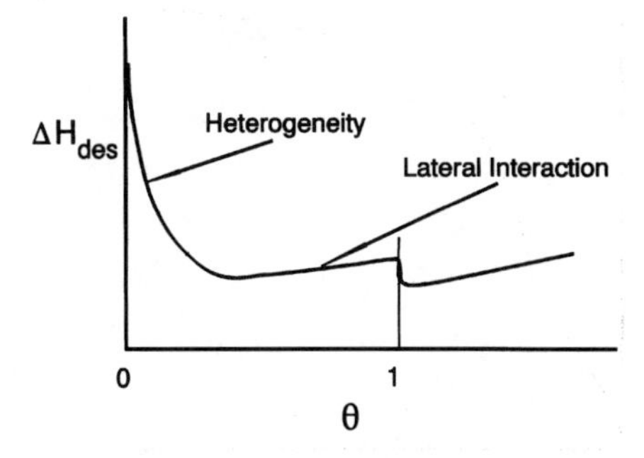

Figure 11.7 Heat of adsorption as a function of adlayer coverage for a slightly heterogeneous surface, showing effects due to high energy sites at low coverage, increasing heat of adsorption due to lateral interactions as the monolayer is approached, and a drop at the onset of second layer formation.

Another way to look at the question of surface heterogeneity in real systems is to measure the heat of desorption as a function of coverage. Here the sites of highest binding energy will dominate the measured value of ΔH_{des} at low coverage, for the reasons discussed, and a large value will be observed for ΔH_{des}. At higher coverages, the sites of weaker binding, which are usually present in much higher number, dominate the desorption process. Consequently, the measured value of ΔH_{des} will decrease as coverage increases, as shown in Figure 11.7. The other effect that appears in this figure is a gradual rise in ΔH_{des} as monolayer coverage is approached, due to presumed attractive lateral interactions, and a small but fairly sharp decreases in ΔH_{des} at monolayer coverage as the second monolayer starts to fill.

For the case of physical adsorption, the net effect of surface heterogeneity on the adsorption isotherm that was developed for uniform surfaces using the two-dimensional gas model is to increase the coverage at any given value of pressure, and to decrease the sharpness of the step associated with two-dimensional condensation. Figure 11.8 shows the difference between a homogeneous surface and a slightly heterogeneous surface for a temperature slightly above the critical temperature for two-dimen-

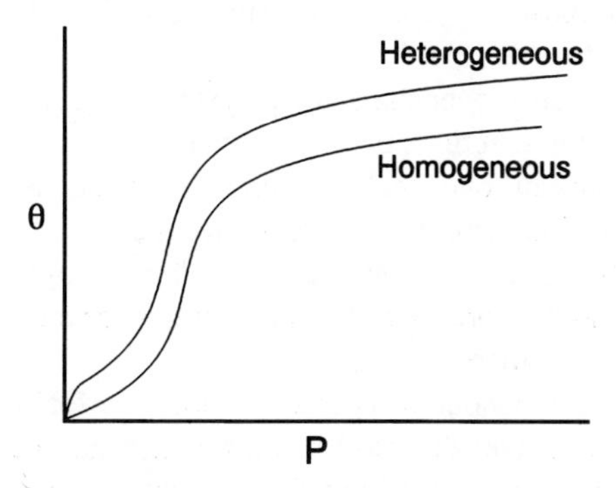

Figure 11.8 Adsorption isotherms for a slightly heterogeneous surface and a homogeneous surface at a temperature above T_{C2}.

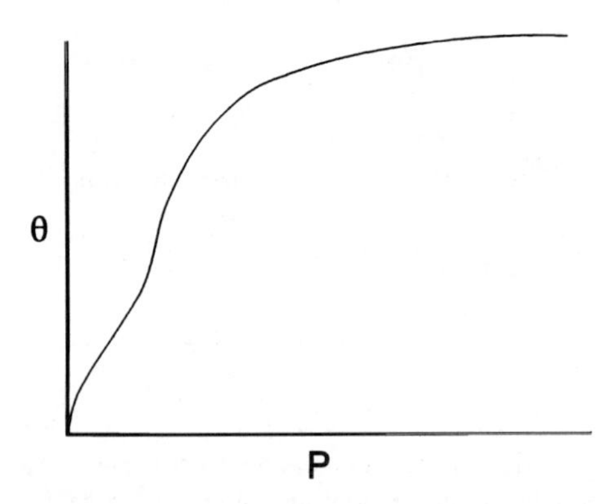

Figure 11.9. Adsorption isotherm for a highly heterogeneous surface.

sional condensation. A more extreme case of surface heterogeneity would lead to an isotherm such as that shown in Figure 11.9. In this case, all traces of the two-dimensional condensation process are hidden by the effects associated with surface heterogeneity. This latter isotherm explains why two-dimensional condensation was not observed in early studies of physical adsorption. The required uniform surfaces are relatively difficult to prepare, especially on high specific surface area samples. Thus, the observation of sharp steps in the adsorption isotherm is the exception rather than the rule.

Chapters 12 and 13 will discuss the question of surface heterogeneity again, in the context of its effect on the processes of chemisorption and surface reaction. Because of the stronger binding forces involved in these cases, the adsorption and catalytic behavior of the system may be dominated by sites of high binding energy such as ledges and kinks.

Problems

11.1. Consider a system in which a gas having a molecular weight of 30 adsorbs on a surface with a heat of desorption ΔH_{des} = 5000 cal/mol on the clean surface. Assume $\tau_o = 10^{-13}$ sec, $n_o = 10^{15}$ cm^{-2}, T = 200 K.

a. Plot the isotherm, θ vs. p, for the pressure range from 0 to 250 Torr, assuming that the system obeys Henry's law.

b. Sketch the isotherm, over the same pressure range, assuming van der Waals behavior, with attractive interactions between molecules.

c. Sketch the isotherm, over the same pressure range, assuming van der Waals behavior, with repulsive interactions between molecules.

11.2. Develop the equation of state relating Π, the two-dimensional pressure, σ, the area per molecule, and T for a two-dimensional gas for the case where the adsorption follows a Langmuir isotherm.

11.3. Plot the expected value of Π as a function of σ for an adsorption process in which an oleic acid monolayer is adsorbed on a water surface and the layer is then compressed by moving a barrier to reduce the available surface area. Assume that the layer follows a two-dimensional van der Waals equation of state, with $a_2 = 2 \times 10^{-28}$

dyn-cm^3/molec^{-2}, $b_2 = 10^{-14}$ cm^2/molec. The temperature is 300 K. The initial surface area is 100 cm^2 and the amount of oleic acid adsorbed is 10^{15} molecules.

Bibliography

Alexander, AE. Simple techniques for measuring force-area curves. In: *Surface Chemistry*. New York: Interscience Publishers, 1949;124.

Birgenau, RJ, and Horn, PM. Two-dimensional rare gas solids. *Science* 1986; 232:329–335.

Brunauer, S, and Copeland, LE. Physical adsorption of gases and vapors on solids. In: *Symposium on Properties of Surfaces*. Philadelphia: ASTM, 1963;59–79.

Klein, ML, and Venables, J, eds. *Rare Gas Solids*. New York: Academic Press, 1977.

Ries, HE, and Kimball, WA. Monolayer structure as revealed by electron microscopy. *J. Chem. Phys*. 1955;59:94–95.

Ross, S, and Olivier, JP. *On Physical Adsorption*. New York: Interscience Publishers, 1964.

Thomy, A, Duval, X, and Regnier, J. Two-dimensional phase transitions as displayed by adsorption isotherms on graphite and other lamellar solids. *Surface Sci. Reports* 1981;1:1.

CHAPTER **12**

Chemisorption

The process of adsorption for systems in which the forces binding the adatom or molecule to the surface are strong chemical forces, such as those involved in covalent, ionic, or strong polar bonding, is generally referred to as *chemisorption*. It differs from physical adsorption both in the strength of the bond between the adsorbed species and the surface and in terms of the specificity of the binding forces. That is, the van der Waals type forces that are primarily involved in physical adsorption exist for all combinations of adsorbed species and surface species. The stronger binding forces involved in chemisorption are more specific, both in terms of which adspecies–surface species combinations lead to chemisorption, and in terms of the preferred positions of the adspecies relative to the surface atoms and to each other. This chapter will look at the chemisorption process in terms of the energetics, the structures formed, and the kinetics of the adsorption and desorption processes.

Energetics of Chemisorption

The energetic situation in systems in which chemisorption takes place may be understood by looking at how the system potential energy changes as the adsorbing species is brought close to the surface, much as has been done previously for the cases of gas scattering and physical adsorption. In the present context, however, the range of possible situations is much more complex. A number of possible situations will be considered. The least complex behavior is the presence of a potential well arising solely from van der Waals interactions between the adspecies and the surface. This case is shown schematically in Figure 12.1a. Here, the depth of the potential well is relatively shallow, on the order of a few kilocalories per mole, and the distance of the minimum from the surface is relatively large. A well of this sort will be present in all cases, irrespective of whether or not chemisorption can occur.

In cases in which chemisorption can occur, the form of the potential energy curve will depend on whether the incoming molecule is adsorbed intact, without the breaking of any intramolecular bonds, or whether the molecule must dissociate in order for its component atoms to reach the chemisorbed state. The simpler case of nondissociative adsorption is shown in Figure 12.1b. The case of dissociative adsorption is shown, for the particular case of a homonuclear diatomic molecule, in Figure 12.1c. In the first of

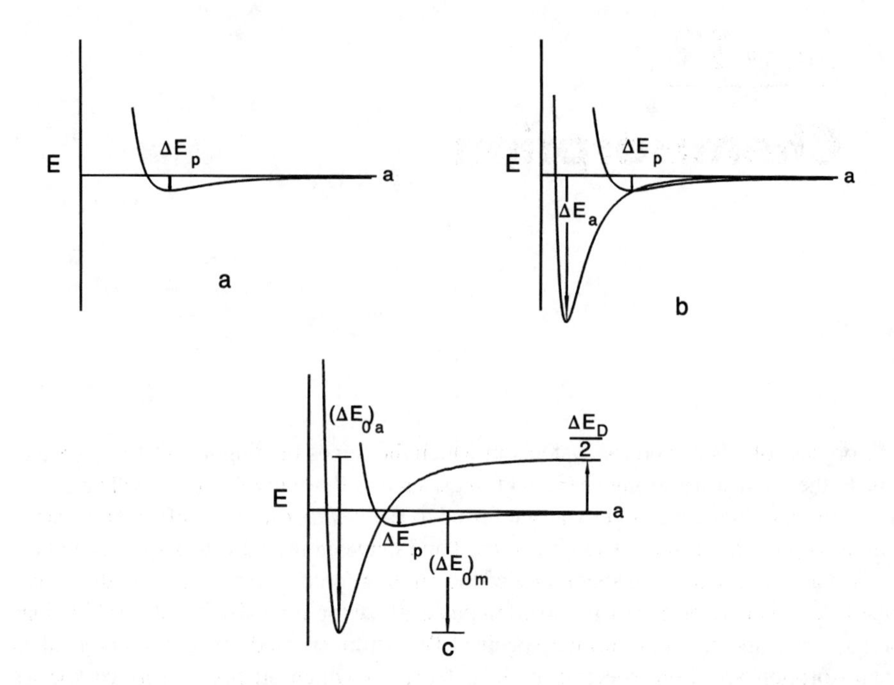

Figure 12.1 Potential energy curves for adsorption showing physical adsorption only (a), nondissociative chemisorption (b), and dissociative chemisorption of a homonuclear diatomic molecule (c).

these cases, the observed energy of desorption will be approximately equal to the well depth relative to the zero of potential energy, ΔE_a in the diagram. In the case of dissociative adsorption one must consider both the possibility of desorption as individual atoms, for which the desorption energy is approximately $-(\Delta E_o)_a$ per atom, and the possibility of recombination to reform the diatomic molecule and subsequent desorption of this species, for which the desorption energy is approximately $-2(\Delta E_o)_m$ per molecule. These two desorption energies are related by

$$(\Delta E_o)a = (\Delta E_o)m - \frac{\Delta E_D}{2}, \qquad 12.1$$

where ΔE_D is the dissociation energy of the diatomic molecule. (Recall that, as they have been defined, adsorption energies are inherently negative; dissociation energies are inherently positive.)

The case of heteronuclear diatomic molecules that adsorb dissociatively is even more complex. Figure 12.2 is the diagram appropriate to CO adsorption on a metal surface, in the case where the adsorption is nonactivated. This case again shows a potential well for the molecularly adsorbed species, but here one must show a single potential well for the combination $(C_a + O_a)$, as the potential wells for the adsorbed carbon and oxygen atoms will not necssarily coincide. Moreover, one must use as a reference for the desorp-

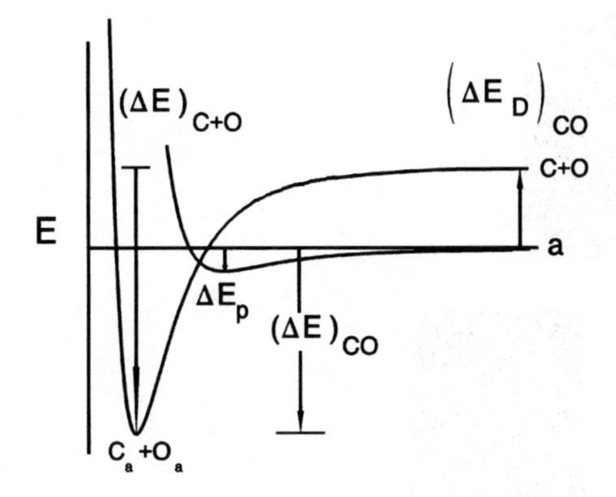

Figure 12.2 Potential energy curves for the dissociative adsorption of a heteronuclear diatomic molecule.

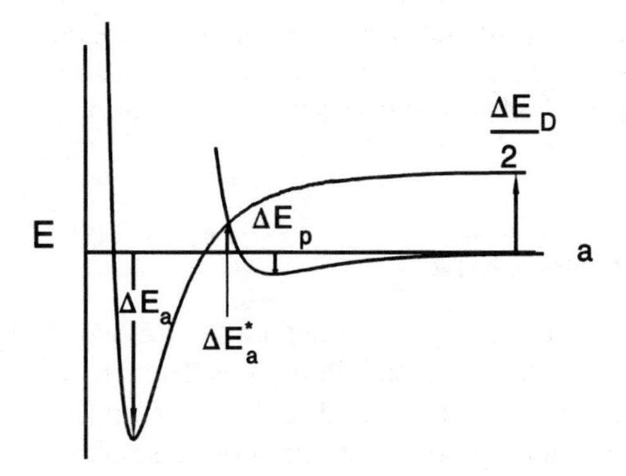

Figure 12.3 Potential energy curves for the dissociative adsorption of a homonuclear diatomic molecule, for the case of a finite activation barrier to chemisorption.

tion of $(C + O)$ as atoms the dissociation energy of CO. Additional possible outcomes for the desorption process also arise in this case, as it is possible that O_2 or C_2 or CO_2 species as well as CO could desorb from a layer containing adsorbed carbon and oxygen. This case thus takes one over the boundary between simple adsorption processes and surface chemical reactions. Chapter 13 will cover processes of this sort in detail.

A final complication that may arise in the case of either molecular or dissociative chemisorption is one in which the crossover between the curves for the physisorbed and chemisorbed states occurs at an energy greater than that of the zero of energy, as shown in Figure 12.3. In this case, the chemisorption process will be activated, with a finite activation energy being required to surmount the energy barrier leading to the chemisorbed state. The kinetics of the adsorption process may show a complicated temperature dependence in this case, and the adsorption rate may depend on the kinetic energy of the incident gas molecules. The section on activated adsorption, later in this chapter, will consider the rate equations for this process.

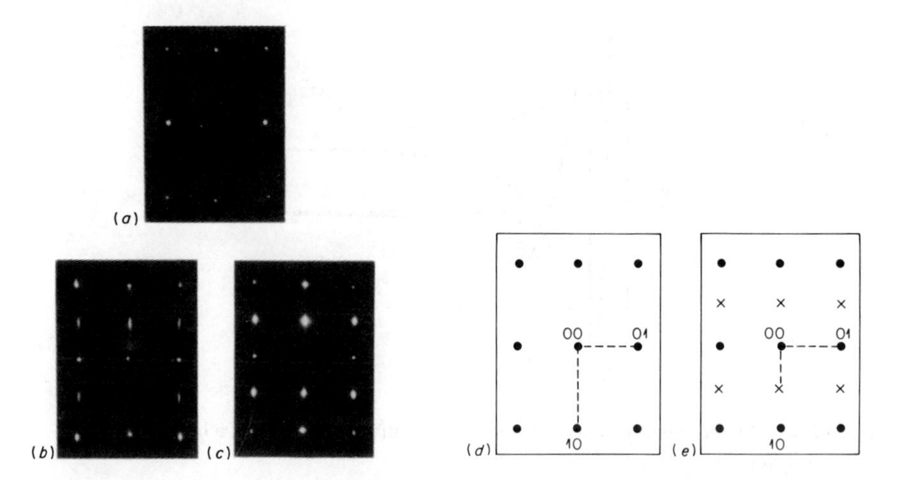

Figure 12.4 LEED patterns showing the ordered structure developed in the course of oxygen chemisorption on Cu (110): (a) (1 × 1) structure of clean surface; (b) (1 × 2) structure after exposure to 1 L of O_2; (c) (1 × 2) structure after exposure to 100 L of O_2 (1L = 10^{-6} Torr-sec.); (d) schematic of clean surface pattern; (e) schematic of (1 × 2) pattern. (Courtesy of John Noonan.)

Adlayer Structure

Consider next the structures formed on the surface in the course of chemisorption. Three aspects of this question will be considered, namely, the long-range periodicity of the structures formed, order–disorder transformations in the adsorbed layer, and the detailed positions of adsorbate and substrate atoms.

Consider first the question of long-range periodicity. Chapter 1 mentioned this phenomenon in passing in the discussion of surface structures as observed by LEED. These LEED studies of chemisorbed layers show that the formation of ordered adlayers takes place in most systems in which chemisorption is observed. Moreover, many systems show more than one adlayer arrangement, depending on the adlayer coverage and temperature. Typical examples of LEED patterns showing ordered adlayers of oxygen on copper and CO on tungsten surfaces are shown in Figures 12.4 and 12.5.

The majority of the chemisorption structures observed show periodicities that are a small multiple of the surface unit mesh. This implies that the nearest neighbor adatom interaction energy is repulsive. This result can be rationalized on the basis of dipole–dipole repulsion between adjacent adatoms, or on the basis that electronic charge withdrawn from the surface to form the chemisorptive bond results in weaker bonding possibilities at adjacent sites. It is also observed, however, that in many systems ordered structures can be seen at quite low total coverages, as shown in Figure 12.4. This implies that there is a net attractive force between second and farther neighbors. Otherwise, ordered structures would only appear when the total coverage was so high that ordering would be the only way to avoid having adatoms on nearest neighbor sites. Note that the LEED measurements also provide some information on the size of

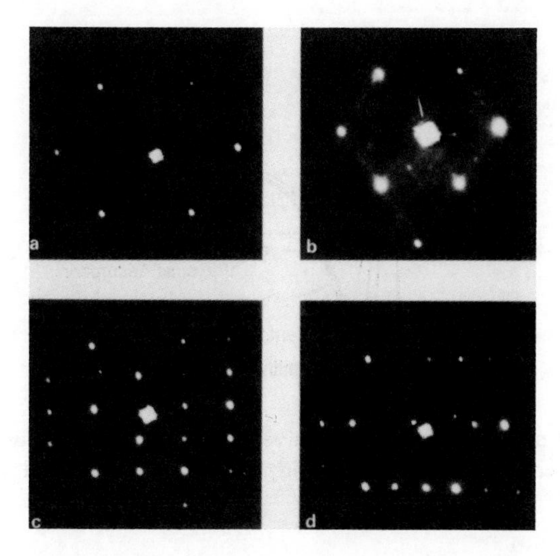

Figure 12.5 LEED patterns showing the ordered structures formed in the course of CO adsorption on W (110): (a) clean surface, (b) $p(2 \times 7)$ structure (based on a rhombic surface unit cell) observed at $\theta \approx 0.2$; (c) $c(4 \times 1)$ structure at $\theta \geq 0.3$; (d) $p(3 \times 1)$ structure at $\theta \approx 0.5$. (Source: Ch. Stienbruchel and R. Gomer, *Surface Sci.* 1977;67:21. Reprinted with permission.)

the ordered regions. Islands of ordered adsorbate structure smaller than the coherence length of the LEED primary electron beam (about 10 nm in practical instruments) will give rise to LEED spots that are diffuse, with the diffuseness increasing as island size decreases. Larger ordered regions, relative to the coherence length, give rise to sharp diffraction spots that do not change in sharpness with further increase in island size. This effect is similar to the particle size broadening seen in x-ray diffraction.

The relation of the unit mesh of these ordered adlayers to the substrate mesh is, in most cases, relatively simple, such as a 2×2 or 2×1. In some other cases, especially at high coverages, more complex orientation relations are observed. In the case of chemisorption, especially for those cases in which the adatom is large compared to the substrate atom, one often observes what is called a *coincidence net structure*, where the lattice points of the adlayer structure and the surface unit mesh coincide at fairly widely spaced intervals. For example, sulfur on Ni (111) forms an $8\sqrt{3} \times 2$ arrangement at coverages in the vicinity of 0.5 monolayer. Finally, in the case of physically adsorbed layers, where the lateral interaction energy is comparable to the site-to-site energy difference parallel to the surface, one occasionally finds incoherent ordered adlayers, in which the atomic arrangement in the adlayer depends only on the atomic size and lateral interaction energy of the adsorbate.

It is also quite frequently observed, as seen in Figure 12.5, that a given system will show several different arrangements of atoms in the adlayers, as the coverage or temperature is changed. As another example, in the case of oxygen chemisorption on Ni(100), one observes a $p(2 \times 2)$ structure at coverages less than $\theta = 0.25$. In the range between $\theta = 0.25$ and $\theta = 0.5$ a $c(2 \times 2)$ structure develops. Further oxygen adsorption

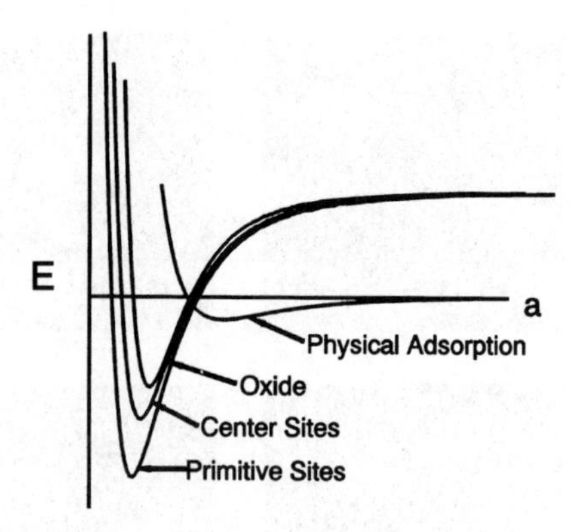

Figure 12.6 Schematic potential energy curves for oxygen adsorption on Ni (100), showing the relative binding energies of oxygen atoms in $p(2 \times 2)$ and $c(2 \times 2)$ adlayer structures and in NiO.

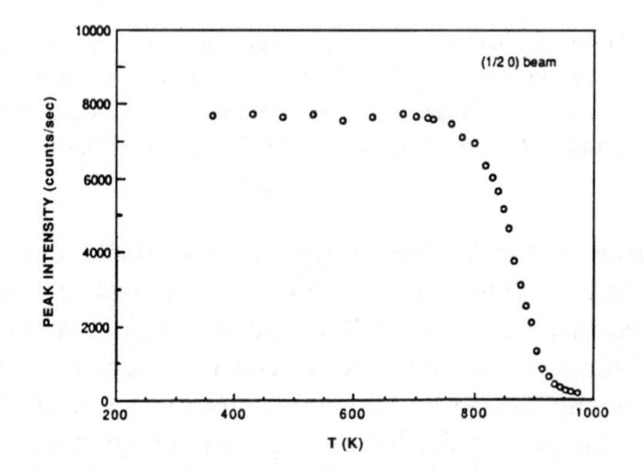

Figure 12.7 Plot of LEED spot intensity as a function of temperature for the (1/2 0) beam of the 2×1 oxygen structure on W (112). The abrupt drop in overlayer spot intensity is due to the transition from an ordered adsorbed phase to a two-dimensional gas. (Source: J-K. Zuo and G-C. Wang, *Phys. Rev.* 1989;B41:7078. Reprinted with permission.)

leads to the formation of a bulk NiO layer. This behavior is consistent with measured values of the heat of adsorption as a function of coverage and implies an energy diagram of the type shown in Figure 12.6.

Just as the observed adlayer structure changes with changes in surface coverage, one may also see changes with surface temperature at a given coverage. The most commonly observed change is a disappearance of order when the temperature is increased. This phenomenon is similar to the transition between a two-dimensional condensed phase and a two-dimensional gas that was discussed in Chapter 11. It is a result of the fact that the order arises from the presence of lateral interaction energies, which will not sustain the ordered layer at temperatures such that kT is large compared to the interaction energy.

Experimentally, this disordering process can be observed in a LEED experiment by measuring the intensity of the fractional order spots associated with the ordered

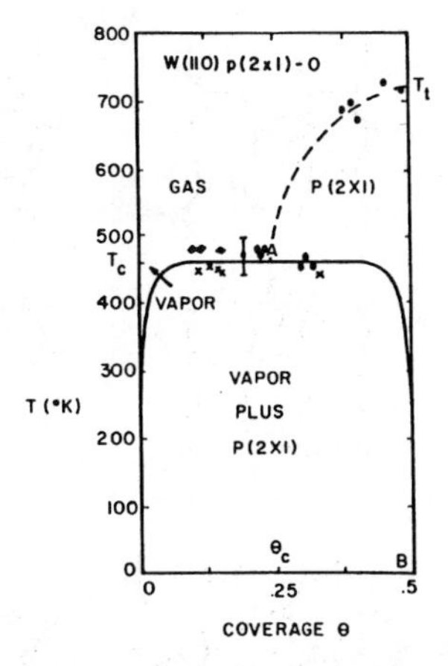

Figure 12.8 Two-dimensional phase diagram for oxygen adsorbed on W (110), showing transitions between the ordered 2 × 1 phase and a disordered two-dimensional gas. (Source: T-M. Lu and M.G. Lagally, *Surface Sci.* 1980;92:133. Reprinted with permission.)

adlayer as a function of temperature. Typically, behavior such as that shown in Figure 12.7 is observed. The intensity of the spots associated with the substrate mesh decreases somewhat with increasing temperature due to the Debye–Waller effect (a decrease in diffracted intensity due to the larger thermal motion amplitude of the atoms in the lattice). The spots associated with the overlayer, however, drop precipitously in intensity over a relatively short temperature range as disordering takes place. This process is reversible, and the adlayer spots will reappear as the temperature is reduced (assuming the temperature was not raised so high that the adsorbate either desorbed or dissolved in the bulk of the crystal).

One can characterize these order–disorder and multiple surface phase phenomena in terms of a two-dimensional phase diagram. This construction is analogous to the conventional phase diagram for a three-dimensional system, in which the regions of stability of various phases or phase mixtures are plotted for a range of temperatures and compositions. In the two-dimensional case, one obtains a plot of adlayer coverage vs. temperature such as is shown in Figure 12.8, showing regions of stability for various two-dimensional gaseous or condensed phases. Diagrams of this sort have been developed for a number of simple adsorbate systems and show many of the features of three-dimensional phase diagrams, including first-order and second-order phase changes, and one- and two-phase regions. One can look at these diagrams as being analogous to the phase diagram for a two-component, three-dimensional system. In the two-dimensional case the two "components" are adatoms and vacant adsorption sites.

As a final topic under adlayer structure, one may consider what is known about the equilibrium adatom positions relative to the positions of the underlying substrate atoms. On any given crystal face are several possible binding sites, involving one or more bonds between surface atoms and the adsorbate atom or molecule. Because the bonding in chemisorption in many cases involves covalent bonds, which have strong

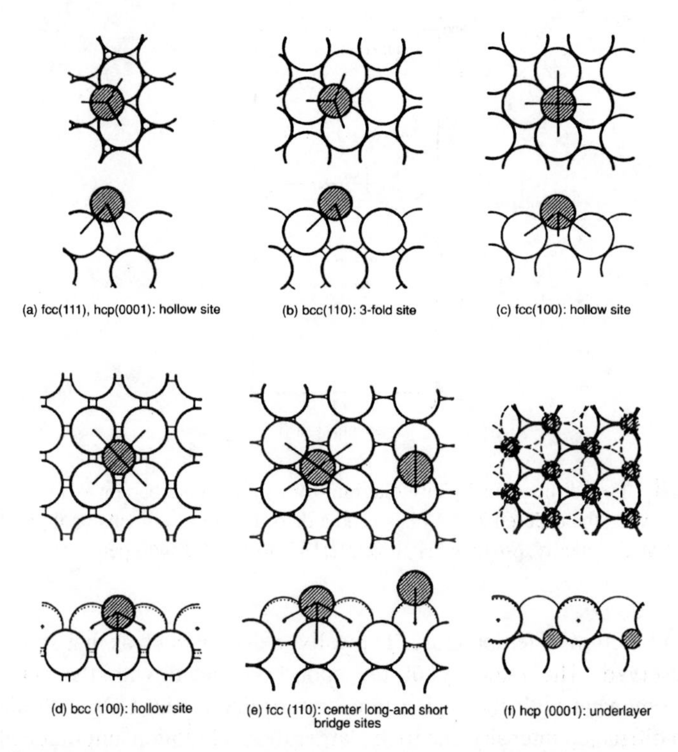

Figure 12.9 Top and side views (above and below in each diagram) of adsorption geometries on various metal surfaces for monatomic adsorbates. Dotted lines show displacements of surface atoms due to adsorption. (Reprinted from G.A. Somorjai, *Chemistry in Two Dimensions*. Copyright © 1981 by Cornell University. Used by permission of the publisher, Cornell University Press.)

directional preferences, the energetically favored site may or may not correspond to the site that provides the largest number of nearest neighbors. For the case of monatomic adsorbates, Figure 12.9 shows a number of possible configurations. In all of the examples shown, the adatom is multiply bonded to the substrate atoms but not always by the maximum possible number of bonds.

In the case of adsorbed polyatomic species, the situation becomes even more complicated, as one must consider the added questions of which atom or atoms of the adsorbed molecule are bonded directly to the surface and what is the orientation of the molecule relative to the surface. Figure 12.10 summarizes a number of the configurations that have been deduced for relatively simple molecules. These involve both cases of adsorption without dissociation and of adsorption with partial dissociation or rearrangement and show cases of bonding both to a single surface atom and to multiple surface atoms.

A number of techniques have been used to provide experimental information on the question of adatom bonding geometry. In the case of monatomic adsorbed species, the most commonly used techniques have been LEED and ion scattering spectrometry

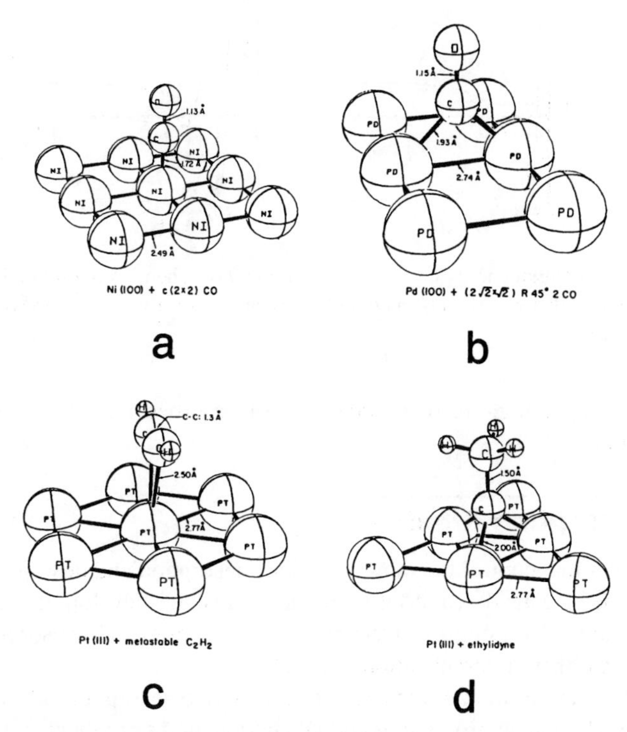

Figure 12.10 Adsorbate configurations for a number of common cases of molecular chemisorption. (Reprinted from G.A. Somorjai, *Chemistry in Two Dimensions*. Copyright © 1981 by Cornell University. Used by permission of the publisher, Cornell University Press.)

(ISS). In the case of LEED, as mentioned earlier, information on atomic positions can be obtained only from analysis of the spot intensity vs. voltage curves. This is a tedious procedure and, consequently, has been used for only a limited number of structures. The various ion scattering spectrometries, which will be discussed in Chapter 15, provide information that requires less interpretation but is more difficult to obtain experimentally. In those few cases in which both LEED and ion scattering have been applied, there is general agreement on atomic positions but differences in detailed values of interatomic spacing.

In the case of polyatomic adsorbates, the experimental techniques applied have been primarily spectroscopic. Ultraviolet photoelectron spectroscopy (UPS), with angle-resolved detection of the emitted electrons, has been used to determine bonding geometry. High-resolution electron energy loss spectroscopy (HREELS), which will be discussed in Chapter 14, has been used to characterize the adsorption process through measurement of the vibrational structure of adsorbed species, as has infrared spectroscopy. Some insight has also been gained by comparing adsorption of a given species on a metal surface with the bonding geometry observed in metal–organic compounds Finally, electron-stimulated desorption (ESD) and photon-stimulated desorption (PSD), which will be discussed in Chapters 14 and 16, respectively, have been

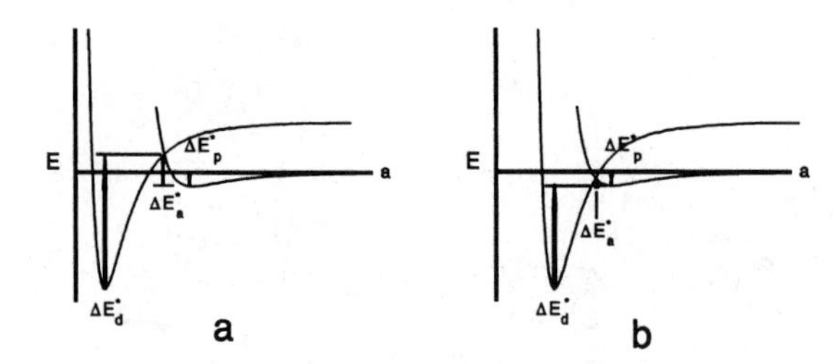

Figure 12.11 Potential energy curves for the dissociative chemisorption of a homonuclear diatomic molecule with a weakly adsorbed molecular precursor, for systems in which $k_p >> k_a$ (a), and $k_a >> k_p$ (b).

used, especially with angle-resolved detection of desorbed ions, to illustrate surface bonding symmetry.

Chemisorption Kinetics

The factors that control the rates of various adsorption and desorption processes have already been considered briefly in the course of developing the Langmuir isotherm in Chapter 10. Looking at the more general case of chemisorption kinetics, one sees that a number of complications can arise.

As an illustration, look at the case of dissociative adsorption of a homonuclear diatomic molecule. The appropriate potential energy curves are those in Figure 12.1c. Potential energy curves for two species are involved: the relatively weakly adsorbed molecular species and the strongly adsorbed atomic species. The energy terms that will be of importance in describing the rate of the adsorption process are the adsorption energy in the molecular state, ΔE_p, the activation energy for the recombination process, $\Delta E_d{}^*$, and the dissociation energy of the diatomic molecule, ΔE_D.

The overall process for the case of precursor-mediated adsorption can be described in terms of a chemical reaction sequence thus:

$$(A_2)_g \underset{k_p}{\overset{I\alpha}{\rightleftarrows}} (A_2)_p \underset{k_d}{\overset{k_a}{\rightleftarrows}} 2(A)_a. \qquad 12.2$$

In this expression I is the impingement rate from the gas phase, α is the accommodation coefficient, or trapping probability, into the weakly adsorbed state, which may be a function of adlayer coverage, and the k_i are rate constants for the various dissociation and desorption processes involved, as described in Figure 12.11. These k_i can generally be expressed in terms of absolute reaction rate theory as the rate constants for first- or second-order reaction steps, with a preexponential term and an exponential in the appropriate activation energy divided by kT.

Formulation of the overall rate equations for the adsorption and desorption process is complex and generally requires that some assumptions be made in order to obtain a specific result. For example, in the present case, if one assumes that chemisorption occurs on a fixed number of sites, that there are no lateral interactions among adatoms, that there is a maximum of one chemisorbed atom per site, that the weakly adsorbed molecular or "precursor" species behaves the same over both filled and empty chemisorption sites, and that the surface lifetime in the molecular state is short (so that the coverage in this state is always small and the population in this state will always be at its steady state value), then one can write for the overall rate of the desorption reaction that

$$r_d = \frac{-d\theta}{dT} = k_p\left[(A_2)_p\right] = \frac{k_p k_d \theta^2}{k_p + k_a(1-\theta)^2}, \qquad 12.3$$

where θ is the fractional coverage in the chemisorbed state. The rate of the adsorption reaction can be written

$$r_a = \frac{\dfrac{k_a(1-\theta)^2 \alpha I}{n_o}}{k_p + k_a(1-\theta)^2}. \qquad 12.4$$

When adsorptive equilibrium has been reached, r_a and r_d will be equal, so that

$$r_a = r_d \qquad 12.5$$

$$\frac{\dfrac{k_a(1-\theta)^2 \alpha I}{n_o}}{k_p + k_a(1-\theta)^2} = \frac{k_p k_d \theta^2}{k_p + k_a(1-\theta)^2}. \qquad 12.6$$

This equation may be solved, to yield for the equilibrium coverage in the chemisorbed state

$$\theta_e = \frac{\kappa p^{1/2}}{1 + \kappa p^{1/2}}, \qquad 12.7$$

where

$$\kappa = \left\{ \left(\frac{k_a \alpha}{k_p k_d n_o} \right) \left[\frac{1}{2\pi mkT} \right]^{1/2} \right\}^{1/2} \qquad 12.8$$

This may be compared to the result obtained in Chapter 10 for the Langmuir iostherm,

$$\theta_e = \frac{\chi p}{1 + \chi p}, \qquad 12.9$$

indicating that Equation 12.7 describes Langmuir-like behavior for the case of dissociative adsorption.

Alternatively, rather than combining the adsorption and desorption rate equations to determine the point of equilibrium, one may look at the rates of the two processes separately. This process is of interest for a number of reasons. First, for many chemisorption systems, the binding energy in the chemisorbed state is so high that $\theta_e \approx 1$ at ambient temperature, even for very low gas phase pressures. Consequently, measurement of the equilibrium coverage does not provide any information on the details of the adsorption process. In this case, kinetic measurements of the rate of approach to equilibrium offer the only means of determining the actual reaction sequence and reaction order and, in turn, the values of the various rate constants and their associated energy terms. As a practical matter, the major importance of chemisorption is in relation to the rate of surface processes such as crystal growth, corrosion and catalysis. The rate of chemisorption is in many cases the rate controlling step for the overall process.

Consider as an example the adsorption rate for the case of dissociative adsorption of a homonuclear diatomic molecule as discussed above. In order to describe the rate of adsorption in such a process, one generally talks in terms of the *sticking coefficient* for chemisorption, defined as

$$S(\theta) = \frac{\frac{dn_a}{dt}}{I} = \frac{r_a}{I}, \qquad 12.10$$

that is, the unit collision adsorption probability. Clearly, the upper limit on S is unity. In practice, S is often much less than unity and in general will change as adlayer coverage changes. For the case considered here it has already been shown that

$$S = \frac{r_a}{I} = \frac{\frac{k_a(1-\theta)^2\alpha}{n_o}}{k_p + k_a(1-\theta)^2}. \qquad 12.11$$

One may consider the form of this equation in terms of the relative values of k_a and k_p. If $k_a >> k_p$, as shown in Figure 12.11b, the surface lifetime in the molecular state will be long compared to the time required for the dissociation process, leading to

$$S = \frac{k_a(1-\theta)^2\alpha}{k_a(1-\theta)^2} = \alpha, \qquad 12.12$$

and the sticking coefficient will be independent of coverage right up to the point where $\theta \to 1$.

It is also useful to look at the effect of various ratios of the rate constants on the overall rate of the desorption process. Consider the case in which $k_a << k_p$. For this case

$$r_d = \frac{k_p k_d \theta^2}{k_p + k_a(1-\theta)^2} \approx k_d \theta^2, \qquad 12.13$$

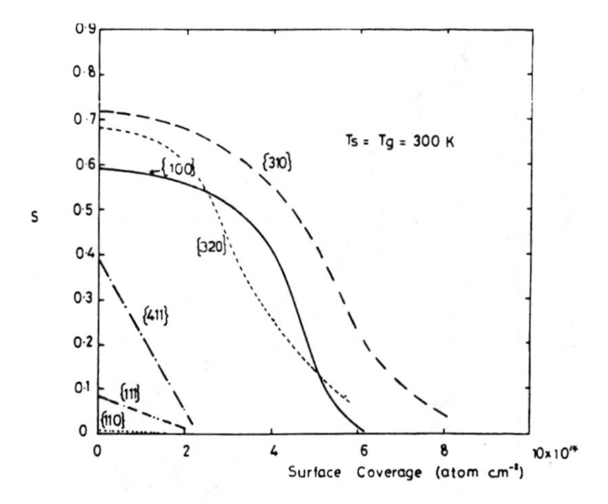

Figure 12.12 Sticking coefficient for the chemisorption of nitrogen on various tungsten single crystal planes. (Reprinted with permission from D.A. King, *CRC Crit. Rev. Solid State and Matls. Sci.* 1977;7:167. Copyright CRC Press, Inc., Boca Raton, Florida.)

and the desorption process will follow second-order kinetics. Alternatively, if $k_a >> k_p$,

$$r_d = \frac{k_p k_d \theta^2}{k_p + k_a(1-\theta)^2} \approx \left(\frac{k_p k_d}{k_a}\right)\left[\frac{\theta^2}{(1-\theta)^2}\right], \qquad 12.14$$

which would show a significant departure from second-order kinetics, especially at large values of θ. Further examples of this type will be found in Chapter 13, where surface reaction kinetics are considered in detail.

The observed behavior of the sticking coefficient as a function of coverage in a range of systems covers the whole range of possibilities suggested by the foregoing calculations. Typical examples are shown in Figure 12.12 for the case of adsorption of nitrogen on various tungsten surfaces. It can be seen that both the initial value of the sticking coefficient and the variation of the sticking coefficient with coverage differ greatly from one crystal face to another. Note that all faces have sticking coefficient values less than unity even at zero coverage, indicating either that α is less than unity or that there is a finite probability of desorption from a precursor state rather than passage into the chemisorbed state. Some of the faces show a range of essentially constant sticking coefficient, suggesting the importance of precursor states in the adsorption process. More detailed measurements, over a range of temperatures, would be necessary in order to fully characterize the adsorption process in this system.

Activated Adsorption

Consider, finally, the kinetic behavior observed in systems in which the crossover between the molecularly physisorbed and dissociatively chemisorbed states occurs at an energy greater than zero, as shown in Figure 12.3. In such cases, a finite

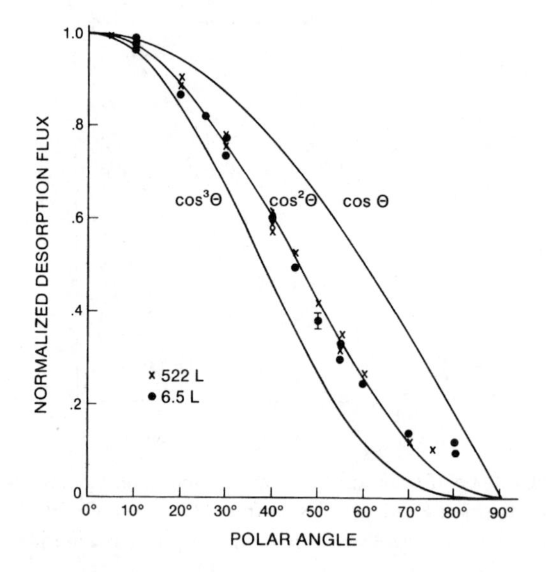

Figure 12.13 Angular distribution of hydrogen desorbing from Fe (110), indicating excess kinetic energy of desorbing molecules. (Source: E.A. Kurz and J.B. Hudson, *Surface Sci.* 1988;195:31. Reprinted with permission.)

barrier is present for both the adsorption and desorption processes. This both influences the temperature dependence of the adsorption rate and gives rise to two new phenomena, namely, the desorption of molecules having excess kinetic energy and a dependence of the adsorption rate on the kinetic energy of the incident molecules.

Consider first the dependence of the adsorption rate on substrate temperature. In terms of the precursor mechanism discussed previously, since $\Delta E_a^* > \Delta E_p^*$, the ratio of the rate constant k_p to k_a, which is

$$\frac{k_p}{k_a} \approx \exp\left(\frac{-\Delta E_p^* + \Delta E_a^*}{kT}\right), \qquad 12.15$$

will decrease with increasing temperature, leading to an increasing chemisorption probability. The alternative case, $\Delta E_p^* > \Delta E_a^*$, leads to a decreasing chemisorption probability with increasing temperature.

In the case of activated adsorption, it is also possible to have direct desorption from the chemisorbed state, without a finite residence time for the desorbing molecule in the precursor state. This leads to the desorption of molecules having excess energy equivalent to the barrier height shown in Figure 12.3. This energy may be carried off either as translational kinetic energy or as excess energy in the internal modes of the molecule. Excess translational kinetic energy can be observed either by a time of flight measurement of the desorbing molecules or as a departure from the expected cosine spatial distribution of the desorbed species. An example of this latter phenomenon is shown in Figure 12.13 for the case of H_2 desorption from Fe (110). The experimental

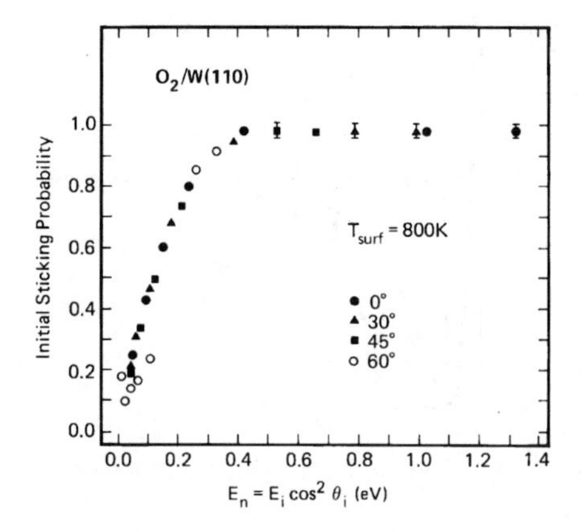

Figure 12.14 Sticking coefficient for chemisorption of oxygen on tungsten, as a function of incident oxygen molecule translational kinetic energy. (Source: C.T. Rettner, L.A. DeLouise, and D.A. Auerbach, *J. Chem. Phys.* 1986;85:1131. Reprinted with permission.)

results show an approximate $\cos^2 \theta$ distribution, indicative of excess kinetic energy normal to the surface.

Direct adsorption from the gas phase is also possible in the case under discussion, and measurement of the adsorption probability as a function of the kinetic energy of the incident molecules provides a means of measuring the height of the barrier to chemisorption. If one considers the behavior of an impinging molecule in terms of the concept of the classical turning point discussed in Chapter 9, one sees that an impinging molecule having an incident kinetic energy greater than the barrier height will have a classical turning point beyond the location of the barrier and can reach the chemisorbed state directly, in a single collision, without trapping in the precursor state. This behavior has been observed in a number of systems by using seeded free jet expansions to produce fluxes of adsorbate molecules of known, high kinetic energies and measuring the chemisorption probability as a function of this energy. Figure 12.14 shows the results of such a study for the case of O_2 chemisorption on W (110). Here the sticking probability, S_o, increases rapidly with increasing molecular translational energy, implying a barrier height on the order of 0.2 to 0.3 eV. The fact that the sticking coefficient scales as the normal component of the incident kinetic energy implies that the barrier is essentially one dimensional. The detailed study of direct adsorption processes such as this is an active area of current research.

Problems

12.1. A gas is chemisorbed on the (111) face of a tungsten single crystal at a coverage of 0.1 monolayer. The heat of desorption of the gas as single atoms is 33 kcal/mol of atoms. The heat of desorption of the gas as diatomic molecules is 26 kcal/mol of molecules. The activation energy for formation of the diatomic molecule on the surface is 28 kcal/mol of molecules. The binding energy of the

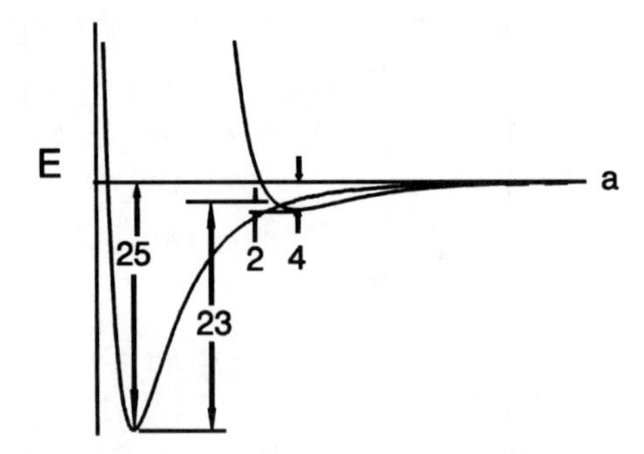

Figure 12.15 Potential energy diagram for the chemisorption of a hypothetical diatomic molecule on nickel.

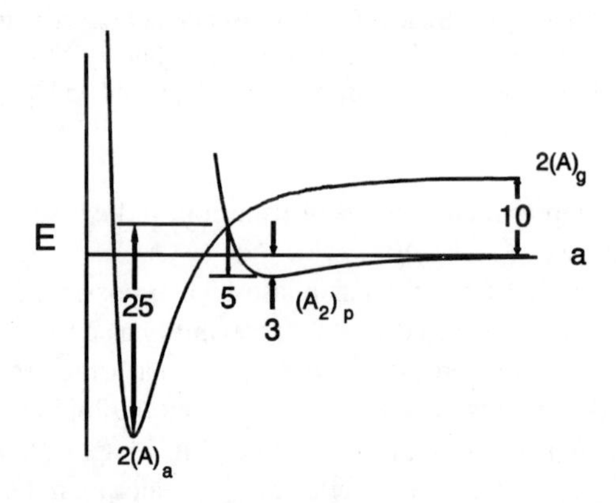

Figure 12.16 Potential energy diagram for the chemisorption of a hypothetical diatomic molecule.

diatomic molecule is 40 kcal/mol. Calculate the desorption rates of atoms and diatomic molecules at 500 K and 1000 K, assuming that precursor effects are negligible and that the preexponential terms in all rate constants are 10^{13} sec^{-1}.

12.2. The adsorption of a diatomic molecule on nickel occurs according to the potential energy diagram shown in Figure 12.15, where all energies are in kilocalories per mole. Develop an expression for the sticking coefficient for chemisorption as a function of coverage in the chemisorbed state, assuming that the temperature is low enough that desorption from the chemisorbed state does not occur and that chemisorption is limited to one monolayer. (Hint: assume that at any instant the precursor state concentration has its equilibrium value.)

12.3. The interaction energy between a homonuclear diatomic molecule, A_2, and a surface is described by the potential energy diagram shown in Figure 12.16, where all energies are in kilocalories per mole. Assume that adsorption into the loosely bound state is molecular physisorption obeying Henry's law, that adsorption into the tightly bound state is dissociative chemisorption, and that sites occupied by chemisorbed atoms are unavailable for physisorption.

a. Develop an expression for the rate of change of coverage in the chemisorbed state.

b. Determine the coverage in the chemisorbed state for the case $T = 100$ K, $t = 100$ sec, $p = 10^{-3}$ Torr. You may assume that $n_o = 10^{15}$ sites/cm^2 and that the preexponential term in all rate constants is 10^{13} sec^{-1}.

12.4. Oxygen chemisorbs molecularly on a metal with a heat of desorption of 25 kcal/mol. There is also a molecularly physisorbed state having a heat of desorption of 4 kcal/mol. The barrier to passage from the physisorbed state to the chemisorbed state is 2 kcal/mol.

a. Sketch the potential energy vs. distance curves for this system.

b. Diagram the reaction sequence leading from O_2 in the gas phase to O_2 in the chemisorbed state.

c. Calculate the equilibrium concentrations in the two adsorbed states, subject to the following assumptions: all $k_o = 10^{13}$ sec^{-1}, $n_o = 1.3 \times 10^{15}$ sites/cm^2, $\alpha = 1$, $T = 400$ K, $p = 10^{-3}$ Torr.

12.5. Calculate the initial sticking coefficient for chemisorption in the system described by the potential energy diagram of Figure 12.16 at $T = 300$ K, subject to the following assumptions:

1. All molecules that strike the surface with total kinetic energy greater than or equal to the barrier height for direct chemisorption are directly chemisorbed.
2. All other molecules are accommodated into the molecularly physisorbed state.

Bibliography

Christman, K. Interaction of hydrogen with solid surfaces. *Surface Science Reports* 1988;9:1–3.

Comsa, G, and David, R. Dynamical parameters of desorbing molecules. *Surface Science Reports* 1986;5:4.

King, DA, and Woodruff, DP. *The Chemical Physics of Solid Surfaces and Heterogeneous Catalysis*. Amsterdam: Elsevier, 1982;2 and 3.

Rhodin, TN, and Ertl, G, eds. *The Nature of the Surface Chemical Bond*. Amsterdam: Elsevier, 1979.

Ricca, F, ed. *Adsorption-Desorption Phenomena*. New York: Academic Press, 1972.

Schmidt, LD. Precursor intermediates in adsorption, desorption and reaction. In: U. Landman, ed. *Aspects of Kinetics and Dynamics of Surface Reactions*. New York: American Institute of Physics, 1980;83–96.

Somorjai, GA. *Chemistry in Two Dimensions*. Ithaca, New York: Cornell University Press, 1981.

Tompkins, FC. *Chemisorption of Gases on Metals*. New York: Academic Press, 1978.

CHAPTER **13**

Surface Chemical Reactions

This chapter expands the treatment begun in Chapter 12 for the process of chemisorption to the more general case of chemical reactions at surfaces. All surface reactions proceed by a sequence of steps, the first of which is usually chemisorption. After this initial step, a number of surface processes can take place, leading eventually to the formation of one or more new chemical species.

Types of Reaction

For purposes of classification, surface reactions may be divided into three categories:

1. *Corrosion reactions.* In corrosion reactions, material from the gas phase interacts with the surface to produce new chemical species that include atoms from the surface. This class can in turn be subdivided into *volatilization reactions*, in which the product species returns to the gas phase and the surface is progressively consumed, and *corrosion layer formation*, in which a nonvolatile surface compound is formed. Examples of volatilization reactions include:

$$H_2O + C \rightarrow CO + H_2 \qquad 13.1$$

$$Cl_2 + Ni \rightarrow NiCl_2 \qquad 13.2$$

$$30_2 + 2Mo \rightarrow 2Mo0_3 \,. \qquad 13.3$$

Examples of reactions that produce a solid corrosion layer include:

$$O_2 + Fe \rightarrow FeO_x \qquad 13.4$$

$$S + Ni \rightarrow NiS \,. \qquad 13.5$$

2. *Crystal growth reactions.* In crystal growth reactions material is deposited on the surface, with or without a decomposition reaction, to extend the surface or to form the solid phase of a new material. Reactions of this type include *physical vapor deposition*, for example

$$(Ag)_v \rightarrow (Ag)_s \,, \qquad 13.6$$

molecular beam epitaxy

$$2(\text{Ga})_v + (\text{As}_2)_v \rightarrow 2\text{GaAs}, \tag{13.7}$$

and *chemical vapor deposition*

$$(\text{NiCl}_2)_g \rightarrow (\text{Ni})_s + (\text{Cl}_2)_g. \tag{13.8}$$

This last type of reaction is essentially the reverse of the volatilization reaction, Equation 13.2.

3. *Catalytic reactions*. In catalytic reactions, material from the surface is not directly involved in the species synthesized or decomposed in the reaction. The surface serves as a site at which the reaction rate is enhanced relative to its rate in the gas phase. Catalytic reactions can be further subdivided into categories such as *exchange reactions*,

$$(\text{H}_2)_g + (\text{D}_2)_g \rightleftarrows 2(\text{HD})_g, \tag{13.9}$$

recombination reactions

$$\text{H}_a + \text{H}_a \rightleftarrows (\text{H}_2)_g\,, \tag{13.10}$$

unimolecular decomposition reactions

$$(\text{N}_2\text{O})_g \rightleftarrows (\text{N}_2)_g + \text{O}_a \tag{13.11}$$

$$(\text{CHOOH})_g \rightleftarrows \text{CO} + \text{CO}_2 + \text{H}_2 + \text{H}_2\text{O}\,, \tag{13.12}$$

and *bimolecular reactions*

$$2(\text{CO})_g + (\text{O}_2)_g \rightleftarrows 2(\text{CO}_2)_g \tag{13.13}$$

$$\text{CO} + 2\text{H}_2 \rightleftarrows \text{CH}_4 + \text{H}_2\text{O}\,. \tag{13.14}$$

This chapter will look in detail at several catalytic reactions and one volatilization reaction. Crystal growth processes will be covered in Chapter 17.

Surface Reaction Sequences

All classes of surface reactions can be thought of in terms of a sequence of steps. In its most general form, the sequence involves adsorption from the gas phase into a molecular precursor state, chemisorption, surface migration to the reaction site, the actual reaction step, surface diffusion away from the reaction site, return to a physically adsorbed state, and finally desorption into the gas phase. Figure 13.1 diagrams this process. Not all of these steps will be involved in any given reaction, and in many cases only one of these steps will control the overall reaction rate. In principle, however, all of these steps must be considered in formulating the detailed reaction mechanism.

A number of these processes have already been discussed in detail, such as accommodation into an adsorbed state, physisorption and chemisorption, and surface migration. The only new step in the sequence is the actual reaction step itself. In discussing this step, it is useful to introduce some new terms that describe this step in various cases.

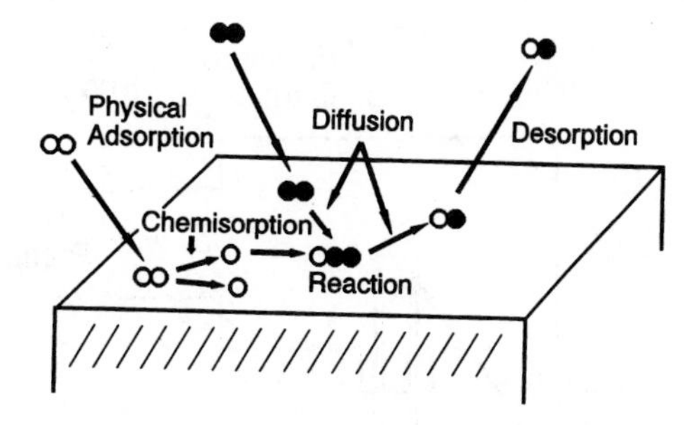

Figure 13.1 Schematic view of the possible steps involved in a surface chemical reaction between two homonuclear diatomic molecules.

Reactions are often referred to as being *structure sensitive* or *structure insensitive*, depending on the way in which the reaction rate changes as a function of surface perfection, or the particle size of the catalytic material. Reactions whose rate depends only on the amount of surface present, and not on its structure, are said to be structure insensitive. Those that show a rate dependent on surface defect structure are said to be structure sensitive.

Depending on the importance of chemisorption before reaction, reactions are classified as following either a *Langmuir–Hinshelwood* or an *Eley–Rideal* mechanism. A bimolecular reaction that procedes by chemisorption of both reactants before the reaction step is said to follow a Langmuir–Hinshelwood mechanism. A bimolecular reaction between one chemisorbed species and a second species that impinges directly from the gas phase is said to follow an Eley–Rideal mechanism.

A term that is frequently used as a measure of reactivity in catalytic systems is the *turnover number*. This is defined as the number of product molecules formed per available surface site per unit time. A related term is the *unit collision reaction probability*, which is simply the ratio of the number of product molecules formed to the total number of reactant molecules striking the surface. Both of these terms give a measure of the efficiency of the catalytic surface.

There is also frequent mention in the catalysis literature of *active sites for reaction*. This is a term that developed in the early days of the study of catalytic reactions, as an explanation of the observed structure sensitivity of many reactions. Much current research in the field is aimed at elucidating the nature of surface configurations that provide especially favorable sites for the surface reaction step. To date, positive correlations have been found between surface reactivity and the presence of ledge or kink sites on a crystal surface, but a detailed understanding of their effect on reactivity is lacking.

Experimental Studies of Surface Reactions

The overall rates of surface catalysed reactions have been studied for some time by catalytic chemists. These studies have, for the most part, been carried out under such conditions and using such techniques that an understanding of the surface reaction at

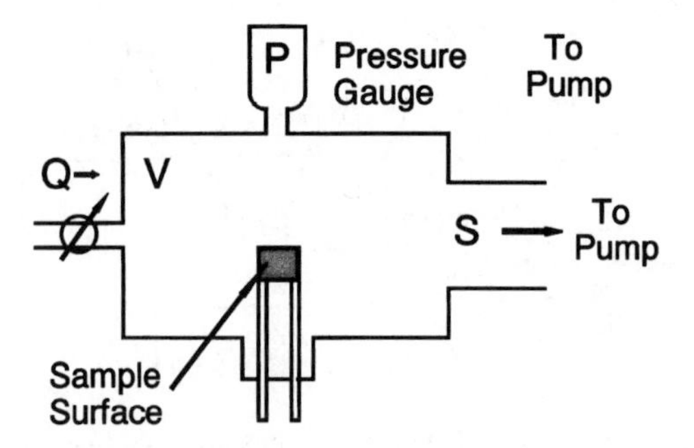

Figure 13.2 Schematic diagram of the experimental setup used for thermal desorption spectroscopy.

the molecular level was not possible. For example, typical studies included measurement of the turnover number for a reaction over a particular catalyst, or of the overall rate of buildup of a corrosion product layer, or of the overall rate of a volatilization process by determining the net rate of weight gain or loss by the sample. Studies of this sort have provided much useful practical information on the overall rates of surface processes and on the macroscopic variables influencing these rates. They do not, however, provide insight into the details of the surface processes at the molecular level.

In order to study the rate of surface processes at the molecular level, a number of techniques have been developed in recent years. Before discussing the kinetics of specific surface reactions, it will be useful to describe two of these techniques in some detail, namely, *thermal desorption spectroscopy* (*TDS*), along with the related technique of *temperature programmed reaction spectroscopy* (*TPRS*), and the more recently developed *molecular beam relaxation spectroscopy* (*MBRS*).

Thermal Desorption Spectroscopy

Consider first thermal desorption spectroscopy. This technique was developed originally as a means of studying chemisorption, and has more recently been extended to use in chemical reaction studies. The basic experimental setup used in TDS is shown in Figure 13.2. The surface to be studied is mounted in an ultrahigh vacuum system in such a way that it can be heated and cooled in a controlled fashion. The vacuum chamber is provided with a total pressure gauge or a small mass spectrometer, and provision is made for leaking the gas or gas mixture of interest into the chamber at a controlled rate.

If the surface is cleaned and then exposed to the test gas at a constant pressure and constant surface temperature, then at steady state

$$p = p_{eq} \tag{13.15}$$

$$KQ = K\,S\,p_{eq}\,, \tag{13.16}$$

where $K = 3.27 \times 10^{19}$ molec/Torr-liter, Q is the leak rate of the gas into the system (Torr-liter/sec) and S is the system pumping speed (liter/sec). If the sample is now heated to decrease the surface lifetime of any gas adsorbed on the surface, at any instant

$$Ar_d + KQ = KSp + KV\left(\frac{dp}{dt}\right), \tag{13.17}$$

where A is the sample surface area and

$$r_d = \frac{-dn_a}{dt} \tag{13.18}$$

is the desorption rate in molecules per square centimeter second.

Equations 13.16 and 13.17 may be combined to yield

$$\frac{dp^*}{dt} + \frac{p^*}{C} = ar_d, \tag{13.19}$$

in which

$$p^* = \left(p - p_{eq}\right), \quad a = \frac{A}{KV}, \quad C = \frac{V}{S}. \tag{13.20}$$

Thus, the value of p^* as a function of time is controlled by the fixed system parameters, A, V, S, Q, and the desorption rate, r_d. The desorption rate is, in turn, a function of the surface lifetime, τ_a, which was defined in Chapter 10. That is,

$$\begin{aligned} r_d &= \frac{-dn_a}{dt} = n_a^m\left(\frac{1}{\tau_a}\right) \\ &= \nu_m n_a^m \exp\left(\frac{-\Delta H_{des}}{RT}\right), \end{aligned} \tag{13.21}$$

where m is the order of the desorption process.

All that remains in order to evaluate $p^*(t)$ is an assumption about the behavior of $T(t)$. The simplest assumption, and the condition used most commonly in practice, is a linear increase of T with time. That is,

$$T = T_0 + \beta t\,. \tag{13.22}$$

Assuming for the moment that the desorption energy is independent of coverage and that there are no effects due to precursor species, Equations 13.19 and 13.21 may be solved to find T_p, the temperature of maximum desorption rate, as

$$\left(\frac{\Delta H_{des}}{RT_p^2}\right) = \left(\frac{\nu_1}{\beta}\right)\exp\left(\frac{-\Delta H_{des}}{RT_p}\right) \quad \text{for } m = 1. \tag{13.23}$$

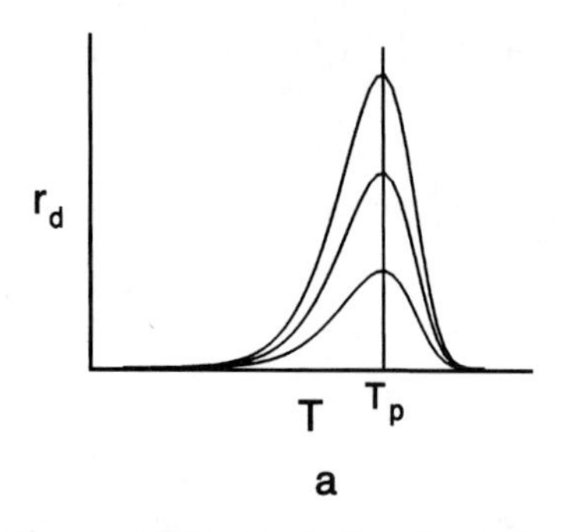

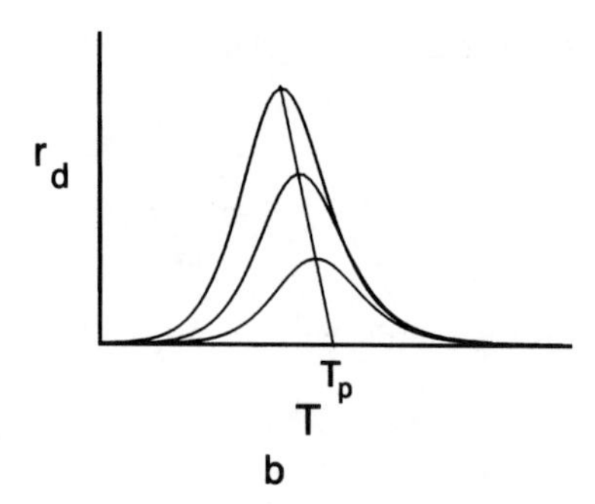

Figure 13.3 Typical results of thermal desorption spectroscopy measurements, showing a first order desorption process (a) and a second order desorption process (b).

$$\frac{\Delta H_{des}}{RT_p^2} = \frac{n_i \nu_2}{\beta} \exp\left(\frac{-\Delta H_{des}}{RT_p} \right) \quad \text{for } m = 2. \qquad 13.24$$

In Equation 13.24, n_i is the initial surface coverage before the temperature increase. A knowledge of T_p, β, and n_i, coupled with an assumed value of ν_1 or ν_2 thus enables one to determine the desorption energy. The equilibrium coverage before the temperature sweep can also be determined from the area under the desorption transient.

The major limitation inherent in the above procedure is that it requires assuming a value for ν_1 or ν_2. Customarily, most studies have assumed $\nu_1 \approx 10^{13}$ sec^{-1}. Actual values, measured by other techniques such as MBRS, however, may differ from this value by as much as 10^5. There are also the uncertainties implicit in the assumptions of coverage-independent desorption energy and assumed order of reaction.

To some extent, these limitations may be avoided by characterizing the desorption process in terms of the shape of the whole desorption rate vs. time curve. To do this, the expression for desorption rate must be integrated. The results of this integration process are

$$\ln\left[\frac{(r_d)_p}{r_d} \right] = \left(\frac{\Delta H_{des}}{R} \right)\left(\frac{1}{T} - \frac{1}{T_p} \right) + \left(\frac{T}{T_p} \right)^2 \exp\left[-\left(\frac{\Delta H_{des}}{R} \right)\left(\frac{1}{T} - \frac{1}{T_p} \right) \right] - 1 \qquad 13.25$$

for the case of first-order kinetics, and

$$\frac{(r_d)_p}{r_d} = \frac{1}{4}\left\{ \exp\left[-\left(\frac{\Delta H_{des}}{2R} \right)\left(\frac{1}{T} - \frac{1}{T_p} \right) \right] + \left(\frac{T}{T_p} \right)^2 \exp\left[-\left(\frac{\Delta H_{des}}{2R} \right)\left(\frac{1}{T} - \frac{1}{T_p} \right) \right] \right\}^2 \qquad 13.26$$

for the case of second-order kinetics. The form of the desorption rate curves for the two cases is given in Figures 13.3 a and b for first- and second-order desorption, respectively. The major differences between the two cases are that the peak temperature is independent of coverage in first-order kinetics but decreases with increasing coverage in the second-order case, and that the desorption curve is assymetrical for first-order desorption but symmetrical in the second-order case.

From the desorption curves, one can determine that

$$\Delta H_{des} = \left[\frac{\varepsilon(r_d)_p}{n_i\beta}\right](RT_p^2) \tag{13.27}$$

and

$$v_1 = \left[\frac{\varepsilon(r_d)_p}{n_i}\right]\exp\left[\frac{\varepsilon(r_d)_p T_p}{n_i\beta}\right] \tag{13.28}$$

for first-order kinetics and that

$$\Delta H_{des} = \left[\frac{4(r_d)_p}{n_i\beta}\right](RT_p^2) \tag{13.29}$$

and

$$v_2 = \left[\frac{4(r_d)_p}{n_i^2}\right]\exp\left[\frac{4(r_d)_p T_p}{n_i\beta}\right] \tag{13.30}$$

for second-order kinetics, where

$$\varepsilon = \frac{n_i}{(n_a)_p}. \tag{13.31}$$

Thus, if one has a means of measuring r_d as a function of time, in the course of a linear temperature sweep, the reaction order may be determined, and ΔH_{des} and v_i may be calculated without making any assumptions about the value of v_i. Typically, r_d may be measured directly if one can make a line-of-sight measurement from the sample surface to the detector, assume cosine law desorption, and differentiate between desorbing material and background gas. More often, one measures $p^*(t)$ rather than r_d. This introduces a complication into the measurement as, recalling Equation 13.19,

$$\frac{dp}{dt} + \frac{p^*}{C} = ar_d, \quad C = \frac{V}{S}, \tag{13.32}$$

one sees that the relation between p^* and r_d also contains terms that depend on other system parameters, primarily the ratio of system volume to pumping speed. In the extreme case that the pumping speed is very low, $S \to 0$, $C \to \infty$, and

$$\frac{dp^*}{dt} = ar_d. \tag{13.33}$$

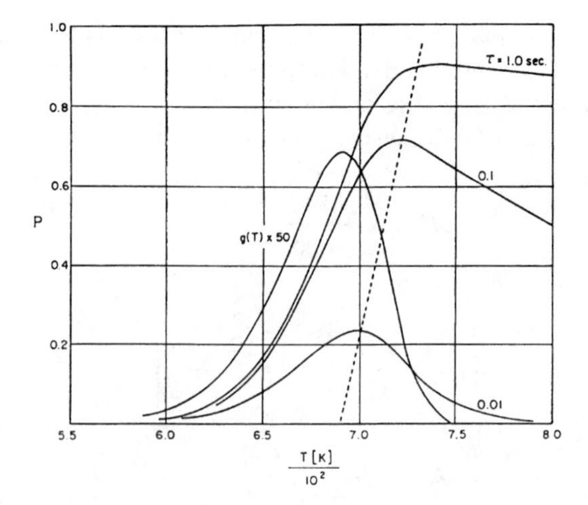

Figure 13.4 Curves showing the pressure rise as a function of temperature in a thermal desorption spectroscopy measurement, for various values of the ratio of system volume to system pumping speed. (Reprinted with permission from P.A. Redhead, *Vacuum* 1962;12:203. Copyright 1962, Pergamon Press plc.)

In this case, if one records $p = f(t)$, the derivative of this curve yields r_d. At higher pumping speeds, the shape of the curve is controlled by the combined effects of r_d and C. The variation of the p^* vs. t curve with pumping speed, for a given $r_d(t)$ is shown in Figure 13.4.

Note that the above treatment still does not account for either variations in ΔH_{des} with coverage, due to lateral interactions or surface heterogeneity, or kinetic effects due to the presence of a loosely bound precursor. Some modifications to the above treatment have been developed to account for either or both of these effects, but in general the results do not give a unique description of the desorption process.

The technique of temperature programmed reaction spectroscopy uses an experimental configuration that is essentially identical to that described above for TDS. In this case, the surface is exposed to a reactive gas or gas mixture at a temperature low enough that the reactants will adsorb but not react. The temperature is then increased, and desorption rate curves are obtained mass spectrometrically for all volatile species formed in the reaction. This can be done either by repeated experiments, monitoring one mass peak each time, or by multiplexing the mass spectrometer output to monitor several masses simultaneously. The desorption peaks obtained may be analyzed as explained above to determine the activation energies for the various desorption steps and to determine the relative amounts of the various product species arising from the surface reaction. This is a useful technique, in that it is relatively simple experimentally and can provide a great deal of information in a relatively short time. However, it does have two major limitations. It is often not possible to determine whether a given desorption peak arises from a process whose rate is limited by the surface reaction step or from the desorption of a product species produced by surface reaction at a lower temperature. The second limitation is that the technique inherently involves sampling the reaction over a

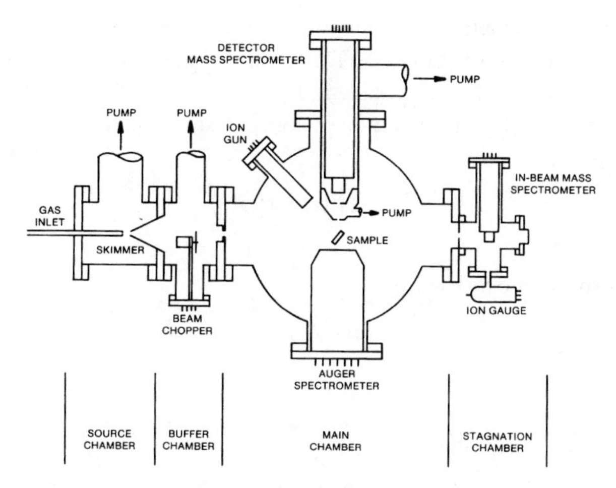

Figure 13.5 Schematic diagram of a typical system used for molecular beam relaxation spectroscopy studies. (Source: D.A. Hoffman and J.B. Hudson, *Surface Sci* 1987;180:77. Reprinted with permission.)

wide temperature range and with continuously changing temperature. If there is a change in mechanism with temperature, the only process that will be observed with TPRS is the mechanism operative at the lower temperature. In spite of these limitations, this technique has been used extensively in surface reaction studies. One such study, the surface catalyzed decomposition of formaldehyde, will be discussed later in this chapter.

Molecular Beam Relaxation Spectroscopy

The final technique for surface reaction rate measurement to be considered in detail is molecular beam relaxation spectroscopy (MBRS). A typical experimental setup for this technique is shown in Figure 13.5. The reactant gas or gas mixture is formed into a molecular beam and allowed to strike the surface on which the reaction takes place. Volatile reaction products are detected by the mass spectrometer after leaving the surface. The reactant beam is usually mechanically modulated, so that the flux to the surface as a function of time is essentially a square wave, although other wave forms, such as a sharp pulse, are also used. The resulting product signal, which generally retains at least partial memory of this modulation, is detected at the mass spectrometer either by lock-in detection at the fundamental frequency of the product flux, or by using digital techniques to collect the whole product waveform.

In either of the above cases, the kinetics of the surface reaction can be inferred from the difference between the waveform of the reactant flux and the waveform of the product flux. As an illustration of how this is done, consider the case of a simple first-order surface process such as adsorption–desorption. That is,

$$R(t) \xrightarrow{\alpha} C(t) \xrightarrow{k_d} P(t), \tag{13.34}$$

in which a time-varying reactant flux, $R(t)$, forms a surface intermediate of concentration, $C(t)$, with a reaction probability, α. This intermediate desorbs to produce a product flux, $P(t)$, with a first-order rate constant, k_d.

A mass balance for the surface species concentration yields

$$\frac{dC(t)}{dt} = \alpha R(t) - P(t). \tag{13.35}$$

Because the desorption process is first order,

$$P(t) = k_d C(t). \tag{13.36}$$

Therefore,

$$\frac{dC(t)}{dt} = \alpha R(t) - k_d C(t). \tag{13.37}$$

This is the equation that must be solved in order to find α and k_d, the parameters that characterize the kinetics of the reaction.

The experimental data, composed of the reactant and product waveforms, can be used to determine α and k_d. One can express $R(t)$ in general as

$$R(t) = R_0\, g(t)\,, \tag{13.38}$$

where $g(t)$ is the gating function of the chopper used to modulate the reactant beam. If the reactant and product signals are being detected with a lock-in amplifier, which is a narrow-band detector that responds only to the fundamental component of the waveform detected, then the effective gating function for the reactant flux is

$$g(t) = \bar{g}\exp(i\omega t)\,, \tag{13.39}$$

or

$$R(t) = R_0\, \bar{g}\exp(i\omega t)\,, \tag{13.40}$$

where $\bar{g}$ is the first Fourier coefficient of the gating function ($2/\pi$ for a square wave) and ω is the angular frequency of the chopper. Because the system is linear,

$$C(t) = C\exp(i\omega t)\,. \tag{13.41}$$

Thus, for a first-order reaction, the mass balance equation becomes

$$\frac{d\left[\bar{C}\exp(i\omega t)\right]}{dt} = \alpha R_o g\exp(i\omega t) - k_d\bar{C}\exp(i\omega t). \tag{13.42}$$

This may be integrated to obtain

$$\bar{C} = \frac{\alpha R_o \bar{g}}{i\omega + k_d}, \tag{13.43}$$

or, since

$$\bar{P} = k_d \bar{C}, \tag{13.44}$$

one has

$$\bar{P} = \frac{\alpha R_o \bar{g}}{1 + \dfrac{i\omega}{k_d}}. \tag{13.45}$$

Using the language of linear systems analysis, a parameter known as the transfer function of the system may be defined as

$$\bar{T}(\omega) = \frac{\bar{P}(\omega)}{R_o \bar{g}} = \frac{\alpha}{1 + \dfrac{i\omega}{k_d}}. \tag{13.46}$$

In general, one can also express this transfer function as

$$\bar{T}(\omega) = \varepsilon(\omega) exp[-i\Phi(\omega)], \tag{13.47}$$

leading to

$$\varepsilon = \frac{\alpha}{\left[1 + \left(\dfrac{\omega}{k_d}\right)^2\right]^{1/2}}, \tag{13.48}$$

$$\Phi = \tan^{-1}\left(\frac{\omega}{k_d}\right).$$

The parameters ϵ and Φ are obtained by comparing the lock-in signals for the reactant and product fluxes, as ϵ is equal to the ratio of the lock-in signal amplitude for the product to the lock-in signal amplitude for the reactant, and Φ is the angle by which the product signal lags the reactant signal.

Figure 13.6 is a plot showing ϵ and Φ as a function of ω for given values of α, k_d. Also shown in Figure 13.6 are the ϵ, Φ plots for other possible reaction sequences. It can thus be seen that application of the MBRS technique can provide information both on the surface reaction sequence, as indicated by the form of the transfer function plot, and on the absolute values of the parameters in the rate equation, through the measured values of ϵ and Φ.

Experimental Examples

Specific examples of reactions that have been studied using the techniques described in the preceding section will now be described. Four different reactions that

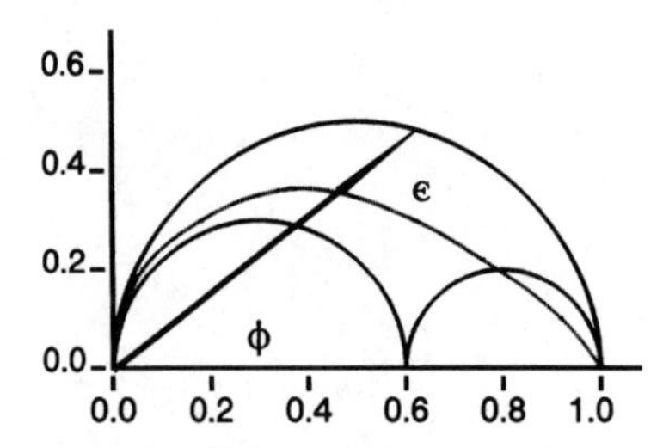

Figure 13.6 Transfer function plots for surface reaction sequences as obtained in molecular beam relaxation spectroscopy. Solid line: first order process; dashed line: branch processes; dotted line: first order reaction with diffusion into the bulk.

cover a wide range of reaction type and a wide range of complexity will be considered: H_2–D_2 exchange, the volatilization of carbon with gaseous O_2, the decomposition of formaldehyde, and the oxidation of CO.

In the case of H_2–D_2 exchange reaction, the overall reaction is

$$(H_2)_g + (D_2)_g \rightleftarrows 2(HD)_g \,. \qquad 13.49$$

In looking at this reaction, one can see why the presence of a catalytic surface is required. In the gas phase, at ordinary temperatures, this reaction does not take place at all, because the energy required to break the H–H bond is 103 kcal/mol. Even if one reactant were present as atoms, the reaction

$$H_2 + D \rightarrow HD + H \qquad 13.50$$

has an activation barrier of 10 kcal/mol.

On many transition metal surfaces, however, the reaction

$$(H_2)_g \rightarrow 2H_a \qquad 13.51$$

is spontaneous, with essentially zero activation energy, and occurs at temperatures as low as 100 K. The adatom binding energy is in the range from 10 to 20 kcal/mol, and the adsorbed atoms can migrate over the surface, encounter one another, and recombine. The resulting weakly adsorbed HD molecules are then easily desorbed into the gas phase. Careful studies of this reaction by MBRS on stepped platinum surfaces vicinal to the (111) reveal that the reaction is structure sensitive. The recombination reaction is much faster (about seven times) at ledge sites than on terrace sites. Figure 13.7 shows the results of this study, in terms of the overall reaction rate as a function of the angle of incidence of the molecular beam relative to the step direction. It is found that the energetics of the reaction are the same on both flat and stepped surfaces. All of the change in overall rate associated with the stepped surface is due to the greater probability of dissociative adsorption of the reactant molecules at the ledge site. The overall reaction sequence is thus

$$\begin{array}{l} (H_2)_g \rightleftarrows (H_2)_p \rightleftarrows 2H_a \\ \qquad\qquad\qquad\qquad\qquad \rightleftarrows 2(HD)_p \rightleftarrows 2(HD)_g \,. \\ (D_2)_g \rightleftarrows (D_2)_p \rightleftarrows 2D_a \\ \qquad\qquad\quad \textit{at ledge} \end{array} \qquad 13.52$$

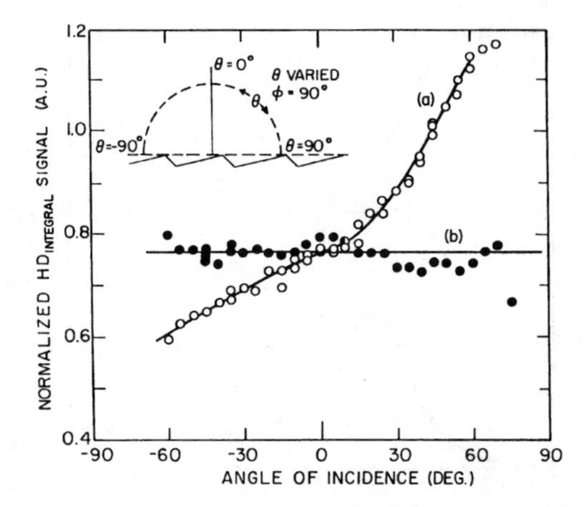

Figure 13.7 Results of a study of the H_2-D_2 exchange reaction on a stepped Pt (111) surface, as measured by molecular beam scattering. HD production as a function of angle of incidence of the molecular beam. (a) Pt (332) (stepped) surface, step edges perpendicular to the incident beam. (b) Pt (332) with step edges parallel to the incident beam. (Source: G.A. Somorjai, *Surface Sci.* 1979;89:496. Reprinted with permission.)

The reason for the increased dissociation efficiency at ledges appears to be that there is essentially zero activation energy for dissociation at the ledge but about a 1 or 2 kcal/mol barrier on the terrace sites. This is consistent with the observed activation barrier for the adsorption and desorption of H_2 on iron described in Chapter 12.

Next consider the reaction of adsorbed carbon with gaseous O_2. In this study, an adsorbed carbon layer was formed by the decomposition of ethylene on an Ni (110) surface, then removed by reaction with O_2 to form CO. This is an example of a so-called *cleanoff* reaction and is similar to volatilization reactions. This reaction was studied using MBRS, with an O_2 molecular beam used as reactant and the gas-phase O_2 and CO fluxes from the surface being measured mass spectrometrically. In addition, the surface carbon and oxygen coverages were measured by Auger electron spectrometry (AES). (AES is an electron spectroscopic technique that will be discussed in Chapter 14. For the present, it is sufficient to know that this technique provides information on surface composition on an atomic basis.)

The results of this study are summarized in Figure 13.8, which contains much kinetic information. If one looks first at the rate at which carbon disappears from the surface, on the basis of the carbon AES signal, one sees that the rate is at first slow, accelerates to a maximum, then drops back to zero as the carbon adlayer is depleted. This same behavior is mirrored in the curve of CO production rate obtained mass spectrometrically. The phase lag of this CO signal initially increases with increasing extent of reaction, then saturates. The surface oxygen coverage remains low until very late in the reaction sequence.

The reaction sequence deduced from these data is shown in Figure 13.9. It involves initiation of the reaction at defect sites in the carbon layer, followed by the growth of "holes" in the adlayer, at which oxygen chemisorption occurs readily. The

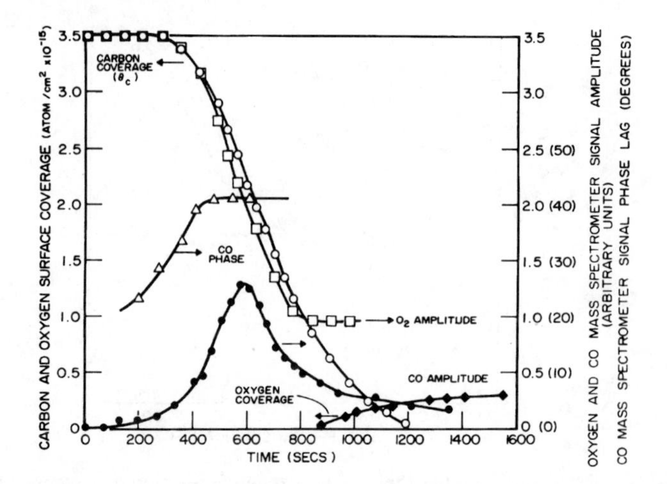

Figure 13.8 Summary of kinetic measurements of the oxidation of a graphitic carbon monolayer on Ni (110). (○) carbon coverage by AES; (□) amplitude of scattered O_2 mass spectrometer signal; (♦) oxygen surface coverage by AES; (●) amplitude and (△) phase lag of product CO mass spectrometer signal relative to scattered, unreacted O_2. (Source: R. Sau and J.B. Hudson, *Surface Sci* 1980;95:465. Reprinted with permission.)

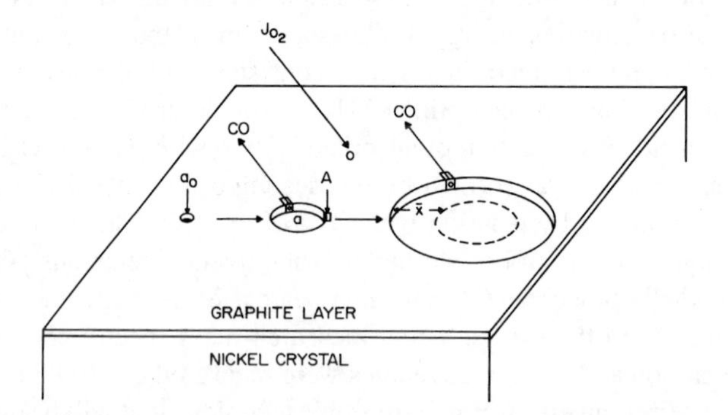

Figure 13.9 Schematic view of the proposed reation mechanism for the oxidation of a graphitic carbon monolayer on Ni (110). At short times: $\Theta_c = \exp(-a_o N_o e^{Kt})$, where $K = 2S_0 J_{0_2} A$. At long times, $\Theta_c = \exp(-K' J_{0_2}{}^2 + 2)$, where $K' = \pi(8AS_o\bar{x})^2 N_0$.

adsorbed oxygen is mobile on the surface, diffuses to the edge of the hole, and reacts to form CO, which is readily desorbed. This reaction takes place quite efficiently initially, until the hole size grows to the point where there is competition for adsorbed oxygen between the surface reaction and dissolution of oxygen into the bulk of the crystal. It is at this point that the phase lag of the product signal saturates. A theoretical rate equation has been developed embodying this reaction sequence. The fit of the theoretical curves to the experimentally measured carbon coverage curve is shown in Figure 13.10.

A third common class of reaction is the unimolecular decomposition process,

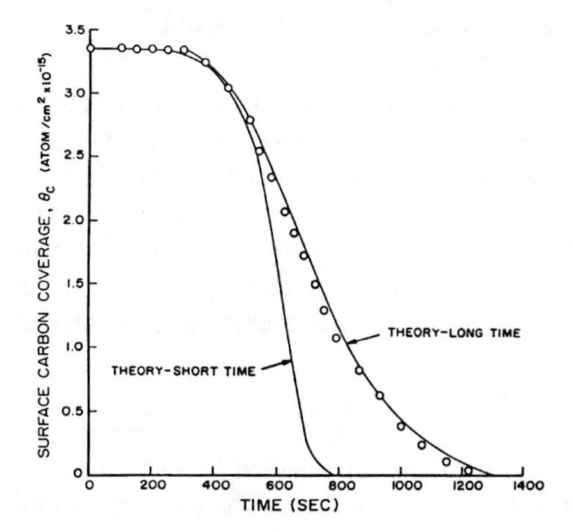

Figure 13.10 Comparison of theoretical predictions of the extent of reaction with the observed carbon coverage by AES, for the oxidation of a graphitic carbon monolayer on Ni (110). (Source: R. Sau and J.B. Hudson, *Surface Sci.* 1980;95:465. Reprinted with permission.)

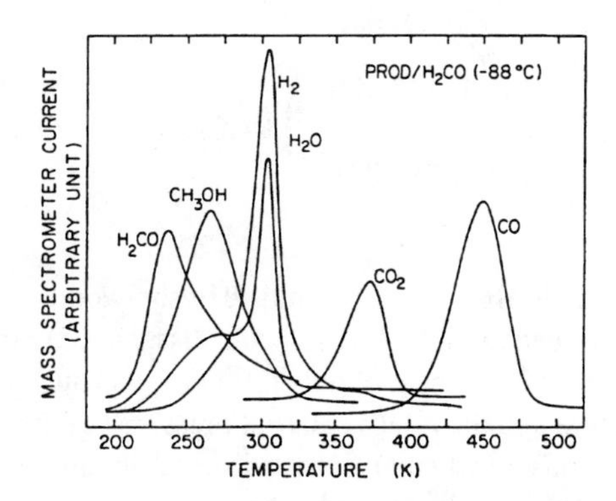

Figure 13.11 Product distribution spectrum following the adsorption of H_2CO on a clean Ni (110) surface at 185 K. (Reprinted with permission from R.J. Madix, In: R. Vanselow, ed. *Chemistry and Physics of Solid Surfaces II*. Boca Raton, Florida: CRC Press, Inc. 1979, p. 63. Copyright CRC Press, Inc. Boca Raton, Florida.)

typified by the decomposition of formaldehyde, H_2CO, on Ni (110). This reaction has been studied by TPRS, leading to the desorption spectra shown in Figure 13.11. A fraction of the adsorbed H_2CO, which increases with increasing H_2CO dose, desorbs without decomposing. At low doses, a large fraction decomposes to yield H_a and CO_a, as indicated by the fact that H_2 and CO desorb at the same temperature and with the same peak shape as is seen in desorption after dosing with pure H_2 or CO. At higher H_2CO doses, the H_2 peak has a first-order shape, indicating that it arises from a surface decomposition process. Methanol, CH_3OH, was also formed at high H_2CO doses. The mechanism of this reaction was deduced by coadsorption of H_2CO and D_2, to provide an independent source of adsorbed D. The observation that this procedure led only to the formation of CH_3OH and CH_3OD indicates that the hydrogen transfer leading to the CH_3 group must take place by an intermolecular reaction involving two H_2CO molecules, while the formation of the OH or OD group can involve the participation of surface H_a or D_a species. The small amount of CO_2 observed desorbs not at the temperature characteristic of CO_2 desorption from adsorbed CO_2 but at the same temperature as was observed for CO_2 desorption in the course of the decomposition of HCOOH, formic acid, indicating a common surface intermediate for these two systems.

Finally, consider a typical bimolecular reaction, the oxidation of CO by the overall reaction

$$(CO)_g + (O_2)_g \rightleftarrows (CO_2)_g. \qquad 13.53$$

This reaction has been extensively studied by many investigators using a variety of techniques on a wide range of surfaces. The major questions that these studies have tried to answer are whether the reaction procedes by a Langmuir–Hinshelwood or an Eley–Rideal mechanism; the effects of varying either the gas-phase pressure or the surface adlayer populations of the two reactants; and the effect of surface structure on the reaction rate and mechanism. The overall reaction scheme deduced is

$$\left.\begin{matrix} \nearrow\!\!\!\swarrow (CO)_p \searrow\!\!\!\nwarrow \\ (CO)_g \rightleftarrows (CO)_a \\ \\ (O_2)_g \rightleftarrows 2(O)_a \\ \searrow\!\!\!\nwarrow (O_2)_p \nearrow\!\!\!\swarrow \end{matrix}\right\} \rightleftarrows (CO_2)_a \rightarrow (CO_2)_g. \qquad 13.54$$

The overall mechanism is thus the Langmuir–Hinshelwood mechanism, with both reactants being chemisorbed (and the O_2 molecule dissociated) prior to CO_2 formation. Because of the differences in the heat of adsorption for CO and O_2, the reaction rate as a function of temperature and the relative partial pressures of the two reactants is complex. The rate also depends on whether CO is adsorbed and the surface then exposed to O_2, or vice versa. All of the results obtained, however, are consistent with the reaction scheme presented in Equation 13.54.

Problems

13.1. It has been determined by molecular beam scattering that the surface lifetime of silver atoms on tungsten is 60 sec at $T = 1000$ K. This system has also been studied by thermal desorption spectroscopy, where the desorption process was found to be first order, with $T_p = 1164$ K when the heating rate was 40 K/sec. What are the values of τ_0 and ΔH_{des}?

13.2. The platinum catalyzed decomposition of NO into N_2 and O_2 is found to obey the experimental rate law

$$\frac{dp_{NO}}{dt} = \frac{-kp_{NO}}{p_{O_2}}.$$

Assuming that all adsorbed species follow the Langmuir isotherm, derive the above rate law, starting with some reasonable assumed mechanism for the surface reaction.

13.3. In the catalytic oxidation of CO to CO_2 on a metal surface, the overall reaction is

$$(O_2)_g + 2(CO)_g \rightarrow 2(CO_2)_g.$$

The following information relative to the surface reaction has been obtained experimentally:

O_2 chemisorbs as atoms with $S_0 = 0.1$.
CO chemisorbs as molecules with $S_0 = 1.0$.
$n_0 = 3 \times 10^{15}$ sites per cm^2 for both gases.
Neither O_2 nor CO chemisorb on occupied sites.
O adatoms recombine and desorb as molecules at 1100°C.
CO desorbs at 400°C.
ΔE^* for the surface reaction between adsorbed O atoms and adsorbed CO molecules is 20 kcal/mol.
CO_2 desorbs at 100°C.

It may be assumed that $\tau_0 = 10^{-13}$ sec for all adsorption–desorption processes and that the rate of surface diffusion is negligible.

a. Write out the detailed course of the reaction, including all surface steps.

b. Develop a general expression for the rate of CO_2 production.

c. Solve this equation for the case of $p_{O_2} = 10^{-4}$ Torr, $p_{CO} = 10^{-5}$ Torr, and $T =$ 300 °C.

13.4. It is desired to remove adsorbed carbon from a metal (111) surface by exposing the surface to O_2 gas at a pressure of 10^{-6} Torr and a temperature of 1200°C. Develop a general expression for the rate of this reaction and estimate the initial rate at the pressure and temperature given, consistent with the information given below:

O_2 is dissociatively adsorbed on the metal with a sticking coefficient of 0.1 and desorbs as molecules at a temperature of 1100°C.
Carbon does not spontaneously desorb at any temperature below the melting point of the metal.
CO desorbs as molecules at a temperature of 400°C.
The major product of the reaction is observed to be CO.
The activation energy for the surface reaction between adsorbed O and adsorbed C is 10 kcal/mol.
The initial C coverage is 0.5 monolayer.
Oxygen does not adsorb on sites covered with adsorbed C.
$n_0 = 3 \times 10^{15}$ sites per cm^2 or all species.

Bibliography

Campbell, CT, Ertl, G, Kuipers, H, and Segner, J. A molecular beam study of the catalytic oxidation of CO on a Pt (111) surface. *J. Chem. Phys.* 1980;73:5862–5873.
Ceyer, ST, Gladstone, DJ, McGonigal, M, and Schulberg, MT. Molecular beams: probes of the

dynamics of reactions on surfaces. In B.W. Rossiter, J.F. Hamilton, R.C. Baetzold, eds. *Physical Methods of Chemistry*, 2nd ed. New York: Wiley, 1988.

D'Evelyn, MP and Madix, RJ. Reactive scattering from solid surfaces. *Surface Sci. Reports* 1985;3:8.

Jones, RH, Olander, DR, Siekhaus, WJ, and Schwarz, JA. Investigation of gas-solid reactions by modulated molecular beam mass spectrometry. *J. Vac. Sci. Technol.* 1972;9:1429–1441.

King, DA, and Woodruff, DP, eds. *The Chemical Physics of Solid Surfaces and Heterogeneous Catalysis*, Vols 1–4. Oxford: Elsevier, 1982.

Landman, U, ed. *Aspects of Kinetics and Dynamics of Surface Reactions*. New York: American Institute of Physics, 1980.

Madey, TE, Yates, JT, Jr., Sandstrom, DR, and Voorhoeve, RJH. Catalysis by solid surfaces. In N.B. Hannay, ed. *Treatise on Solid State Chemistry*. New York: Plenum Press, 1976;6B:1–124.

Madix, RJ. Molecular transformations on single crystal metal surfaces. *Science* 1986;233:1159–1166.

Somorjai, GA. Catalysis and surface science. *Surface Sci.* 1979;89:496–524.

Wise, H, and Oudar, J. *Material Concepts in Surface Reactivity and Catalysis*. San Diego: Academic Press, 1990.

PART III

Energetic Particle–Surface Interactions

CHAPTER **14**

Electron–Surface Interactions

This chapter and Chapter 15 will consider in detail the processes that take place when a surface, either clean or adsorbate covered, is exposed to a flux of charged particles. This chapter will consider electron impact; Chapter 15, ion impact. The majority of the techniques that have been developed in recent years for the characterization of surface structure and surface reactions are based on these various charged particle interactions.

Surface Electronic Structure

The equilibrium structure of surfaces was discussed in Chapter 2 in terms of occupied and unoccupied electron states. The picture developed, for the case of a clean metal surface at a finite temperature, is recalled in Figure 14.1. Note that at 0 K, all of the allowed electron energy states are filled, up to a level known as the Fermi level. At finite temperatures, states slightly above the Fermi level have a finite probability of occupancy, and those slightly below the Fermi level have a finite probability of being unoccupied. The energy difference between the Fermi level and the vacuum level is known as the work function, Φ. This chapter will consider the ways in which electrons, either free or associated with an atom or molecule from the gas phase, can interact with the allowed electronic states of the surface, either filled or unfilled, to modify the picture presented in Figure 14.1, and in the process provide information on the nature of the surface.

Work Function Changes

Consider first processes that can change the value of the work function of the clean surface. These involve interactions between the surface and atoms or molecules from the gas phase, such as those diagrammed in Figure 14.2a and b. Figure 14.2a shows the results of bringing an atom, having an occupied electronic state closer to the vacuum level than the work function of the surface involved, close to the surface. As the atom comes close enough to the surface that the electron charge densities of the bulk and atomic electronic states overlap (that is, when adsorption takes place), there will be either partial or complete transfer of the electron from the atom to the surface, so that the Fermi level is uniform throughout. This will result in either a partial or com-

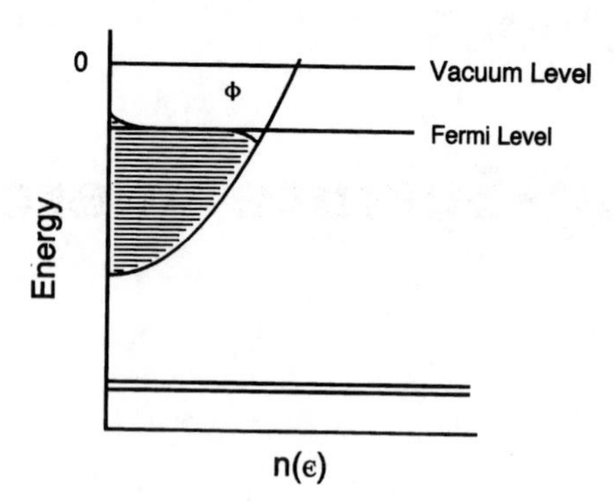

Figure 14.1 Density of filled and unfilled electronic states in a typical metal.

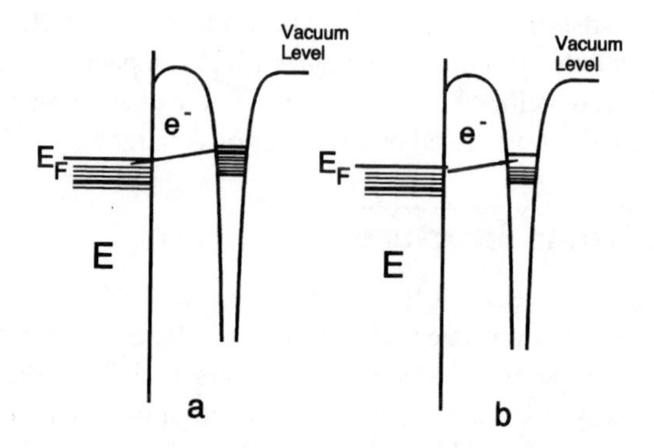

Figure 14.2 Charge transfer processes between an adsorbed atom and a metal surface, showing transfer to the surface from an adsorbed atom having a filled electronic state slightly above E_F (a), and transfer from the surface to an adsorbed atom having an unfilled state slightly below E_F (b).

plete net positive charge on the adatom and a consequent reduction in the work function. In the alternative case, shown in Figure 14.2b, the adsorption of an atom having an unoccupied electronic state slightly below the Fermi level of the metal will result in net electron charge transfer to the adsorbed species and a corresponding increase in the work function.

A similar phenomenon is observed in the case that a polar molecule is adsorbed on a surface, even without electron transfer from the surface. In this case, polar molecules that adsorb with the positive end of the dipole away from the surface will reduce the work function and vice versa.

In either case, measurement of the work function change accompanying adsorption provides significant information on the details of the adsorption process. The sign of the work function change indicates the direction of the electron transfer associated with formation of the adsorbate–substrate bond, or the orientation of the adsorbed species in the case of adsorbed polar molecules. To first order, the magnitude of the

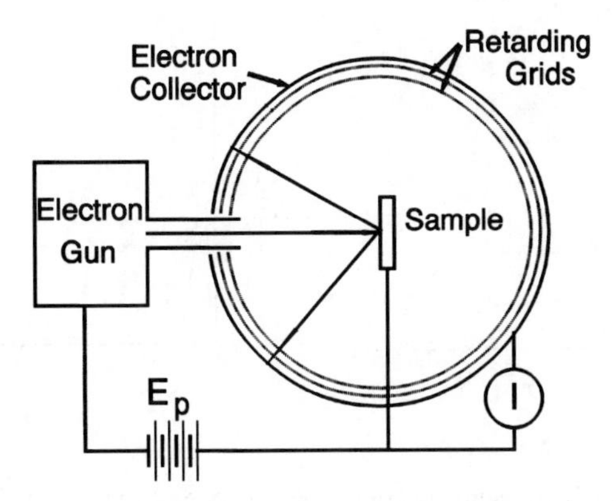

Figure 14.3 Schematic view of a system for the study of secondary electron emission. (Note the similarity to an LEED apparatus.)

work function change is proportional to the amount of material adsorbed, at least at total coverages small enough that lateral interactions can be ignored. Thus, the application of any of the techniques for work function measurement described in Chapter 2 that do not seriously perturb the situation at the surface can be used to study the adsorption process.

Secondary Electron Emission

When a surface is bombarded by a flux of electrons, over a wide electron energy range, a number of phenomena take place. Many of these processes lead to the emission of electrons from the surface. The electrons generated by these processes are known collectively as secondary electrons. Figure 14.3 shows a typical system for studying the secondary electron generation process. A source of electrons, of known energy and current, produces a direct flux of electrons to the surface. The energy of interest can range from a few electron volts to many thousand electron volts. The electrons scattered at the surface, or produced in the course of the interaction between the primary electron beam and the surface, are collected by the spherical collector surface. High transparency grids between the surface and the collector can be biased electrically to permit determination of the energy distribution of the secondary electrons. (Note that this is essentially the apparatus that was seen in the discussion of LEED.)

The results of a typical experiment of the type described are summarized in Figure 14.4. Figure 14.4a shows the total current collected by the collector as a function of the bias applied to the retarding grid structure. If the retarding grids are at the same potential as the surface, then all electrons leaving the surface will reach the collector. As the grids are raised to an increasingly negative potential, to repel electrons having an energy lower than the potential difference between the sample surface and the grids, the total current decreases and drops to zero when the retarding potential becomes greater than the primary electron energy. The details of the secondary electron energy distribution become much more apparent if, rather than plotting electron current vs.

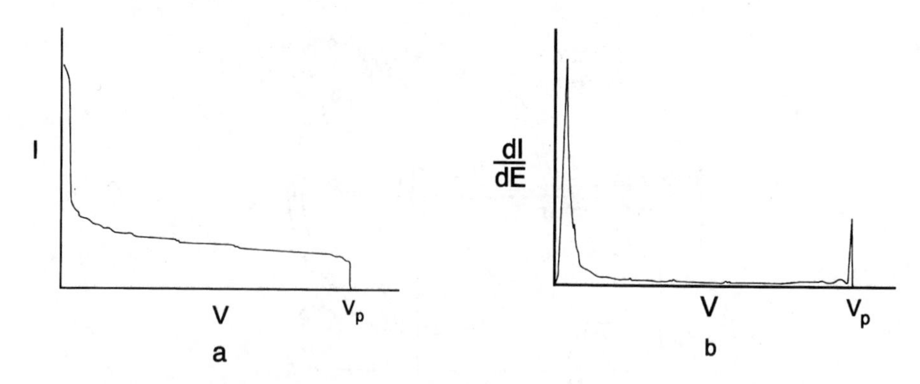

Figure 14.4 Results of a measurement of the secondary electron current from a surface being bombarded by an electron beam having an energy E_p showing collector current as a function of retarding potential (a), and the derivative of curve *a* with respect to energy, $n(E) = dI/dE$ (b).

retarding voltage, one plots the derivative of this curve, the so-called $n(E)$ vs. E curve, as shown in Figure 14.4b. Plotting the secondary electron energy distribution in this fashion reveals at least four different classes of secondary electrons. The sharp peak at the energy of the incident electron beam consists of *elastically scattered primary electrons.* The features at energies immediately below the primary peak represent incident electrons that have undergone *discrete energy losses* due to interactions with other electrons, either in the surface or associated with adsorbed species. The broad peak at very low energies is due to the so-called *true secondary electrons*, which arise from inelastic interactions with the primary beam. The small sharp peaks spread throughout the energy range arise from so-called *Auger neutralization processes* and are called *Auger electrons.*

Each of these four classes of secondary electrons will be examined in turn. Consider first the elastically scattered primary electrons. These are electrons that struck the surface as part of the incident electron beam and were scattered without any loss of energy. Their behavior is similar to specularly reflected light or to x-rays. They are the electrons that are involved in electron diffraction, as they produce diffraction effects according to Bragg's law, as discussed in Chapter 1.

The inelastically scattered primaries again are electrons that entered with the primary beam but have lost energy in the course of interaction with the surface. The sharp peaks in this part of the energy distribution represent specific loss processes, and study of these peaks can provide much information about the nature of the surface and about species adsorbed on the surface. In the case of clean surfaces, these peaks arise primarily from plasmon excitation of the electrons within the solid. Plasmon excitations require energies in the 10 eV range. One generally observes a so-called *bulk plasmon* and in addition a *surface plasmon.* On the basis of a classical calculation, one can determine that the frequency of the bulk plasmon is given by

$$\omega_p = \left(4\pi n_o\right)^{1/2}. \tag{14.1}$$

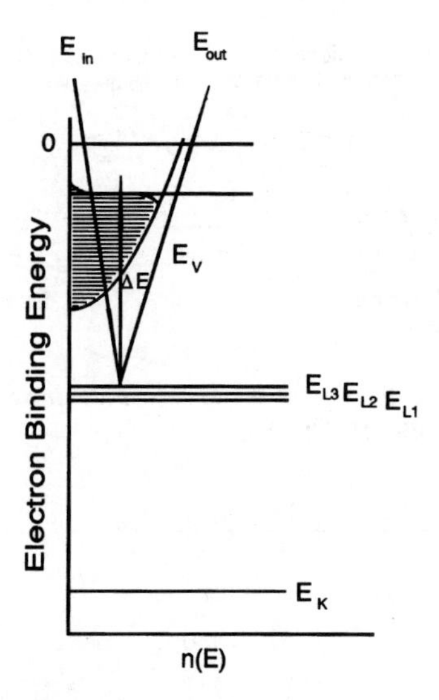

Figure 14.5 Excitation of an electron from a normally filled state to a normally empty state by interaction with an incident electron. This process leads to detection of an inelastically scattered primary electron.

That is, the frequency, and consequently the energy, depend only on n_0, the number density of free electrons. Typical values for metals are from 10 to 20 eV. In the case of metals, the surface plasmon frequency (and consequently its energy) are related to the bulk plasmon frequency by

$$\omega_{sp} = \frac{\omega_p}{\sqrt{2}}. \quad 14.2$$

These plasmon excitations will give rise to features in the inelastic loss region of the secondary electron distribution at the energies given by the primary energy minus the plasmon energy.

One can, in some cases, also observe loss peaks due to the excitation of an electron from the solid to an allowed state that is normally empty. This process is diagrammed in Figure 14.5. The electron causing the excitation will be observed in the secondary electron distribution at an energy given by the primary energy minus the energy gained by the excited electron, ΔE in the figure.

Similarly, one can observe loss features associated with the excitation of vibrational modes of either the surface atoms or of atoms or molecules adsorbed on the surface. In this case, the excitation energies involved are much smaller, generally a fraction of an electron volt. Consequently, the separation between the primary electron energy and the energy at which the loss feature is observed is small, and a detection technique having very high energy resolution is required in order to observe these features. The experimental technique developed for studying these loss processes is known as *high resolution electron energy loss spectroscopy* (*HREELS*). The experi-

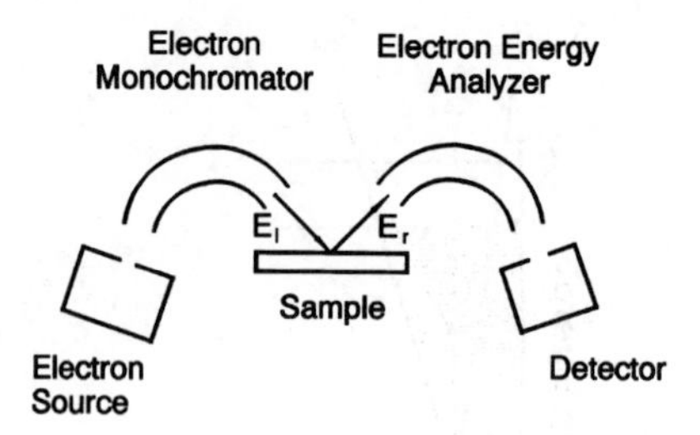

Figure 14.6 Schematic diagram of the experimental arrangement for high resolution electron energy loss spectroscopy (HREELS).

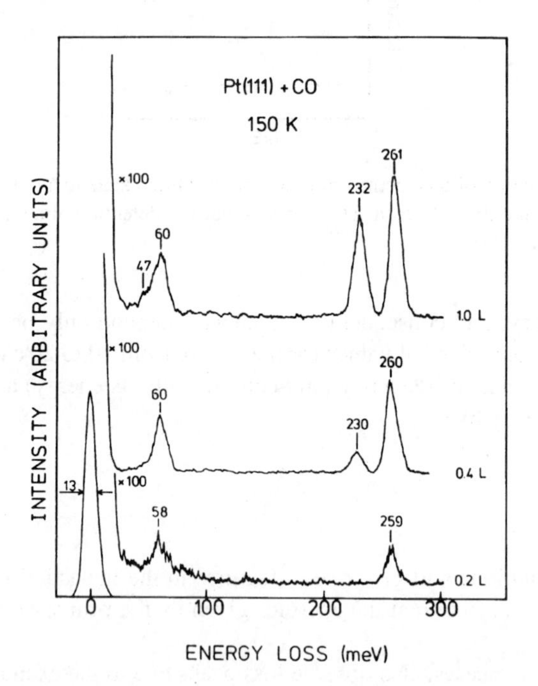

Figure 14.7 HREELS spectra of CO adsorbed on Pt (111) for various CO coverages. The surface was saturated with CO at 300 K, then heated to the temperatures shown. (Source: H. Hopster and H. Ibach, *Surface Sci.* 1978;77:109. Reprinted with permission.)

mental setup for this technique is shown schematically in Figure 14.6. A monoenergetic beam of electrons, having an energy on the order of 10 eV, is formed by the electron gun–primary analyzer combination and allowed to strike the sample surface. The scattered electrons are detected by a second energy analyzer and a high sensitivity detector. In operation, the pass energy of the detector is varied to display the energy distribution of the scattered electron beam. Figure 14.7 shows a typical result of such a measurement, for the case of a Pt (111) surface covered with varying amounts of adsorbed carbon monoxide. Note that the loss peaks are in all cases very small com-

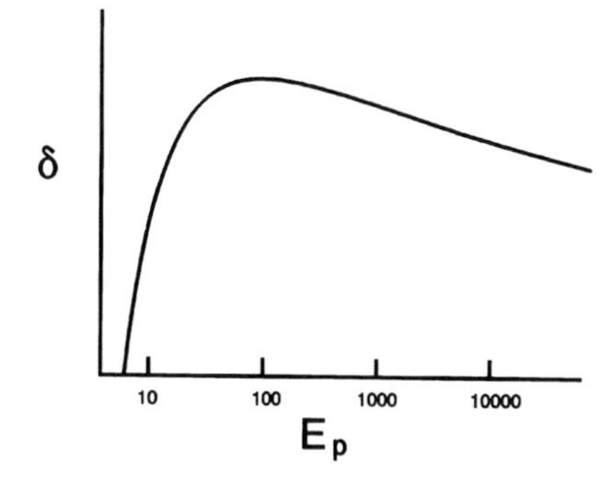

Figure 14.8 The secondary electron emission ratio, δ, as a function of the energy of the primary electron beam.

pared to the height of the elastically scattered electron peak. The loss features shown correspond to vibrational excitation of the C-O bond for linear (200 meV) and bridge bonded (230 meV) CO molecules and to vibrational excitation of the C-Pt bond for linear (60 meV) and bridge bonded (47 meV) molecules. Note that as the temperature is raised, both the total amount of adsorbed CO and the fraction that is bridge bonded change. The vibrational frequencies also change slightly with temperature. The vibrational energies in these measurements are often expressed in wave numbers, cm^{-1}. This is a common convention in optical spectroscopy. The relation between wave numbers and electron volts is

$$100 \text{ cm}^{-1} = 12.41 \text{ meV}. \qquad 14.3$$

Note that the losses observed in HREELS are quite small compared to an electron volt.

The third contribution to the secondary electron energy distribution is the so-called *true secondaries*. These are electrons that were originally in the solid and have received enough energy from interactions with the primary electron beam that they have been raised to states above the vacuum level and consequently ejected from the surface. These electrons make up the smooth curve of the low energy part of the secondary electron distribution. The total contribution of these electrons to the secondary electron flux is plotted in Figure 14.8. Here the flux is plotted as δ, the secondary electron emission ratio, as a function of primary electron energy. The parameter δ is defined as the number of secondary electrons generated per incident primary electron. From Figure 14.8 it is seen that essentially no secondary electrons are generated for primary energies less than the work function. Once this threshold is crossed, the secondary electron yield rises rapidly to a maximum in the 100 eV primary electron energy range, then decreases slowly with increasing primary energy. This overall behavior arises from the combination of increasing excitation probability with increasing primary electron energy, leading to the initial rise, and deposition of the energy of the primary electron deeper within the target as primary energy increases, which produces excited electrons too far from the surface to escape without further scattering.

The question of how far an excited electron will travel within a solid before undergoing a second inelastic scattering event has been looked at in some detail, due to

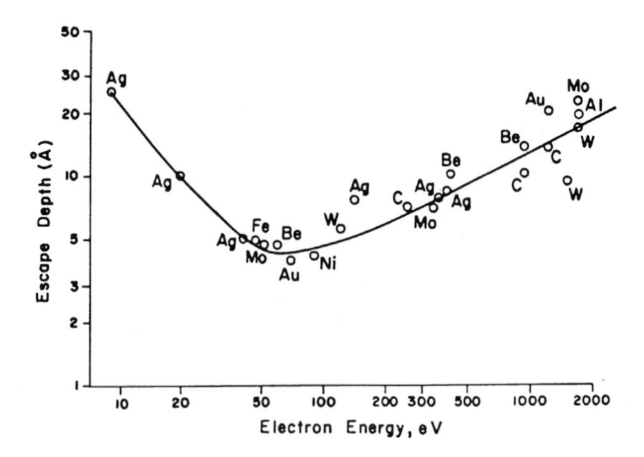

Figure 14.9 The electron inelastic mean free path as a function of initial kinetic energy. This is also known as the electron escape depth and is a measure of how far an electron of given kinetic energy will travel within a solid without undergoing an inelastic collision. (Reprinted with permission from P.W. Palmberg, *Anal. Chem.* 1973;45:549A. Copyright 1973, American Chemical Society.)

its importance in determining the surface sensitivity of various surface electron spectroscopic techniques, such as UPS, discussed in Chapter 2, and Auger electron spectroscopy (AES) to be discussed in the next section of this chapter. The general relation deduced is shown in Figure 14.9. This figure summarizes data obtained for a wide range of materials and is plotted as the electron escape depth, defined as the depth at which the probability of escape from the sample without inelastic scattering is 1/e, as a function of the energy of the excited electron. Note that the curve has a minimum in the 50 to 100 eV range and that the escape depth is less than 2 nm over the whole energy range shown. This will be discussed in greater detail in the discussion of surface analysis by AES in the next section.

As a final point on the subject of the true secondaries, consider the typical value of the secondary emission ratio, δ. This parameter is greater than unity for many materials over a wide range of primary energies. This fact is the basis of the most commonly used detector for low-level electron or ion fluxes, the so-called *secondary electron multiplier*. A typical structure for such a device is shown in Figure 14.10. An electron

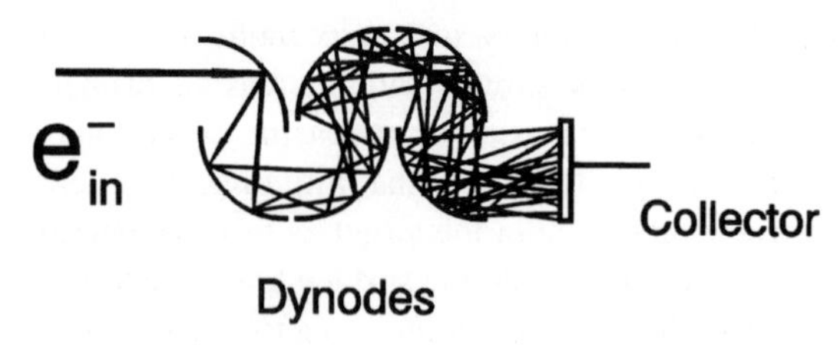

Figure 14.10 Schematic view of a typical secondary electron multiplier structure.

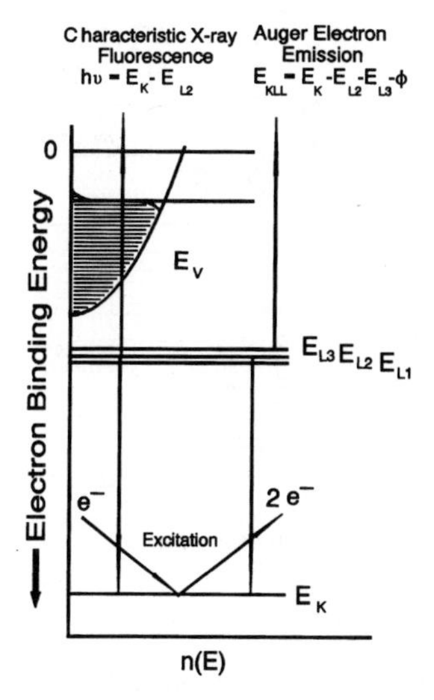

Figure 14.11 Energy diagram showing the processes leading to x-ray fluorescence or Auger electron emission after inner shell excitation of an atom by an incident energetic electron.

incident on the first plate, or *dynode,* will cause the ejection of, on the average, δ secondary electrons. These electrons, in turn, are accelerated into the second dynode, where each of these, in turn, generates an average of δ additional secondaries. This process is repeated down the length of the dynode structure, leading to a current gain, G, of

$$G \approx (\delta)^n, \qquad 14.4$$

where n is the number of dynodes. Practical gains in such structures are in the range of 10^6–10^8 and permit the detection of individual electron impacts on the first dynode. Ions and photons can also be detected by these structures.

Consider finally the contribution to the secondary electron distribution arising from Auger excitation processes. These electrons form the basis of the currently most widely used technique for surface elemental chemical analysis, Auger electron spectroscopy (AES). The process giving rise to these electrons is shown in Figure 14.11. The initial step in the process is the excitation of an atom in the target, in this case by electron impact, in which an electron is removed from a stable core level state. These core level states are allowed electronic states of the solid, derived from the allowed atomic orbitals of the atoms that make up the solid. They are sharply defined in energy, generally localized around a particular ion core, and lie at energies well below those of the states in the conduction band which have been discussed so far. Similar excitations can also be caused by absorption of a photon or by ion impact. Once the atom is excited, it will relax by one of two competing processes shown in Figure 14.11. One alternative is for an electron from a less tightly bound state to drop to the empty core level, with a photon being emitted to carry off the energy difference between the two levels, E_K–E_{L2} in the example shown. This is the process of x-ray fluorescence and predominates for cases in which the excited core level has a high binding energy. The

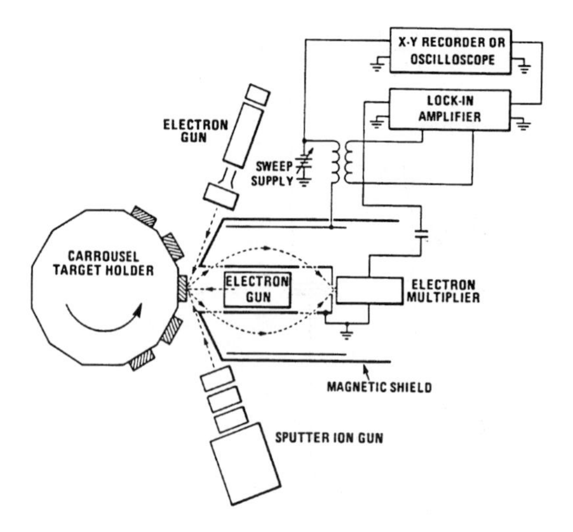

Figure 14.12 Schematic view of a typical apparatus for Auger electron spectroscopy, based on a cylindrical mirror electron energy analyzer (CMA). (Reprinted with permission from P.W. Palmberg, *Anal. Chem.* 1973;45:549A. Copyright 1973, American Chemical Society.)

alternative relaxation process involves the dropping of an electron from a less tightly bound level, with, in this case, the excess energy being carried off by the emission of a second electron from another less tightly bound level by a so-called *radiationless* or *Auger deexcitation process*. The kinetic energy of the emitted electron is given, in the example shown, by

$$E_{KLL} = E_K - E_{L2} - E_{L3} - e\Phi, \qquad 14.5$$

where Φ is the work function of the surface.

Auger Electron Spectroscopy

Because AES is a popular analytical method, it will be looked at in some detail. Figure 14.12 show the most common experimental configuration. The primary electron beam generally has an energy in the 3000 – 5000 eV range. The secondary electrons are energy analyzed in a device called a *cylindrical mirror electron energy analyzer*. This device can be adjusted to pass only a narrow range of electron kinetic energies. As the analyzer voltage is swept over the range of interest, the $n(E)$ vs. E distribution is obtained directly as shown in Figure 14.13a. More commonly, a small ac modulation is superimposed on the sweep voltage, and ac detection techniques used at the detector. In this case, the detected signal is the derivative of the $n(E)$ vs. E curve, as shown in Figure 14.13b. The net result of this process is to greatly improve the signal-to-noise ratio of the analyzer. Most Auger spectrometers are operated in this mode. The Auger spectrum shown in Figure 14.13b contains three kinds of information. The position of the peaks along the energy axis allows qualitative determination of what elements are present at the sample surface, by comparison with reference spectra such as that shown in

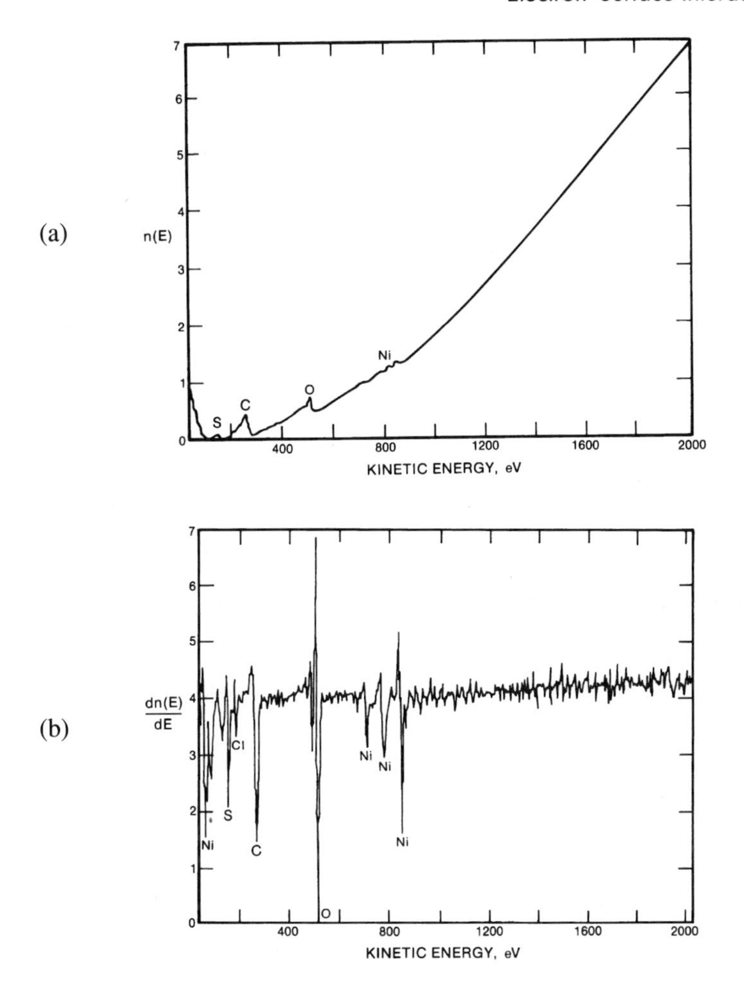

Figure 14.13 Auger spectra of a contaminated nickel surface, showing $n(E)$ vs. E curve (a), and $dn(E)/dE$ vs. E curve (b).

Figure 14.14. The peak-to-peak height of the derivative feature is, to first order, directly proportional to the atomic concentration of the species in the near-surface region of the sample and can be used, with appropriate corrections for relative transition intensities and chemical effects, to provide quantitative information on surface concentrations. It is also possible, in some cases, to obtain chemical state information from the Auger peaks. This is especially true of the peaks arising from transitions involving the valence electron states, as these are the states with energies and relative populations that are most affected by chemical state changes.

Another common use of AES is in the determination of the variation of composition with depth in thin layers. As was pointed out previously, the mean escape depth for Auger electrons is on the order of 1 nm. Thus, information is obtained only from the very near surface region. However, it is possible to slowly remove material from the

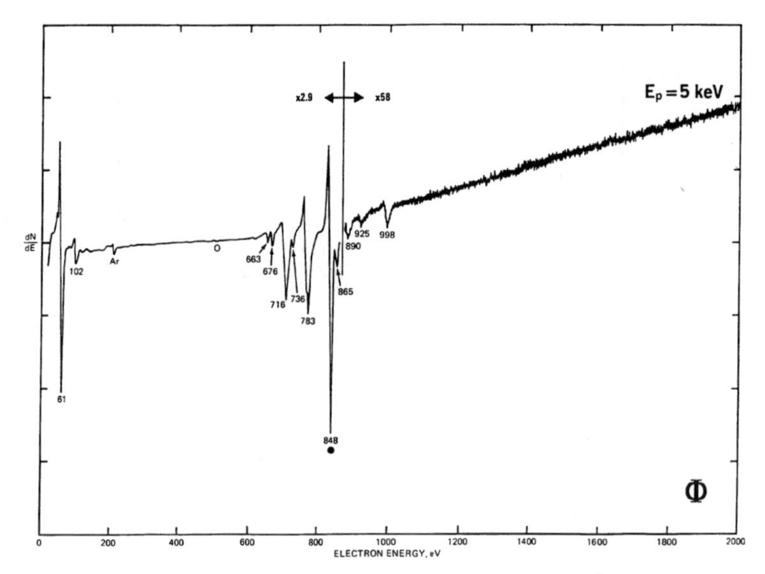

Figure 14.14 Reference spectrum for Auger electron spectroscopy of a pure nickel sample, using a primary electron energy of 5 keV. (Source: L.E. Davis, N.C. MacDonald, P.W. Palmberg, G.E. Raich, and R.E. Weber, *Handbook of Auger Electron Spectroscopy,* 2nd ed. Eden Prairie, Minn.: Physical Electronics Div. of Perkin-Elmer Corp., 1976. Reprinted with permission.)

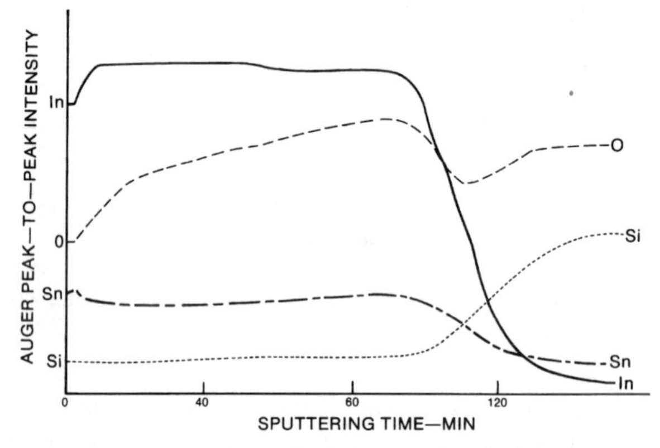

Figure 14.15 Auger depth profile of an In-Sn oxide thin film on glass, showing the way in which the concentrations of the various elements vary with depth below the original film surface.

surface by bombarding the surface with ions. This process is called *sputtering* and will be discussed in more detail in Chapter 15. As the surface is etched away, Auger spectra are taken repeatedly, to determine the way in which the concentrations of the various elements present change with depth. A typical Auger depth profile, for a thin film of In-Sn oxide on glass, is shown in Figure 14.15. Here the peak-to-peak height of the derivative feature for each element is plotted as a function of ion bombardment time, which is directly proportional to depth below the original surface.

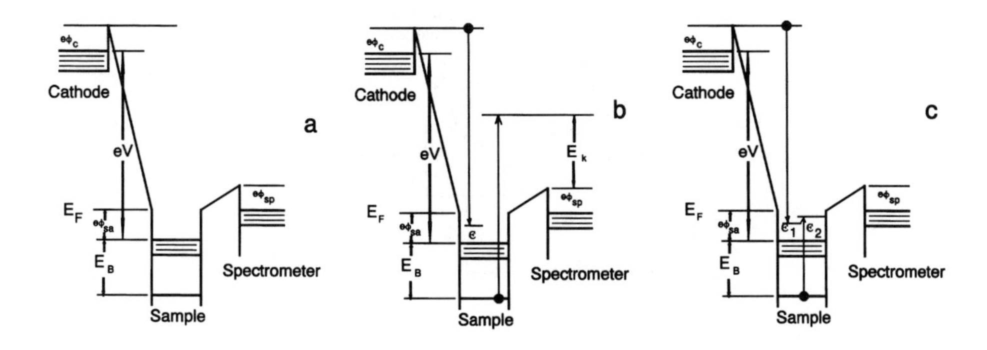

Figure 14.16 Electron energy diagrams representative of various core level spectroscopy experiments. (a) Energy diagram for an electron source, a sample, and an electron energy spectrometer. (b) Energies involved in the inelastic scattering of an electron at a solid surface. (c) Energies involved in the capture of a primary electron and attendant excitation of a core level electron.

Core Level Spectroscopies

Consider again the question of the excitation of core level electronic states in solids by electron impact. Several additional means exist for using these excitations to provide information on the structure of the solid. Collectively, these techniques are referred to as core-level spectroscopies. One can describe these spectroscopies with reference to the energy diagram shown in Figure 14.16a, which shows the energy values for electrons in a system composed of an electron emitting surface (the cathode), a sample, and an energy analyzer (the spectrometer).

In the context of this diagram, the electron energy loss process discussed previously is represented by the transitions shown in Figure 14.16b. An electron from the source is captured in one of the allowed unoccupied electronic states of the sample. A second electron, from the sample, appears in the vacuum space with an energy, as measured by the spectrometer, of

$$E_k = (eV + e\Phi_{cath}) - (E_B - e\Phi_{spect}) - \epsilon. \qquad 14.6$$

Here ϵ is the energy difference between the state occupied by the electron from the cathode and the Fermi level of the sample. The maximum possible value of E_k is thus

$$(E_k)_{max} = (eV + e\Phi_{cath}) - (E_B - e\Phi_{spect}). \qquad 14.7$$

Thus, if one knows V, Φ_{cath}, and Φ_{spect}, one can determine the binding energy of the level, E_B.

Alternatively, one can look at the processes that occur when we have an excitation of the sort shown in Figure 14.16c. In this case, an electron from the source is accommodated into an unfilled level of the sample, and an electron from a core level of the sample is excited to a second unfilled level of the sample. In this case

$$eV + e\Phi_{cath} = E_B + \epsilon_1 + \epsilon_2, \qquad 14.8$$

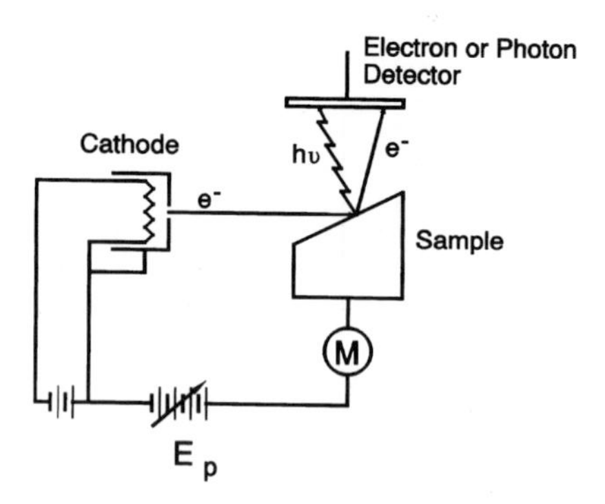

Figure 14.17 Schematic view of the experimental arrangement for soft x-ray appearance potential spectroscopy (SXAPS).

and the critical energy required to excite an electron from the state having E_B to the lowest unfilled state is

$$eV_{crit} = E_B - e\Phi_{cath}. \qquad 14.9$$

That is, eV_{crit} is the minimum energy required for core level excitation. Consider the possible observable effects of this excitation, which does not, in itself produce any new species in the vacuum. Three possibilities exist:

1. When relaxation occurs to fill the core hole, a soft x-ray may be emitted.
2. Alternatively, the core hole may relax by Auger emission.
3. A new channel for removing electrons from the elastic peak appears.

Each of these possibilities is the basis of an analytical technique for studying the electronic structure of the solid. Consider first the case in which relaxation occurs by soft x-ray emission. This is the basis of a technique called *soft x-ray appearance potential spectroscopy* (*SXAPS*). Figure 14.17 shows the experimental setup for this technique. The sample is bombarded by electrons from a source capable of producing a wide range of electron energies. A photon detector of some sort is placed in line-of-sight to the sample, and the total photocurrent is measured as the bombarding electron energy is increased. In this situation, the probability of photon emission is a function of which levels, ϵ_1, ϵ_2 are occupied and which are unoccupied. All other things being equal, the excitation probability is proportional to the self-convolution of the density of unfilled states near the Fermi level. The photoemission probability is also related to the relative probability of deexcitation by x-ray or Auger processes. The net result is that there will be sharp increases in the total photocurrent at energies corresponding to the core-level binding energies, as shown in Figure 14.18. The details of these discontinuities, as revealed by looking at the derivative of the photocurrent vs. electron energy curve, thus provide information on the density of *unfilled* states in the sample. (Note the contrast with ultraviolet photoelectron spectroscopy—that technique provides information on the density of *filled* states near the Fermi level.)

One can, alternatively, look at the case in which the excitation relaxes by an Auger process. In this case, the emitted Auger electrons add to the total secondary elec-

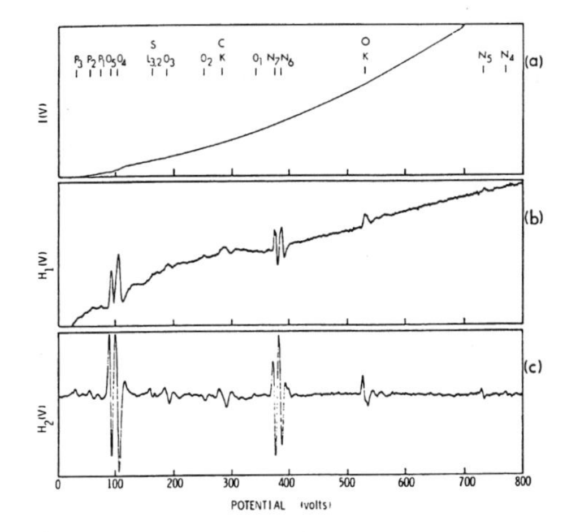

Figure 14.18 Typical results of a soft x-ray apperance potential spectroscopy measurement, showing integral curve, x-ray photocurrent in detector as a function of primary electron energy (a), the derivative of curve a, dI/dE vs. E_p (b), and the second derivative of curve a, d^2I/dE^2 vs. E_p (c). (Source: R.L. Park and J.E. Houston, *J. Vac. Sci. Technol.* 1974;11:1. Reprinted with permission.)

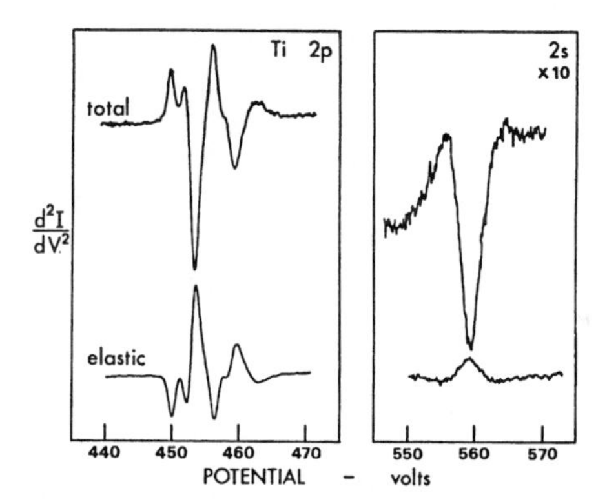

Figure 14.19 Results of an Auger electron appearance spectroscopy experiment, showing the AEAPS and DAPS features arising from the ionization of the Ti 2p and 2s core levels. Note that the AEAPS (total) and DAPS (elastic) curves are essentially mirror images. (Source: M.L. DenBoer, P.I. Cohen, and R.L. Park, *Surface Sci.* 1978;70:643. Reprinted with permission.)

tron current, both by themselves and through any additional true secondaries that are generated by collisions with Auger electrons within the sample. The net result is a sharp increase in secondary electron current as the critical energy for excitation of a core hole is exceeded. The study of the electron current emitted in this case is called *Auger electron appearance potential spectroscopy* (*AEAPS*). In this case, the change in signal with energy, at the excitation edge, is controlled by the density of unfilled states near the Fermi level and the probability of Auger deexcitation. Figure 14.19 shows the result of an AEAPS measurement on titanium.

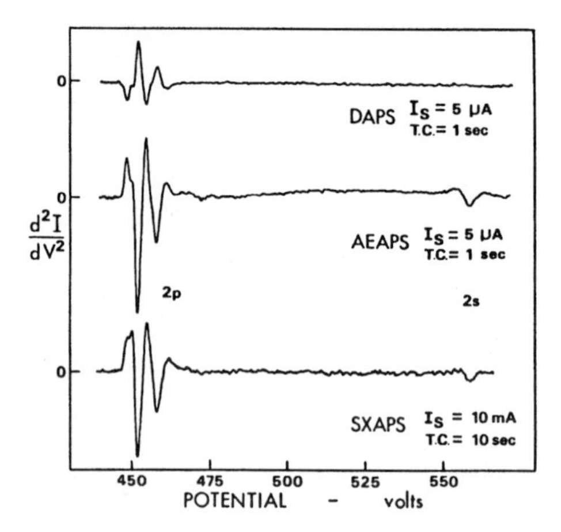

Figure 14.20 Comparison of the results of SXAPS, AEAPS, and DAPS measurements of a single titanium sample. Second derivative spectra are shown in all cases. (Source: M.L. DenBoer, P.I. Cohen, and R.L. Park, *Surface Sci.* 1978;70:643. Reprinted with permission.)

Finally, one may look at the effect of the excitation process on the elastic peak of the secondary electron spectrum. Any process that can result in an inelastic interaction between an incoming electron and the sample will reduce the number of electrons that are elastically scattered. Consequently, if the electron collector is biased to collect only the elastically scattered electrons, there will be a decrease in the collected signal every time the electron energy is raised beyond the critical value for excitation. Again, this technique provides information on the density of unfilled states. This technique, referred to as *disappearance potential spectroscopy* (*DAPS*), is also shown in Figure 14.19. As an example of the results obtained using these three spectroscopies, Figure 14.20 shows DAPS, SXAPS, and AEAPS second derivative spectra for titanium.

Electron-Stimulated Desorption

The interaction of electrons with solid surfaces can also cause the removal of atomic, molecular, or ionic species from the surface. Several mechanisms are involved in these processes, which go by the general name of *electron-stimulated desorption* (*ESD*).

The earliest studies of this phenomenon involved measurement of the production of ions from solid surfaces under the influence of bombardment by electrons in the 100 eV range. This process led to the observation of spurious signals in mass spectrometers, due to the electron-stimulated desorption of species such as O^+, F^+, and Cl^+ from surfaces in the ionization region of the mass spectrometer under bombardment by the electrons used to ionize the gas-phase species in the instrument. More detailed study of this phenomenon revealed that, in addition to positive ions, neutral atoms with high energies (several electron volts) and excited neutral species were also produced in the process. Working independently, Redhead [1964], and Menzel and Gomer [1964] put forth explanations of this phenomenon. The mechanism proposed is shown schematically in Figure 14.21. The initiating step in the process is the ionization or

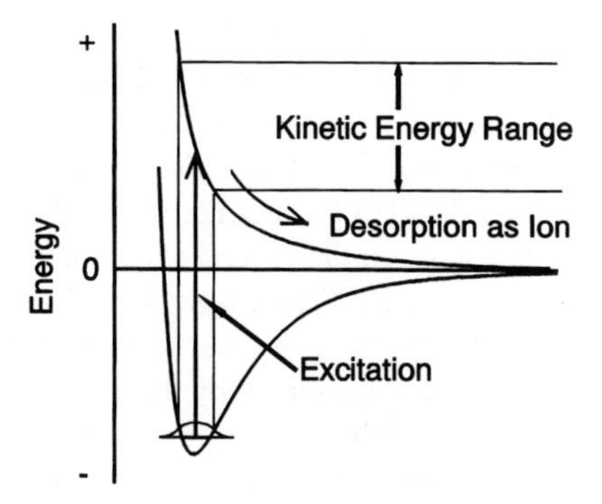

Figure 14.21 Potential energy diagram showing the processes involved in the electron-stimulated desorption of an adsorbed species as an ion, according to the Redhead–Menzel–Gomer model.

excitation of an adsorbed atom or molecule by electron impact. An electron can be removed from one of the stable levels of an adsorbed species, just as was the case for the initiating event in the Auger process discussed in this chapter.

This excitation process results in a charged or excited adspecies, whose interaction energy curve with the solid may be very different from that of the ground state neutral species. In particular, the excited species may have a shallow attractive potential well or no attractive region at all. After the excitation process, the system will relax by the excited species moving away from the surface, thus reducing the system potential energy and imparting equivalent kinetic energy to the excited species. If no further processes intervene, then the ion or excited neutral will appear in the gas phase with a kinetic energy range as shown in the figure.

The generation of ground state neutral species with excess kinetic energy can be explained by a similar process, as shown in Figure 14.22. The initial excitation event is the same as in the previous case. In this case, however, since the excited species is very

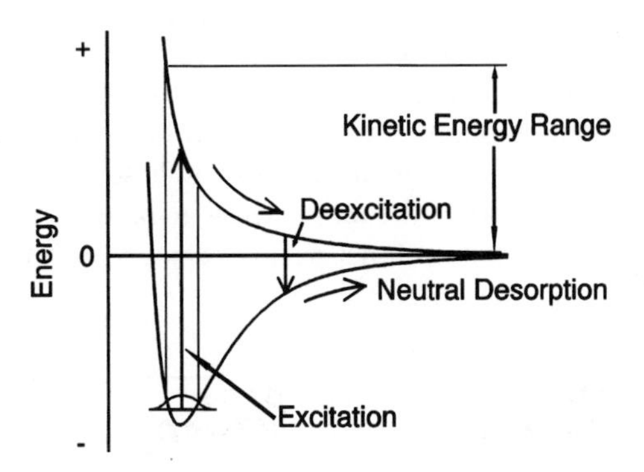

Figure 14.22 Potential energy diagram showing the processes involved in the electron-stimulated desorption of an adsorbed species as a neutral species with excess kinetic energy, according to the Redhead–Menzel–Gomer model.

close to the surface, it has a finite probability of being neutralized by an Auger or resonance neutralization process, in which an electron from the surface tunnels into the vacant electronic level of the excited species. The excess energy from this process can be carried away either by emission of an Auger electron from the solid or by emission of an electron from the excited level of the atom or molecule. The net result of this process is to return the excited species to the ground state potential energy curve but with whatever kinetic energy it had acquired in the time between the excitation and neutralization processes. If this kinetic energy is greater than the potential energy difference between the vacuum level and that represented by the ground state curve at the point of deexcitation, then the atom or molecule can escape from the surface as a neutral species with excess kinetic energy. Because the probability of these neutralization or deexcitation processes is strongly dependent on the distance between the excited species and the surface, these processes will be much more efficient for species that are found close to the surface in the adsorbed ground state. The cross section for emission as ions or excited neutrals will thus be much greater for species bound far from the surface, or those atoms of an adsorbed molecule that are farthest from the surface, than for species bound close to the surface. For example, a large H^+ ion flux can be observed from adsorbed H_2O or NH_3, which are adsorbed with the N or O atom closest to the surface, while very small H^+ ion signals are observed from chemisorbed hydrogen.

Another common observation is that oxides or other insulating materials very often decompose to some extent under the influence of electron bombardment, with the anion being removed preferentially. This is a significant practical problem for high voltage insulators in vacuum, as the reduction of the oxide can reach the point where its insulation value is reduced. This is also a potential problem in analytical techniques such as Auger spectrometry, where the near-surface stoichiometry of a sample may be significantly changed by the bombarding electron current used in the analysis. This phenomenon has been explained by a considerably different mechanism than the desorption of adsorbed species described previously. The so-called *Knotek–Feibelman mechanism* used to describe this process is shown schematically in Figure 14.23. In this case, the initiating event is core-level ionization of one of the metal ions of the oxide. Because the oxide structure is ionic, with the valence electrons of the metal transferred to the oxygen anion, no valence electrons are available to permit relaxation of this core hole. The required electrons are drawn from the oxygen anion, in a so-called *interatomic Auger process*. Normally, this would lead to a neutral oxygen species, which would still be bound to the surface. In a fraction of these processes, however, two electrons are emitted to carry away the excess energy involved in the Auger process. In this case, the oxygen atom is left with net positive charge, surrounded by positive metal ions, and is ejected from the surface.

Electron-Stimulated Desorption Ion Angular Distribution (ESDIAD)

Studies of the angular distribution of the ions or neutral species emitted in the ESD process have revealed that the emission in many cases is not spatially uniform but shows a strong preference for a particular polar and azimuthal angle, or group of

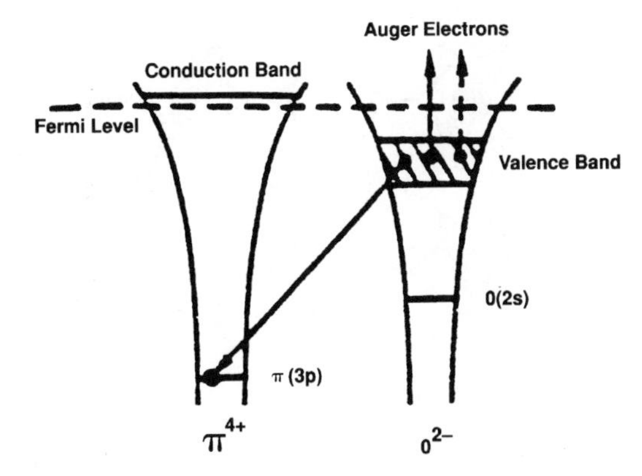

Figure 14.23 Energy diagram detailing the electron-induced desorption of an oxygen ion from a maximal valence oxide, according to the Knotek–Feibelman mechanism. (Source: M.L. Knotek and P.J. Feibelman, *Surface Sci.* 1979;90:78. Reprinted with permission.)

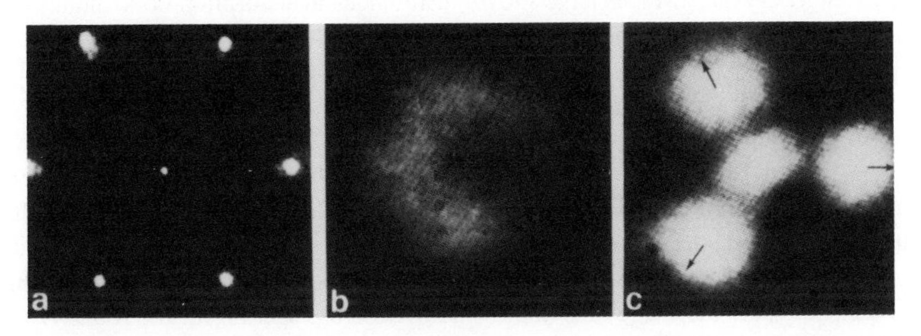

Figure 14.24 ESDIAD study of water adsorption on clean and oxygen-dosed Ni (111) showing LEED pattern of clean surface (a), H^+ ESDIAD from H_2O adsorption on clean surface (b), and H^+ ESDIAD from H_2O adsorption on oxygen-dosed surface (c). (Source: T.E. Madey and F.P. Netzer, *Surface Sci.* 1982;117:549. Reprinted with permission.)

angles. This technique has been used to provide information on the bonding geometry of adsorbed species, a technique known as *electron-stimulated desorption ion angular distribution (ESDIAD)*. The general observation is that ions are emitted more-or-less along the direction of the bond between atom and surface, or the bond direction within the molecule for fragments formed from adsorbed molecules. An example of such a study is shown in Figure 14.24, for the case of H_2O adsorbed on an Ni(111) surface on which a fractional monolayer of oxygen had been preadsorbed. The emitted species in this case is the H^+ ion. The angular distribution of the emitted ions indicates that the H_2O molecules are adsorbed intact, with the O atom bound directly to the surface. Analysis of the angular distribution pattern indicates that the H_2O molecule is distorted only slightly in the adsorption process and that the adsorbed molecules are azimuthally oriented as shown in Figure 14.25, with the threefold symmetry of the ESDIAD pattern asrising from three differently oriented domains of adsorbed species on the hexago-

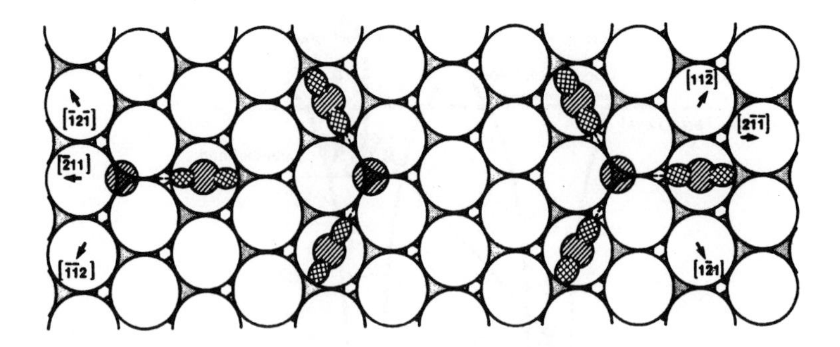

Figure 14.25 Schematic bonding model for O + H_2O on Ni (111); single O atoms are shown influencing the ordering of one, two, or three H_2O molecules. (Source: T.E. Madey and F.P. Netzer, *Surface Sci.* 1982;117:549. Reprinted with permission.)

nally symmetric Ni (111) surface. Similar studies of atomically adsorbed species on stepped single crystal surfaces have shown differences in adsorption bond angles to the surface for atoms adsorbed at terrace and at ledge sites.

Problems

14.1. Sketch the secondary electron energy distribution curve ($n(E)$ vs. E) for bombardment of aluminum by 2000 eV electrons. Include peaks arising from elastic scattering, bulk and surface plasmon losses, and Auger transitions.

14.2. Typically, electron impact ionization cross sections have a maximum at $E/E_B = 3$. Calculate the value of E at the maximum cross section for the K and L shells of magnesium.

14.3. How thick a layer of aluminum would one need in order to reduce the yield of KLL Auger electrons from an underlying layer of magnesium by 90% ?

14.4. Calculate the expected Auger electron kinetic energies for carbon, nitrogen, oxygen, and fluorine, and compare the results to the measured KLL Auger energies. (Tables of electron binding energies and KLL Auger energies can be found in the book by Feldman and Mayer listed in the Bibliography.)

14.5. In a core level spectroscopy experiment on chromium, at what bombarding electron energy would one first see soft x-ray emission from the Cr M shell? At what bombarding electron energy would electrons from the LMM Auger transition first appear?

Bibliography

Brundle, CR, and Baker, AD. eds. *Electron Spectroscopy: Theory, Techniques and Applications.* New York: Academic Press, 1981.

Chopra, DR, and Chourasia, AR. Characterization of semiconductor surfaces by appearance potential spectroscopy. In: G.E. McGuire, *Characterization of Semiconductor Materials.* Park Ridge, N.J.: Noyes Data Corp., 1989.

Czanderna, AW, ed. *Methods of Surface Analysis.* New York: Elsevier Publishing Company, 1975.

Davis, LE, MacDonald, NC, Palmberg, PW, Raich, GE, and Weber, RE. *Handbook of Auger Electron Spectroscopy*. Eden Prairie, Minn.: Physical Electronics Industries, 1976.

Egelhoff, WF, Jr. Core-level binding-energy shifts at surfaces in solids. *Surface Sci. Reports*, 1987;6:4/5.

Feldman, LC, and Mayer, JW. *Fundamentals of Surface and Thin Film Analysis*. New York: North-Holland, 1986.

Honig, RE. Surface and thin film analysis of semiconductor materials. *Thin Solid Films* 1976;31:89–122.

Menzel, D, and Gomer, R. Desorption from surfaces by slow-electron impact. *J. Chem. Phys.* 1964;40:1164.

Palmberg, PW. Quantitative analysis of solid surfaces by Auger electron spectroscopy. *Anal. Chem.* 1973;45:549A–556A.

Ramsier, RD, and Yates, JT, Jr. Electron stimulated desorption: principles and applications. *Surface Sci. Reports,* 1991;12:6–8.

Redhead, P. Interaction of slow electrons with chemisorbed oxygen. *Can. J. Phys.* 1964; 42:886.

Spanjaard, D, Guillot, C, Desjonquers, MC, Treglia, G, and Lecante, J. Surface core level spectroscopy of transition metals: a new tool for the determination of their surface structure. *Surface Sci. Reports*, 1986;5:1/2.

Weissmann, R, and Muller, K. Auger electron spectroscopy—a local probe for solid surfaces. *Surface Sci. Reports*, 1984;1:5.

CHAPTER **15**

Ion–Surface Interactions

This chapter continues the discussion of charged particle–surface interactions begun in Chapter 14. The consideration of the interactions between atomic or molecular ions and solid surfaces will point up a number of parallels to phenomena discussed in Chapter 14. There will be major differences, however, associated to a large extent with the fact that ions are much more massive than electrons and will, in general, interact more strongly with the substrate lattice.

Ion Scattering

First, consider processes in which an incident ion is scattered from a solid surface. Studies of this type have been carried out for a wide variety of ionic species and solid surfaces, over a wide range of ion energies. The general class of behavior observed depends primarily on the ion energy involved. Two energy ranges will be considered in detail: the so-called *low energy range*, in the vicinity of 1000 eV, and the *high energy range*, in the vicinity of 10^6 eV.

In the low energy range, the predominant interaction mechanism is hard sphere scattering, which was considered in the discussion of atom–surface scattering in Chapter 9. The fact that the incident particle is charged makes very little difference. The basic equation for energy transfer in this hard sphere collision is

$$\frac{E_1}{E_o} = \left[\frac{m^2}{(m+M)^2}\right]\left[\cos\theta + \left(\frac{M^2}{m^2 - \sin^2\theta}\right)^{1/2}\right]^2. \qquad 15.1$$

The values m, M, and θ are defined in Figure 15.1. For the case of scattering through an angle of 90°, Equation 15.1 reduces to

$$\frac{E_1}{E_o} = \frac{M-m}{M+m}. \qquad 15.2$$

Because the scattered particle is charged, its energy can be readily measured using an energy analyzer similar to that used in the various electron spectroscopies discussed in

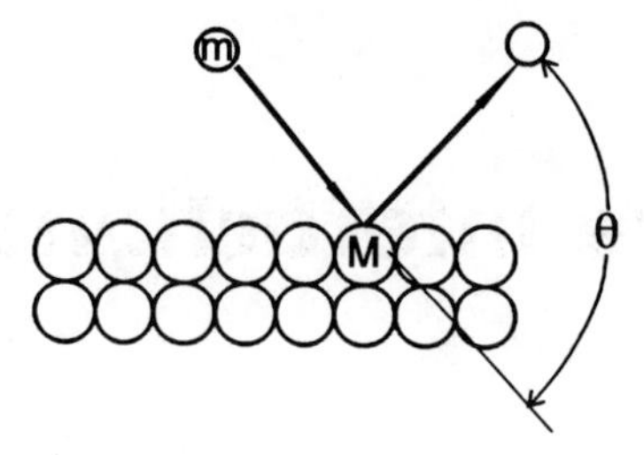

Figure 15.1 Schematic view of ion-surface scattering in the hard sphere regime.

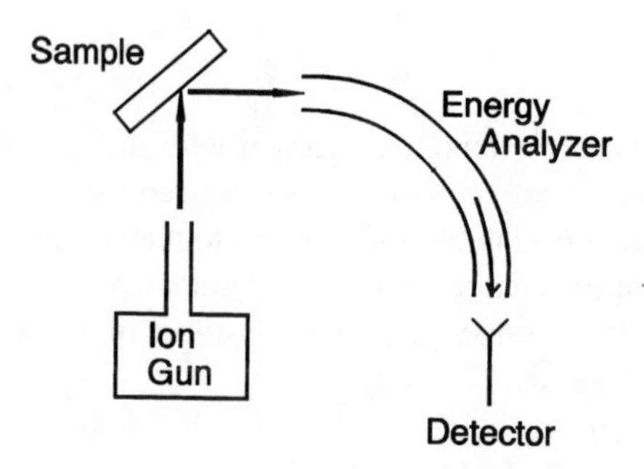

Figure 15.2 Schematic view of an experimental system for ion scattering spectroscopy.

Chapter 14 but biased to the opposite polarity. Energy analysis of the scattered ions, combined with a knowledge of the primary ion species mass, scattering angle, and primary ion energy, provides information on the mass (or masses) of the atomic species in the topmost layer of the scattering surface. This technique is called *ion scattering spectroscopy* (*ISS*).

It can be seen from Equation 15.2 that only collisions in which $m < M$ will yield a loss peak. Consequently, practical ion scattering spectrometers generally use a light bombarding ion. Generally, a chemically inert ion is the primary projectile; ^4He, ^3He, and ^{20}Ne are commonly used gases. Figure 15.2 shows a typical experimental setup for ISS. It is similar to that used in EELS, with an ion gun to provide a monoenergetic primary beam, the sample surface, and an energy analyzer. Figure 15.3 shows a typical result for $^4He^+$ scattered from metallic aluminum and from an aluminum oxide thin film. The relative peak heights for the surface species provide information on the surface stoichiometry, provided appropriate corrections are made for ion neutralization effects.

Ion scattering spectroscopy is also a sensitive technique for the detection of surface impurities, as shown in Figure 15.4. Here, spectra taken of the surface of a gold–nickel alloy after heat treatment in a vacuum show clearly the presence of molybdenum, which has segregated to the surface in the course of heating. Note that in this case spectra have been taken using both He^+ and Ne^+ ions. The increased mass resolution associated with the use of a higher mass probe particle and the consequent improvement in the ability to determine the masses of the surface species are quite apparent.

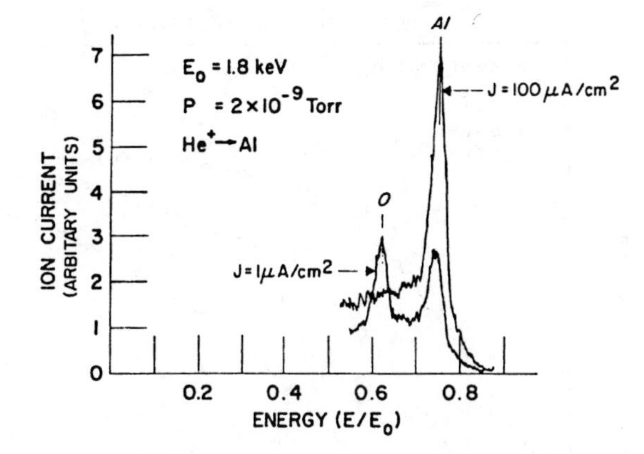

Figure 15.3 Partial spectra of 1800 eV He^+ scattering from a clean aluminum surface and from an Al_2O_3 thin film on that surface. The stoichiometry of the film may be determined from the relative peak heights, after correction for ion neutralization processes. (Source: D.P. Smith, *Surface Sci.* 1971;25:171. Reprinted with permission.)

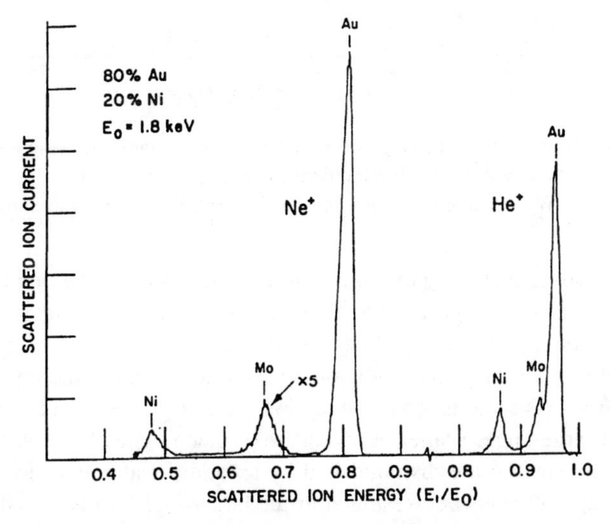

Figure 15.4 ISS spectra from a gold surface contaminated with small amounts of molybdenum and nickel. Note the improved resolution associated with the use of Ne^+ as the scattering species. (Source: D.P. Smith, *Surface Sci.* 1971;25:171. Reprinted with permission.)

Consider next the alternative case of high energy ions. Fluxes of ions in the megaelectron volt energy range can be produced in relatively small particle accelerators. Here again, singly or doubly ionized helium atoms are the most commonly used projectile. The scattering process for ions in this energy range is more complicated than that for the low-energy ions described previously, primarily because their high energy permits them to penetrate some distance into the solid and because the probability of a

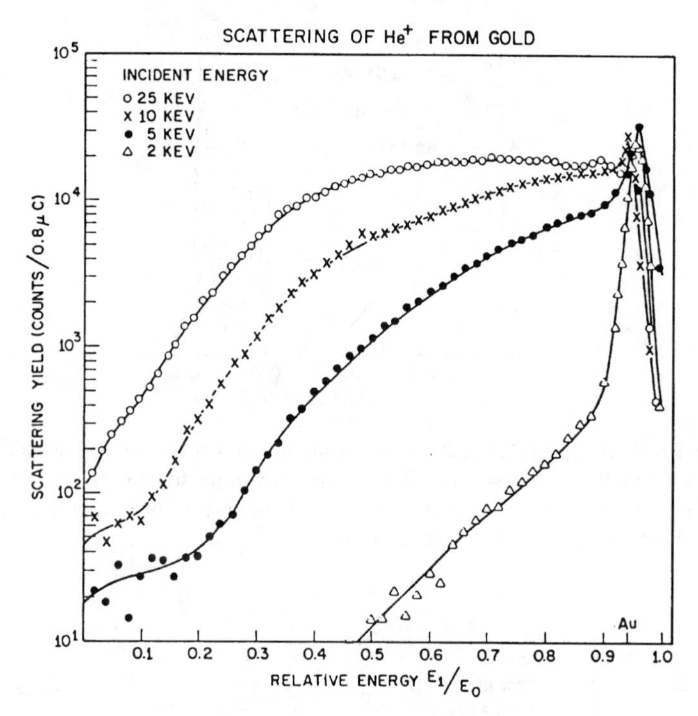

Figure 15.5 Helium ion scattering spectra from a gold surface, showing increasing penetration below the surface with increasing incident ion energy. (Source: D.J. Ball, T.M. Buck, D. MacNair and G.W. Wheatley, *Surface Sci.* 1972;30:69. Reprinted with permission.)

collision that causes scattering through a large angle is small. The trend with increasing energy is illustrated in Figure 15.5, showing energy loss spectra for He^+ scattering from gold. At the lowest energy, 2 keV, essentially all of the scattering is in the sharp peak associated with the initial collision with the topmost atomic layer of the sample (note that the y-axis scale is logarithmic). As the energy increases, more and more of the observed large-angle scattering events take place beneath the surface. Ions scattered below the surface are distinguished by an additional energy loss, due to small-angle scattering collisions before and after the large-angle collision. At energies in the megaelectron volt range, the curve would show a sharp rise at the characteristic loss energy given by the hard sphere model and then be essentially flat down to an energy loss characteristic of the thickness of the sample.

The study of ion scattering in this range is called *Rutherford backscattering spectroscopy* (*RBS*). A typical experimental setup and the results obtained from it are shown in Figure 15.6. The ion beam emerges from the accelerator and strikes the sample. Those ions that are scattered through a large angle, shown as $(\pi - \theta)$ in the figure, are detected by a solid state detector, which measures both particle flux and particle energy, yielding the result shown in the lower part of the figure. For each of the elements present in the sample, a roughly trapezoidal region appears in the spectrum. The high-energy edge of the trapezoid corresponds to ions that have scattered from the outer-

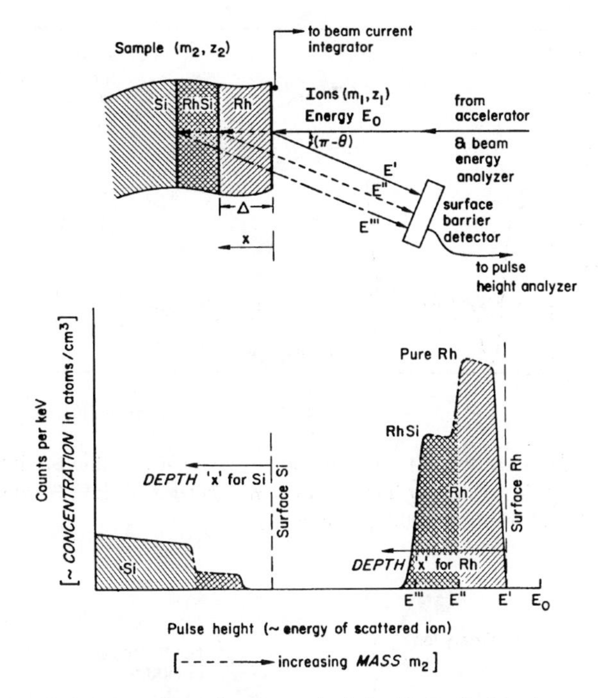

Figure 15.6 Experimental configuration for analysis by Rutherford backscattering spectroscopy (RBS), and the spectrum observed for scattering from an Rh-RhSi film on silicon. (Source: W. Reuter and J.E.E. Baglin, *J. Vac. Sci. Technol.* 1981;18:282. Reprinted with permission.)

most atoms of that species in the sample. In the case of rhodium, these atoms are at the sample surface, and the edge appears at the value calculated for hard sphere scattering. For silicon, the highest energy scattered ions will be at an energy lower than that calculated, due to energy loss by the ions while traversing the rhodium before striking the silicon. The height of each area is proportional to the concentration of the atom at that depth in the sample.

This technique thus provides an in-depth profile of concentration, without the necessity of destroying the sample, as was the case in AES depth profiling. On the other hand, it is often difficult to identify a thin layer sandwiched between two thicker layers, as shown in Figure 15.7, or to analyze for a small amount of a light element in a thick layer of a heavy element.

Sputtering

As was mentioned in the discussion of AES depth profiling in Chapter 14, the impact of ions on surfaces will cause removal of material from the surface. This phenomenon was recognized, named, and used in commercial processes long before it was understood at the atomic level. It is called *sputtering*. The currently accepted mechanism of the sputtering process is shown schematically in Figure 15.8. An energetic ion strikes the surface, penetrates, and after a series of collisions with atoms in the solid,

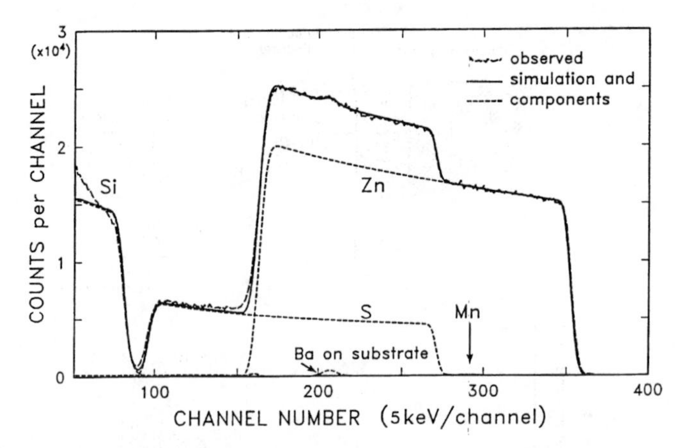

Figure 15.7 RBS spectrum of a thin film sample having a very thin $BaTiO_3$ layer between a silicon surface and a ZnS overlayer film. (Source: W. Reuter and J.E.E. Baglin, *J. Vac. Sci. Technol.* 1981;18:282. Reprinted with permission.)

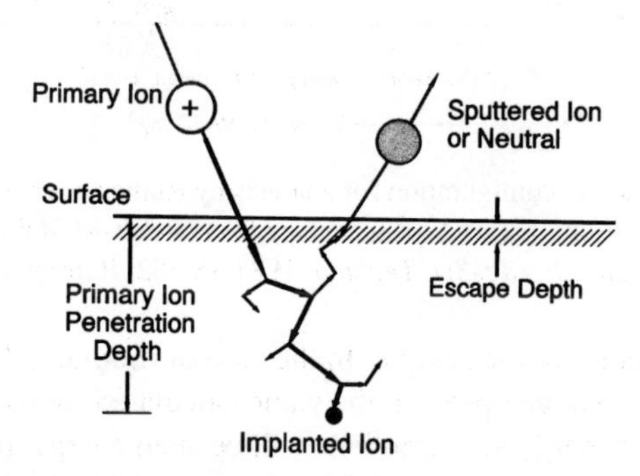

Figure 15.8 Sequence of events following the impingement of an energetic ion on a solid surface. The resulting collision cascade leads to ion implantation and removal of surface species by sputtering.

comes to rest. Certain of these collisions will impart momentum to atoms of the solid in the direction of the surface. Through a chain of collisions, this momentum may be transferred to an atom or atoms at the surface, causing their ejection from the solid as neutrals or as positive or negative ions.

Practical use is made of the sputtering process in three major areas: Materials synthesis, controlled etching, and surface analysis by secondary ion mass spectrometry (SIMS). The materials synthesis application involves subjecting a target material to a flux of ions to cause sputtering. The sputtered particles are then collected by and deposited upon the receiving surface, or substrate, causing the growth of a new solid phase. In many cases, sputtering has advantages over other techniques of crystal growth, such as evaporation–condensation, in that it does not require heating the target to a high temperature in order to achieve a high rate of material transfer.

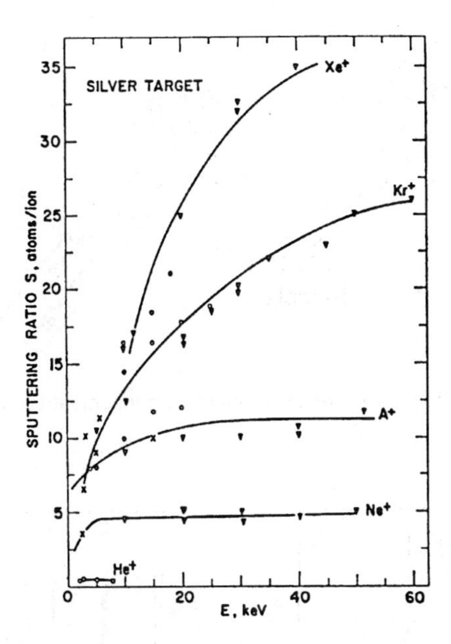

Figure 15.9 Sputter yield as a function of incident ion energy, for various rare gas ions incident on a polycrystalline silver surface. (Source: M. Kaminsky, . *Atomic and Ionic Impact Phenomena on Metal Surfaces*. Berlin: Springer-Verlag, 1965, p. 157. Reprinted with permission.)

The use of sputtering as a controlled etching technique is the basis of several surface and in-depth analytical methods. The overall sputtering rate in any system is a complicated function of the ion used, its mass and kinetic energy, and the target material chemical species and crystallographic orientation. As a general rule, if one defines a sputter yield in terms of the number of atoms removed per incident primary ion, one obtains curves such as those shown in Figure 15.9. Here the sputter yield is plotted as a function of primary ion energy for the various rare gas ions incident on a metal surface. The most obvious trends are that there is essentially no sputtering for ions having energies less than the binding energy of the target atoms. When this threshold is exceeded, the sputter rate for a given ion increases rapidly, then continues to rise, although less rapidly. This behavior is reminiscent of the curve of secondary electron emission ratio vs. primary electron energy and is governed by similar considerations. The other trend visible in Figure 15.9 is the increasing sputter efficiency with increasing primary ion mass, all other things being equal. There has also been considerable research on the relative sputtering rates of various solid materials under a given set of ion bombardment conditions. This parameter is of great interest in the depth profile analysis of multicomponent materials. Any preferential sputtering of one component will result in a surface enriched in the other component and consequently lead to errors in the analysis.

The final application of sputtering to consider is the analysis of the sputtered particles to determine the surface composition, the technique called *secondary ion mass spectrometry* (*SIMS*). A basic setup for SIMS is shown in Figure 15.10. Ions produced by sputtering are collected by a lens system, monochromatized, and injected into a

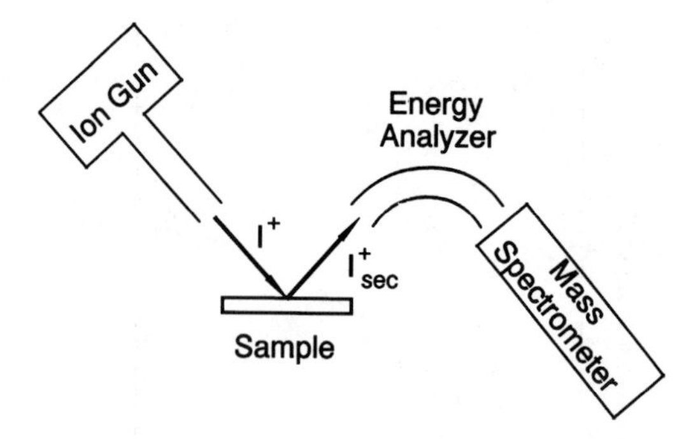

Figure 15.10 Schematic view of the experimental arrangement for secondary ion mass spectrometry (SIMS).

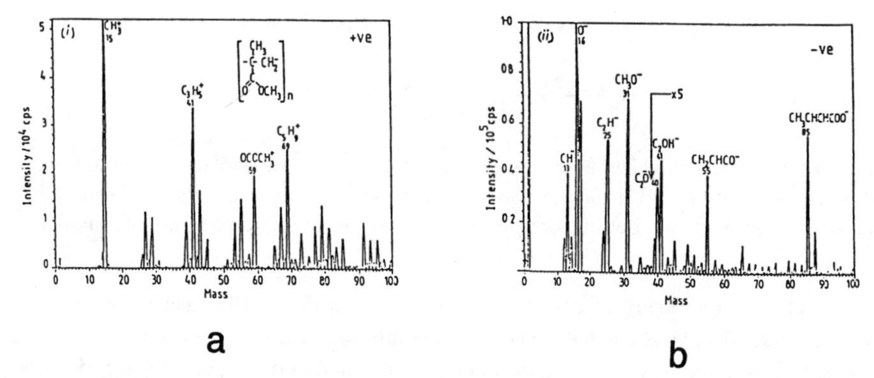

Figure 15.11 Positive and negative ion SIMS spectra from a polymethylmethacrylate (PMMA) sample. (Source: A.J. Eccles and J.C. Vickerman, *J. Vac. Sci. Technol.* 1989;A7:234. Reprinted with permission.)

device that separates the ion beam according to the mass-to-charge ratio of the ions. The mass separated ion beam is then detected with a device such as an electron multiplier. From these data, a secondary ion mass spectrum can be plotted, as in Figure 15.11, which shows the signal intensity vs. mass for both the positive and negative ion currents resulting from ion bombardment of a polymethylmethacrylate (PMMA) sample. Note that the spectra show ions that result from fragmentation of the basic polymer unit in the course of the sputtering process. These fragmentation patterns provide much useful information on the structure of organic materials.

The three principal modes of operation in SIMS are referred to as *dynamic, static, and scanning*. In dynamic SIMS, the ion flux to the surface is large enough that a fairly high rate of sputter etching takes place. Consequently, as the analysis proceeds, information is obtained from greater and greater depths below the original sample surface. In this mode, one may obtain spectra such as those shown in Figure 15.11, or alternatively, one can measure the concentration vs. depth of one or more of the species present in the sample. This SIMS depth profiling is especially useful in determining dopant profiles in

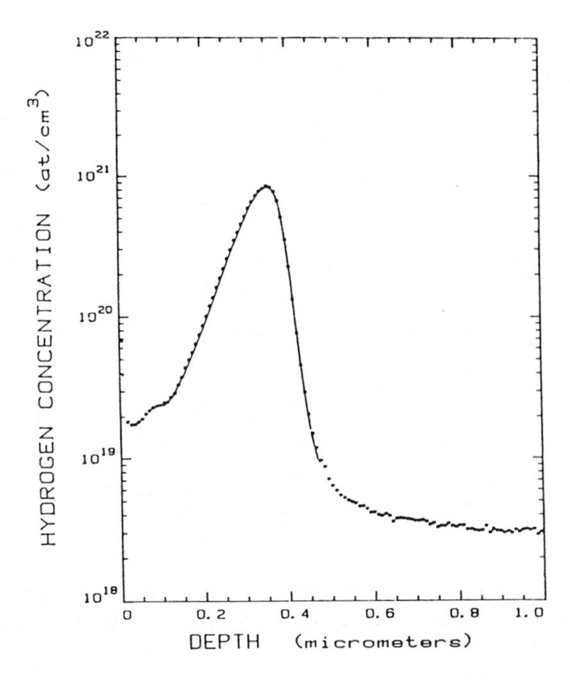

Figure 15.12 Depth profile of a hydrogen ion implant in silicon, obtained using dynamic SIMS with a Cs^+ primary ion beam. (Source: C.W. Magee and E.M. Botnik, *J. Vac. Sci. Technol.* 1981;19:47. Reprinted with permission.)

semiconductors, as the doping levels of interest are usually below the detection limit of other surface analytical techniques such as AES or RBS. A typical implant profile of hydrogen in silicon obtained using dynamic SIMS is shown in Figure 15.12.

In static SIMS, one operates with an ion bombardment rate such that the sputter etch rate is small compared to a monolayer of material during the duration of an experiment. This mode is useful in the study of thin organic layers on metals, as the resulting fragment ions contain a great deal of information on the structure of the organic species. This technique and the related technique of fast atom bombardment, which uses a high energy neutral bombarding beam, are currently undergoing rapid expansion as analytical techniques in the area of organic structure determination.

Figure 15.13 shows the setup used in scanning SIMS. The basic configuration is similar to a scanning electron microscope, in that a high energy particle beam is rastered over the sample surface and that the particles generated by the interaction of the beam with the surface are detected and used to form an image of the surface. In this case, however, the beam is an O_2^+ ion beam, and the detected particles are the secondary ions sputtered from the surface. Since the detector mass spectrometer can be tuned to accept a single ionic species, the output of the instrument is a map of the surface concentration of the species of interest. Successive scans using different mass spectrometer settings can provide a complete picture of elemental distribution over the surface. In current instruments, the lateral resolution is on the order of 1 μm.

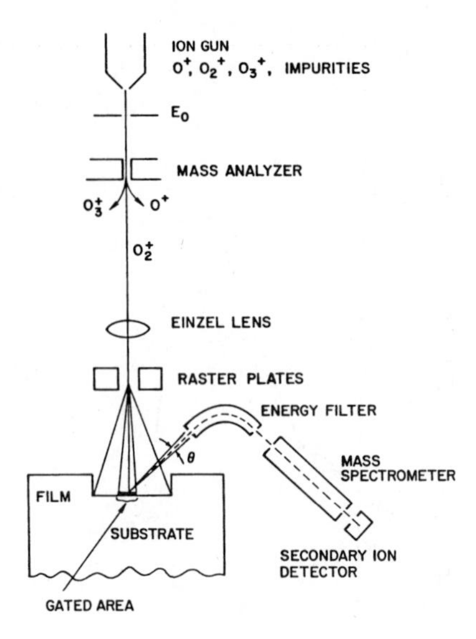

Figure 15.13 Schematic view of the experimental arrangement for scanning SIMS, using a mass-separated oxygen ion beam. (Source: W. Reuter and J.E.E. Baglin, *J. Vac. Sci. Technol.* 1981;18:282. Reprinted with permission.)

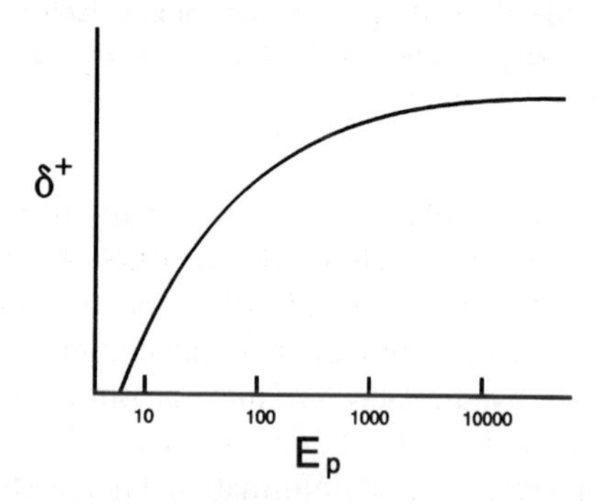

Figure 15.14 Secondary electron yield from a surface under ion bombardment as a function of the energy of the ion beam.

Secondary Electron Generation

The impingement of energetic ions on a surface also causes the generation of secondary electrons by mechanisms similar to those involved in the generation of secondary electrons by electron impact. As in the previous case, the overall rate of secondary electron emission can be characterized in terms of a secondary emission ratio, δ^+. The behavior of δ^+ with primary ion energy is shown in Figure 15.14. Again, this figure is reminiscent of the curve for secondary electron generation by electron impingement, with a threshold at the work function, an initial rapid rise, and a decreas-

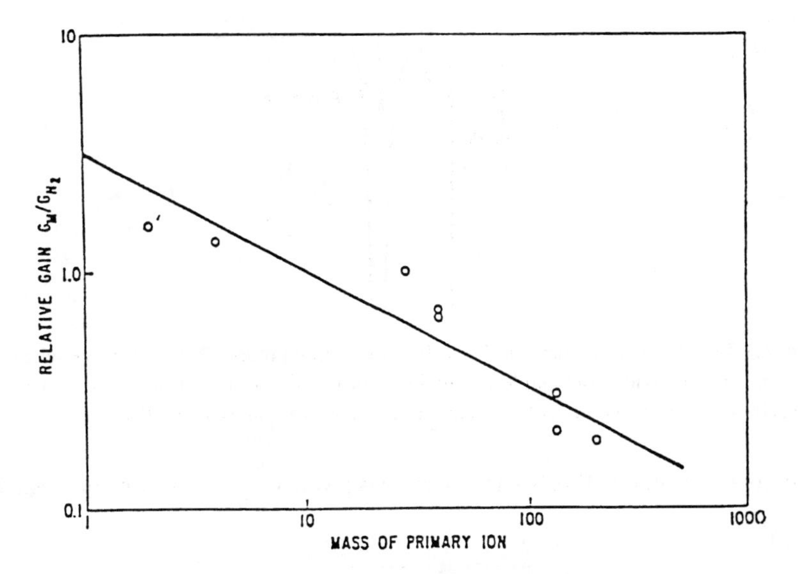

Figure 15.15 Secondary electron yield from a surface under bombardment by beams of various ionic species at a given primary energy, as a function of primary ion mass. (Source: R.E. Grande, R.L. Watters and J.B. Hudson, *J. Vac. Sci. Technol.* 1966;3:329. Reprinted with permission.)

ing rate of rise at higher ion energies. In this case, the curve does not actually turn over at high ion energies due to the lesser penetrating power of the massive ions.

It is also interesting to consider the variation of the secondary emission ratio with ion mass, at constant energy. An experimental plot of this relation, given in Figure 15.15, shows regular decrease in δ^+ with the square root of ion mass. The significance of these two figures is, first, that because ion impingement produces electrons, electron multiplier structures can be used as ion detectors and, second, that the overall gain of the multiplier in this application will depend upon both ion energy and ion mass.

Surface Ionization and Neutralization

A number of processes take place when atoms or ions are brought close to a metal surface and participate in some sort of charge exchange process. These include surface ionization, resonance neutralization, resonance ionization, and surface Penning ionization. All of these processes can be understood in terms of the electron potential energy diagram for the metal surface. The parameter that differs from case to case is the nature or electronic state of the incoming particle.

In surface ionization, the incoming particle is an unexcited neutral atom that has a filled electronic state above the Fermi level of the metal. That is, the ionization potential, I, is smaller than the work function, Φ. This case is shown in Figure 15.16. In this case, the energy of the system will be reduced if an electron leaves the adsorbed atom and occupies an unfilled electronic state near the Fermi level of the metal. When such an ionized adsorbed species desorbs, it may do so as a positive ion. The ratio of ions to

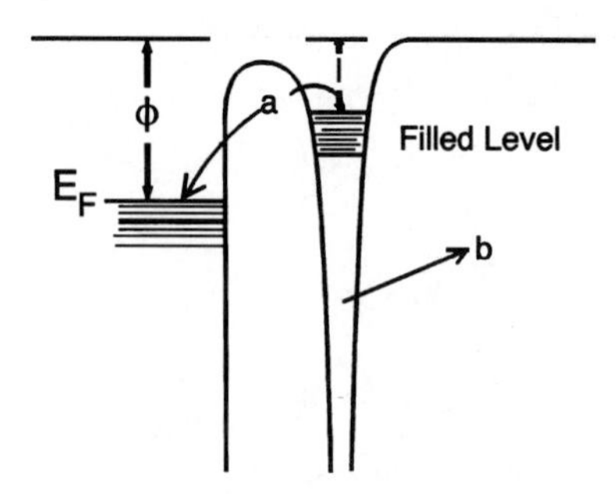

Figure 15.16 Energy diagram for the surface ionization process that takes place when an atom having an ionization potential less than the work function of the surface approaches the surface, showing electron transfer to the solid (a), and desorption as an ion (b).

neutrals in the desorption flux from the surface is given by the Saha–Langmuir equation,

$$\frac{N_+}{N_o} = \exp\left[\frac{-(I - \Phi)}{kT}\right]. \qquad 15.3$$

Note that if $I < \Phi$, the exponential term is positive, giving $N_+ > N_0$ and a ratio that decreases with increasing temperature. If $I > \Phi$, then the reverse is true. Experimentally, it is found that I is significantly less than Φ only for a few situations, primarily the adsorption of the alkali metals on the surfaces of refractory metals such as tungsten and molybdenum. This effect has been used both as a source of alkali ions, by applying a salt of an alkali metal to a refractory metal surface and then heating it in a vacuum, and as a detector for molecular beams of alkali or alkaline salt species, by allowing them to strike a heated tungsten filament and collecting the resulting ions. This last technique provides a very sensitive detector, as it responds only to the readily ionizable alkali atoms.

In the case of resonance neutralization, the incoming particle is an ion, having an ionization potential much larger than the work function. (These experiments are usually carried out with helium, which has $I = 24$ eV). In this case, when the atom is close to the surface, an electron from the conduction band can tunnel into the helium ion, neutralizing it as shown in Figure 15.17. The excess energy associated with this process can be removed from the system by the excitation into the vacuum of a second electron in a process essentially similar to the Auger deexcitation process studied in Chapter 14. The energy of the emitted electron in this case is

$$E_k = I - E_a - E_B. \qquad 15.4$$

Since any of the electrons in the conduction band can cause the neutralization, and any electron in the conduction band can carry away the excess energy, the electron flux produced will cover a range of energies twice the width of the conduction band. Study of the emitted electron flux vs. energy provides information on the density of filled states in the conduction band of the metal. This technique is known as *ion neutralization spectroscopy* (*INS*).

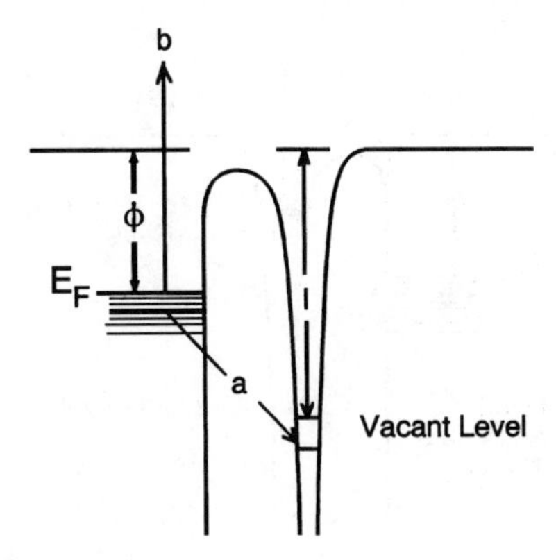

Figure 15.17 Energy diagram for resonance neutralization of an ion having an ionization potential large compared to the work function of the surface it is approaching, showing neutralization of the ion by an electron from the solid (a), and ejection of an Auger electron to carry off the excess energy from process a (b).

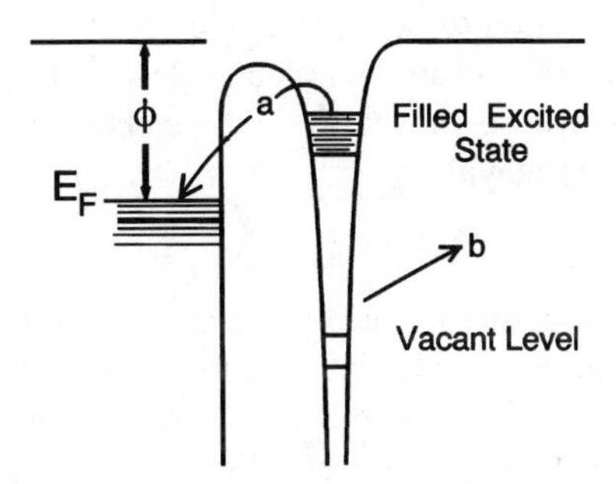

Figure 15.18 Energy diagram for resonance ionization of an excited atom having a filled electronic state above the work function of the solid and an empty electronic state below the work function of the solid, showing electron transfer to the solid (a), and desorption as an ion (b).

Resonance ionization involves an incoming particle that is an excited neutral atom having a filled electronic state above the work function of the metal and an empty state below the work function, as shown in Figure 15.18. In this case, the electron from the excited state can jump to an unfilled state near the Fermi level of the metal. This now leaves the adsorbed atom as a positive ion, with $I > \Phi$, and the resonance neutralization process described above can again take place or desorption as an ion can occur. Again, this technique provides information on the density of filled states in the conduction band of the metal.

The final process to be discussed also involves an excited neutral as a probe particle but in this case on an adsorbate-covered surface as shown in Figure 15.19. In this figure, the density of occupied states of the adsorbed species is shown adjacent to the conduction band of the metal. In this case, because of the presence of the adsorbate, the

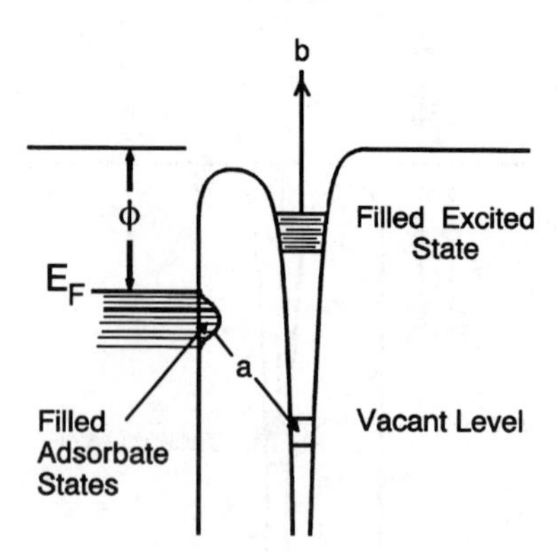

Figure 15.19 Energy diagram for surface Penning ionization electron spectroscopy (SPIES), showing electron transfer from a filled adsorbate electronic state to the vacant level of the excited probe atom (a), and emission of an electron from the excited state of the probe atom (b).

excited probe atom (typically helium) cannot approach close enough to the metal surface for resonance neutralization to take place. However, an electron from the filled adsorbate levels can tunnel into the unfilled ground state of the probe atom. The excess energy from this process can be carried away by ejection of the electron from the excited state of the probe atom, with

$$E_k = I - E_a - I_2 \,. \qquad 15.5$$

Since I and I_2 are known properties of the probe atom, measurement of the distribution of E_k arising from this process provides information on the density of filled electronic states of the adsorbed species. This recently developed technique is known as *surface Penning ionization electron spectroscopy*, or *SPIES*.

Problems

15.1. An ISS experiment is carried out on the surface of a polycrystalline $FeAlO_3$ sample, using 2000 eV $^4He^+$ ions and a scattering angle of 90°.

a. Sketch the expected plot of ion yield vs. energy for the scattered ions.

b. If the sample were a single crystal, rather than polycrystal, how might this affect the peak height ratios in the spectrum?

15.2. Calculate the energy loss for a $^4He^{++}$ ion in the case of a collision with each of the following species, for the case in which the ion is scattered through an angle of 160°:

a. A gold nucleus.

b. An oxygen nucleus.

c. A second helium atom.

15.3. An RBS depth profile experiment is carried out on a sample made up of a 1000 Å layer of PtSi on a silicon substrate, using 2 MeV helium ions, and a scattering angle of 170°. The rate of energy loss by the ions traversing the PtSi is 38 eV/Å, and in traversing silicon is 25 eV/Å. The collision probability with a given atomic species is proportional to the square of the atomic number of the atom being struck. Sketch the expected backscattering spectrum.

15.4. An alloy of 50% Fe and 50% Al is depth profiled by sputter etching with 5000 eV $^{40}Ar^+$ ions. If the relative sputter ratio Al/Fe is 1.3/1.0, what is the steady state surface composition of the alloy as sputtering takes place?

15.5. An INS experiment is carried out using $^4He^+$ ions incident on the surface of a metal having a Fermi energy of 5 eV relative to the bottom of the conduction band and a work function of 3 eV. What is the range of kinetic energies of the emitted electrons?

Bibliography

Bevolo, A. Ion/solid interactions in surface analysis. In: G.E. McGuire, ed. *Characterization of Semiconductor Materials*. Park Ridge, N.J.: Noyes Data Corp., 1989.

Czanderna, AW, ed. *Methods of Surface Analysis*. Amsterdam: Elsevier Publishing Company, 1975.

Feldman, LC, and Mayer, JW. *Fundamentals of Surface and Thin Film Analysis*. New York: North-Holland, 1986.

Garrison, BJ, and Winograd, N. Ion beam spectroscopy of solids and surfaces. *Science* 1982;216:805-812.

Goff, RF, and Smith, DP. Surface composition analysis by binary scattering of noble gas ions. *J. Vac. Sci. Technol.* 1970;7:72.

Kaminsky, M. *Atomic and Ionic Impact Phenomena on Metal Surfaces*. New York: Academic Press, 1965.

Kasi, SR, Keng, H, Sass, CS, and Rabalais, JW. Inelastic processes in low energy ion-surface collisions. *Surface Sci. Reports* 1989;10:1–104.

Morgan, RE. Secondary ion mass spectrometry. In: G.E. McGuire, ed. *Characterization of Semiconductor Materials*. Park Ridge, N.J.: Noyes Data Corp., 1989.

Reuter, W, and Baglin, JEE. Secondary ion mass spectrometry and Rutherford backscattering spectroscopy for the analysis of thin films. *J. Vac. Sci. Technol.* 1981;18:282–288.

van der Veen, JF. Ion beam crystallography of surfaces and interfaces. *Surface Sci. Reports* 1986:5: 5/6.

CHAPTER **16**

Photon–Surface Interactions

This chapter concludes the study of the interaction of energetic particles with surfaces, looking in this case at the processes that take place when surfaces are bombarded with photons over the energy range from a few electron volts to a few thousand electron volts. As in the previous two chapters, it will be seen that this photon bombardment can cause the excitation of both electronic and vibrational states of surface species and the ejection of electrons and ions from the surface.

Types of Interaction

Of all the ways in which photons can interact with matter, the only two that are important in the present context are elastic scattering and photon absorption. Other processes, such as Compton scattering and pair production, are important only at much higher energies than are of interest here. Elastic scattering, in which there is momentum transfer but no energy transfer, leads to the observation of diffraction, as has been seen for the elastic scattering of electrons and atoms. Photon absorption, which is the basis of the photoelectric effect, is the initial event in all of the photon-based spectroscopic techniques to be considered, and the initiating event in laser induced processes such as laser-induced desorption and surface reaction enhancement.

Diffraction

The concept of *diffractive scattering*, in which the interaction of a wave with a regular array of scattering sites leads to maxima and minima in the scattered intensity as a function of angle, was first discovered many years ago with visible light. Other concepts previously introduced, such as the Ewald sphere, were first developed in the study of the diffraction of x-rays from bulk crystals. All of the applications that this book has considered represent extensions of these concepts to systems of lower dimensionality or to other types of waves.

Although the diffraction of x-rays has become a major method for the study of the structure of bulk condensed phases, it has been much less useful in the study of surfaces. The reason for this is that photons generally interact much less strongly with atoms than do electrons or other atoms. The net result of this difference is that the information

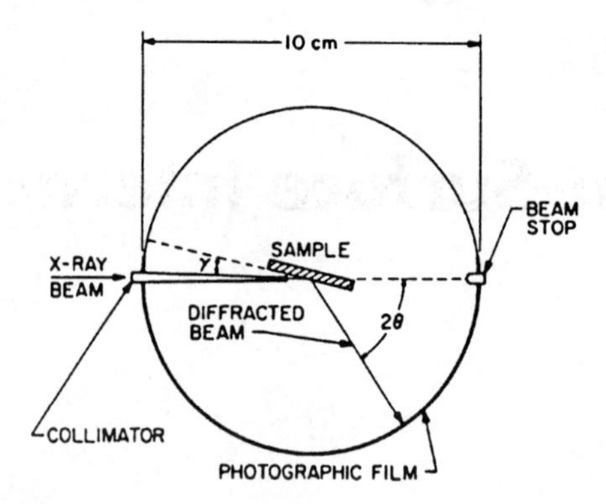

Figure 16.1 Read camera for the determination of thin film structure by low angle x-ray diffraction. (Source: S.S. Lau, W.K. Chu, J.W. Mayer and K.N. Tu, *Thin Solid Films* 1974;23:205. Reprinted with permission.)

obtained in an x-ray diffraction experiment is dominated by diffraction from the bulk of the sample to a depth of several thousand nanometers, rather than the 1 nanometer range sampled by LEED. Some use has been made of x-ray diffraction in the study of thin crystalline films on surfaces, however, for the case of films in the hundred nanometer thickness range. The technique is known as *glancing angle x-ray diffraction*, in which the x-ray beam is incident on the surface at a very low angle, in order to maximize the distance traveled by the beam in traversing the thin film. One common configuration for this technique is shown in Figure 16.1, the so-called *Read camera*. This is similar to cameras used in conventional x-ray diffraction studies of crystalline powders, except that the x-ray beam is incident on the sample at a low angle, γ, typically 5–10°. Figure 16.2 shows an example of the diffraction pattern produced in such a camera for the case of a nickel film on silicon. Diffraction from the film, which is fine-grained polycrystalline, gives rise to the ring pattern. Diffraction from the single crystal silicon substrate gives rise to the spot pattern shown. Measurements of this sort are important in the study of thin films, as they provide information on the phases present and their relative amounts.

The Photoelectric Effect

When photons in the energy range of interest here interact inelastically with electrons, the net result is the absorption of the photon, with a commensurate increase in the energy of the electron. Chapter 2 touched on this process briefly in the discussion of photoelectron emission and ultraviolet photoelectron spectroscopy.

The cross section for the photon absorption process can be calculated using quantum mechanics. The basic equation for the transition probability, W, for the transition of an electron from an allowed initial state to an allowed final state due to the absorption of a photon is

$$W = \left(\frac{2\pi}{\hbar}\right)\rho(E)\langle \Psi_k | H' | \Psi_m \rangle^2 \sec^{-1}, \qquad 16.1$$

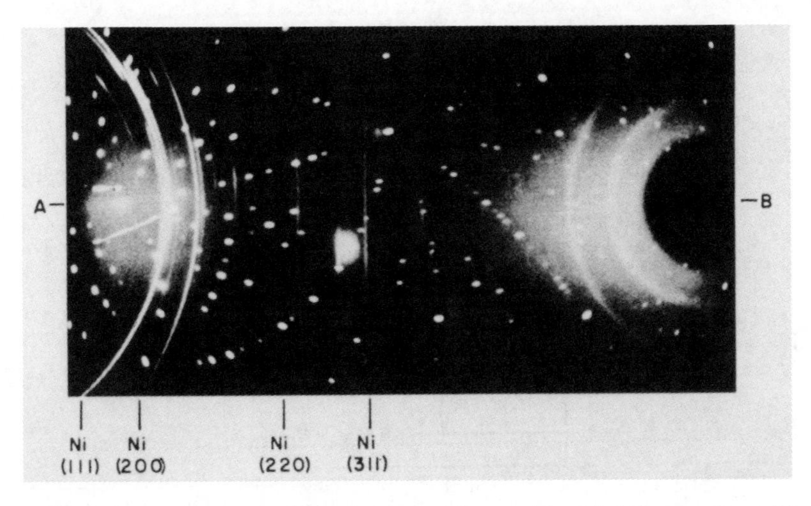

Figure 16.2 Low angle x-ray diffraction pattern from a thin nickel film on silicon. The rings arise from the polycrystalline nickel film, the sharp spots from the single crystal silicon substrate. (Source: S.S. Lau, W.K. Chu, J.W. Mayer and K.N. Tu, *Thin Solid Films* 1974;23:205. Reprinted with permission.)

where $\rho(E)$ is the density of states for the final state and the term in brackets is related to the overlap of the initial and final state wave functions. This expression is known as the *Fermi golden rule*.

For the case of photoemission from a free electron type state, for example, a conduction band state in a metal, solution of Equation 16.1 leads to a cross section for excitation, σ, of

$$\sigma = \left(\frac{8e^2\hbar}{mc}\right)\left(\frac{E_B{}^{3/2}}{E^{5/2}}\right), \qquad 16.2$$

where c is the velocity of light, e and m are the electron charge and mass, respectively, E_B is the binding energy of the initial state of the electron, and $E = h\nu$ is the photon energy.

For the case of a core level electron, bound to an ion core, the corresponding equation is

$$\sigma = \left(\frac{128\pi}{3}\right)\left(\frac{e^2\hbar}{mc}\right)\left(\frac{E_b{}^{5/2}}{E^{7/2}}\right), \qquad 16.3$$

or, putting in values for the physical constants,

$$\sigma = \left(\frac{7.5}{h\nu}\right)\left(\frac{E_B}{h\nu}\right)^{5/2} \text{Å}^2. \qquad 16.4$$

The most important point to note from this equation is the dependence of the cross sec-

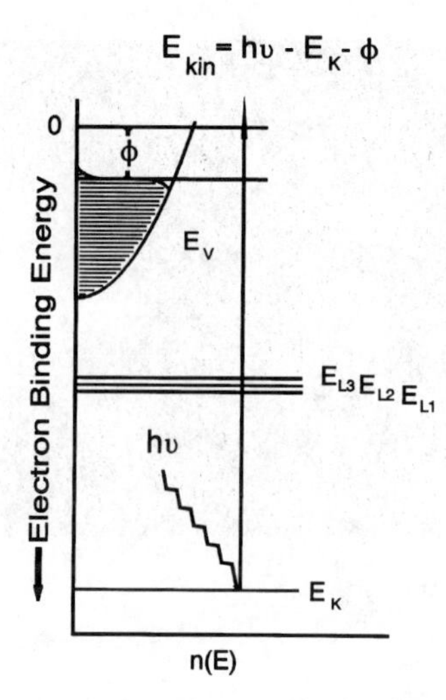

Figure 16.3 Electron energy diagram illustrating the processes involved and the energy relationships in x-ray photoelectron spectroscopy.

tion on the photon energy. The process cannot occur unless $h\nu \geqslant E_B$ and will have a maximum probability at $h\nu = E_B$. The cross section drops off fairly rapidly, as $(1/h\nu)^{7/2}$, as the photon energy is increased beyond the threshold value.

The absolute values of σ are small compared to the values calculated for electron bombardment–induced ionization. Consequently, the signal levels in photon excited processes, such as x-ray photoelectron spectroscopy, to be discussed in the next section, are generally much lower than those for electron excited processes, such as AES. On the other hand, the background levels are also much lower in photon-excited spectroscopies.

X-ray Photoelectron Spectroscopy

The most commonly used surface analytical technique based on the photoelectric effect is *x-ray photoelectron spectroscopy* (*XPS*). This technique is similar in principle to the ultraviolet photoelectron spectroscopy discussed in Chapter 2. The major difference is that, in this case, typical photon energies are in excess of 1000 eV. The basis of the technique is illustrated in Figure 16.3. Here an energy diagram shows the stable filled electronic states of a solid metal. The absorption of an incident photon of energy $h\nu$ causes the emission of an electron having an energy

$$E_{kin} = h\nu - E_i - \Phi, \qquad 16.5$$

where E_i is the binding energy of the electronic level excited by absorption of the photon. Again, note that photoemission will be observed only for those levels for which

$$E_i \leq h\nu + \Phi. \qquad 16.6$$

Because the electron is emitted from a core level having a well-defined energy, due to the absorption of a photon having a well-defined energy, this photoelectron will have a

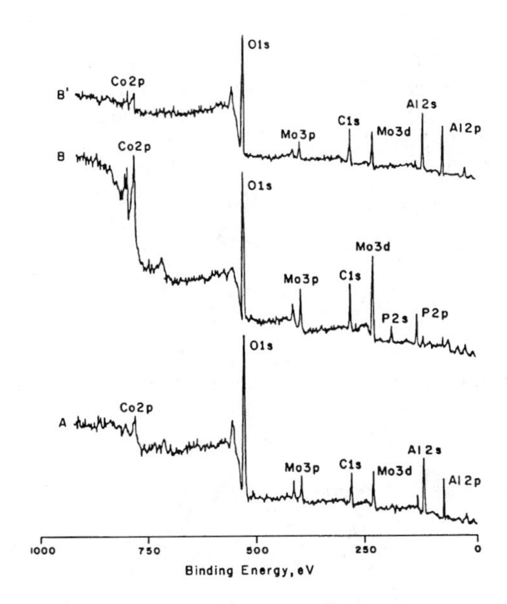

Figure 16.4 XPS survey scans for alumina supported cobalt molybdate catalyst samples, showing high activity catalyst (a) low activity catalyst (b), and bulk composition of catalyst (c). (Source: W.M. Riggs and M.J. Parker, in: A.W. Czanderna, ed. *Methods of Surface Analysis.* Amsterdam: Elsevier Publishing Company, 1975, p. 103. Reprinted with permission.)

well-defined energy characteristic of the atom involved. Moreover, the total photoelectron current arising from any such process will be directly proportional to the concentration of the atomic species involved in the near-surface region of the sample. Energy analysis of the photoelectron flux thus provides both qualitative and quantitative analysis of the region sampled. The x-rays required for excitation are usually obtained either from a metal-target x-ray tube (aluminum and magnesium targets are typically used) or from a synchrotron radiation source. This last type of source has the advantages of high intensity and variable but sharply defined photon energy, thus permitting the photon energy to be "tuned" to maximize the cross section for a transition of interest.

The analytical system used to carry out the energy analysis is similar to that used in AES, consisting of an electron energy analyzer and appropriate signal processing electronics. In this case, because of the low signal levels and the relatively low background level, pulse counting techniques are usually used, rather than the signal averaging and derivative techniques used in AES. Typical wide energy range, or survey, XPS spectra are shown in Figure 16.4, for a series of alumina-supported cobalt molybdate catalyst samples. The three spectra represent different catalyst surface compositions resulting from exposure to different environments. The information obtained from these spectra is similar to that obtained in AES.

A major advantage of XPS is that the energy resolution of practical systems is high enough to resolve the small changes in electron binding energy that accompany changes in the chemical state of the atom being excited. These shifts arise from the fact

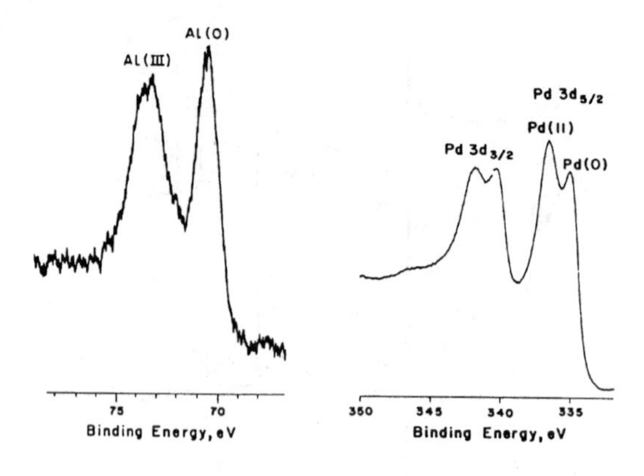

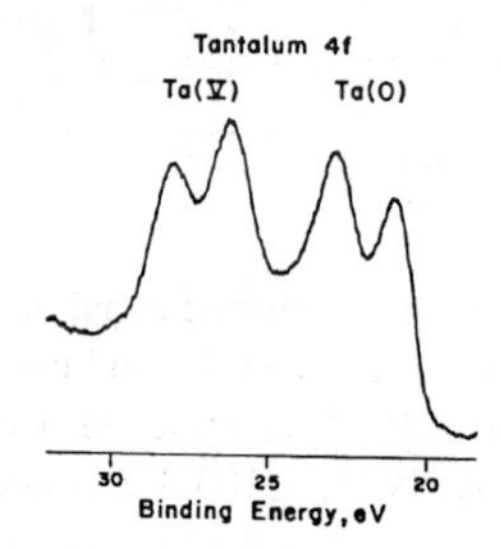

Figure 16.5 High resolution XPS spectra of aluminum, palladium, and tantalum surfaces, showing chemical shifts associated with oxidation. (Source: W.M. Riggs and M.J. Parker, in: A.W. Czanderna, ed. *Methods of Surface Analysis*. Amsterdam: Elsevier Publishing Company, 1975, p. 103. Reprinted with permission.)

that as electronic charge is added to or removed from a given atom in the process of forming a chemical bond, the resulting change in the screening of the core electrons results in small changes in the energies of the allowed electronic states. The magnitude of these changes is on the order of 1 to 5 eV. Because of this ability to resolve these so-called *chemical shifts,* this technique is often referred to by the alternative name *electron spectroscopy for chemical analysis* (*ESCA*).

Examples of these chemical shifts are shown in Figure 16.5 for the case of the shifts that accompany the oxidation of aluminum and palladium and tantalum. In the case of palladium, the level excited is the 3d level, which is inherenty a doublet—two closely spaced sublevels separated by about 5 eV. The case of tantalum is similar, but in this case the doublet is associated with the 4f level.

Information on concentration or chemical state as a function of depth can be obtained either by ion etching, as was the case for AES, or by varying the angle between the sample and the detector. At low polar angles (detector angle far from the surface normal), a larger fraction of the observed signal comes from the near-surface region of the sample, due to the limitation on depth sampled associated with the electron escape depth, as discussed in Chapter 14. This effect is shown in Figure 16.6. Here

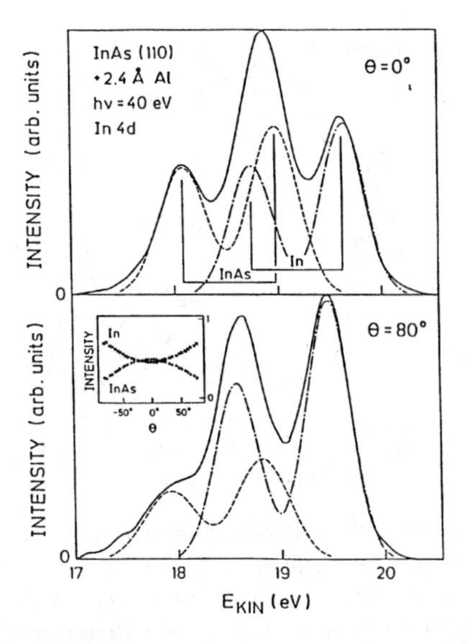

Figure 16.6 High resolution XPS study of an InAs sample, on which a thin layer of aluminum has been deposited. Measurements of the metallic and arsenide contributions to the In 4d XPS peak as a function of detector takeoff angle indicate that the metallic contribution arises from indium diffusion into the aluminum layer. (Source: H.W. Richter, J. Barth, J. Ghijsen, R.L. Johnson, L. Ley, J.D. Riley and R. Sporken, *J. Vac. Sci. Technol.* 1986;B4:900. Reprinted with permission.)

the In 4d peak is shown, from a sample consisting of an InAs crystal covered with a thin layer of aluminum. This peak, again a closely spaced doublet, shows about a 1 eV shift between metallic indium and the indium in InAs. The increased height of the metallic contribution at low detector polar angles indicates that this contribution is a near-surface one and arises from diffusion of indium into the aluminum overlayer.

XPS is also very useful in the analysis of organic layers, as the shifts in the carbon, oxygen, or other peaks present can help identify the chemical nature of the material. In the example shown in Figure 16.7, a polymer having the structure shown at the top of the figure yielded the carbon and oxygen peak shapes shown in the lower part of the figure. The two different oxygen atom environments and the three different carbon environments present in the structure lead to clearly resolved contributions to the observed XPS peaks. Moreover, in the case of carbon, the contribution from the six equivalent carbon atoms in the aromatic ring is essentially three times as large as the contributions from the two carbon atoms each in the COO and O–CH_2 environments.

Photodesorption

The desorption of adsorbed species can be induced by photon bombardment, just as it can by electron bombardment or surface heating. The study of the phenomena involved in photon-induced desorption has taken place largely since high energy, high intensity lasers and synchrotron radiation photon sources have become available, as the cross sections for photodesorption processes have been found to be small.

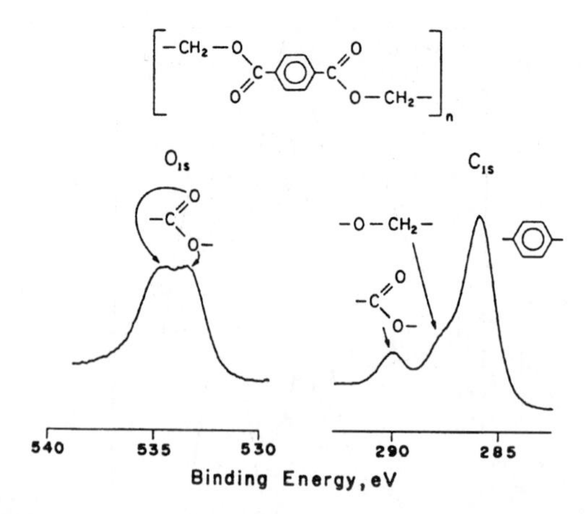

Figure 16.7 High resolution XPS study of a nylon-type polymer, showing contributions to the oxygen 1s and carbon 1s peaks from atoms in different chemical environments. (Source: W.M. Riggs and M.J. Parker, in: A.W. Czanderna, ed. *Methods of Surface Analysis*. Amsterdam: Elsevier Publishing Company, 1975, p. 103. Reprinted with permission.)

The desorption of surface species as a result of photon bombardment can arise from several different mechanisms. These may be categorized as direct heating of the surface by photon bombardment; indirect, or resonant, heating associated with resonant absorption of photons to cause vibrational excitation of adsorbed species; and photon-stimulated desorption processes, where the excitation process leads diretly to the desorption of an atom or molecule.

Consider first the direct excitation process. When studies using high energy, high intensity photon sources were first begun, it was hoped that surface analogs to gas-phase photochemistry could be developed. In the case of gas-phase photoexcitation, absorption of a photon leading to either vibrational or electronic excitation of a molecule is observed to cause fragmentation or ionization of the excited species. Because there is a sharp maximum in the photon absorption cross section for cases in which the photon energy is equal to the energy difference between a normally occupied and an excited state, specific vibrational modes or electronic excited states can be selectively excited by matching the photon energy to the excitation energy. Because different isotopes of a given atom lead to different vibrational frequencies when that atom is part of a larger molecule, it was hoped that laser excitation could be used to selectively desorb one isotopic species in preference to others, leading to a technique for isotope separation. For the case of adsorbed molecules, this hope was ill-founded. The problem is that, in this case, a molecule close to a solid surface has many pathways by which the excitation generated by photon absorption can be rapidly dissipated into the bulk of the solid, for example by phonon or plasmon excitation or by the generation of electron–hole pairs in the solid. In a few studies, true direct photodesorption arising from vibrational excitation has been observed, but these usually involve systems in which an inert "spacer" layer, for example, an adsorbed inert gas, has been used to impede energy transfer from the excited adsorbed species to the substrate.

The situation is somewhat different for the case of electronic excitation of surface species. In this case, the exciting radiation is usually synchrotron radiation in the soft x-ray range (10–100 eV). Desorption of oxygen ions from oxides has been observed to result from a mechanism similar to that involved in electron-stimulated desorption of ions from oxides, namely the Knotek–Feibelman mechanism. In both cases, the initiating event is the excitation of a surface atom by either electron- or photon-induced core level ionization. Relaxation of this initial core level vacancy by various Auger processes can lead to formation of an oxygen species that is unbound and which can leave the surface as a free ion. By analogy to the case of electron-stimulated desorption (ESD), this process is known as *photon-stimulated desorption* (*PSD*). Again, by analogy with the electron excited case, it is possible to obtain information on surface bonding geometry by measuring the angular distribution of the desorbed ions, in this case the technique is known as *photon-stimulated desorption ion angular distribution* (*PSDIAD*).

A much larger fraction of the results of photodesorption experiments can be explained in terms of the indirect, or resonant, heating mechanism. In this case, the initiating event is resonant photon absorption by a vibrational mode of an adsorbed molecule. Because the cross section for this process depends on matching the photon energy to the vibrational mode frequency, the energy input into the system has sharp maxima at the resonant frequencies. Thus, the process shows the strong photon frequency dependence typical of gas-phase photochemical processes. However, because of the rapid redistribution of the energy of the initial excitation into the phonon modes of the substrate, the desorption events that follow the initial excitation are essentially thermal desorption events driven by the local heating.

Desorbed species arising from these processes generally have kinetic energy distributions that are Maxwellian but at a temperature somewhat lower than the maximum temperature rise calculated from the measured energy input. The reasons for this temperature difference are not completely clear but may be associated with the dynamics of the desorption process or may simply be an indication that desorption occurred before the time of the maximum in surface temperature.

Laser-Induced Desorption

In a large majority of the studies involving the desorption of surface species under the influence of intense photon bombardment, the desorption process is due to direct heating of the substrate by absorption of the incident photons. Photon energies in these studies are generally in the infrared range and are readily absorbed at metal surfaces.

The desorption process in these cases, essentially a very rapid thermal desorption process, can be described analogously to the thermal desorption spectroscopy discussed in Chapter 13. The major difference in the present case is the heating rate, which can be as high as 10^{10} K/sec. This rapid heating rate has two major consequences. First, as mentioned for the case of indirect heating, the temperature of the desorbed species, as measured by time of flight techniques, is generally not the same as the calculated surface temperature rise. Second, the product distribution observed in the

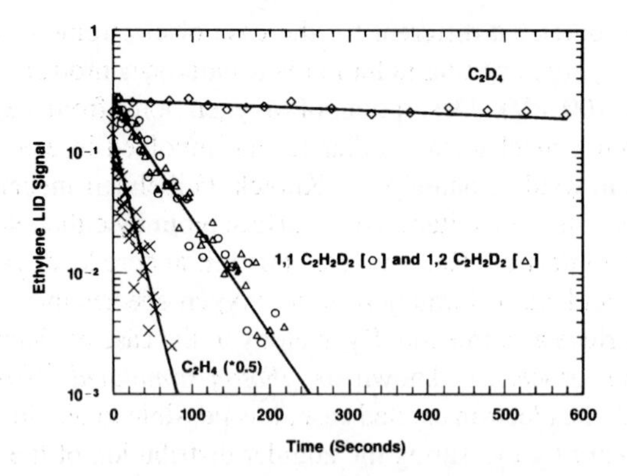

Figure 16.8 Laser-induced thermal desorption study of the rate of ethylene decomposition on an Ni (110) surface, showing the effect of isotopic substitution on reaction rate. (Source: F. Zaera and R.B. Hall, *Surface Sci.* 1987;180:1. Reprinted with permission.)

laser-induced desorption case may be very different from that observed in conventional thermal desorption measurements. The general result is that laser-induced desorption is much more likely to result in the desorption of the intact adsorbed molecule, as opposed to the desorption of molecular fragments or new species produced in a surface reaction commonly seen in conventional TDS or TPRS experiments. This last difference is associated with the relative speed with which intermolecular or intramolecular processes take place involving the adsorbed species. The general observation is that intramolecular rearrangement and the formation of new species by surface reaction are relatively slow, compared to breaking of the adsorptive bond between the adspecies and the surface.

The rapid desorption of adsorbed species due to direct heating of the surface is the basis of an emerging technique for the study of surface processes known as *laser-induced thermal desorption* (*LID* or *LITD*). The utility of this technique is based on the observation that the rapid heating associated with laser irradiation is most likely to produce desorption of the intact adsorbed species, as opposed to the enhancement of surface decomposition processes. This fact enables LID to be used as a probe of the surface concentration of a reactant species as a function of time. This technique has been used both to study the rate of surface decomposition reactions that do not produce gas-phase reaction products and to study the rate of surface diffusion of adsorbed species.

In the case of surface reaction kinetics, the system is set up so that a finely focused laser beam irradiates a small spot on the surface, on the order of 100 μm in diameter. The system is arranged so that the laser beam can be moved relative to the surface, so that a fresh spot is sampled with each laser pulse. If the surface is heated by conventional means to a temperature at which the reaction of interest is relatively rapid, the amount of reactant desorbed by successive laser pulses, as measured by a mass spectrometer in the system, gives a measure of the degree of reaction that has occurred. Figure 16.8 shows the results of such a study, for the case of the decomposition of various ethylene isomers on an Ni (110) surface. The decrease in the ethylene

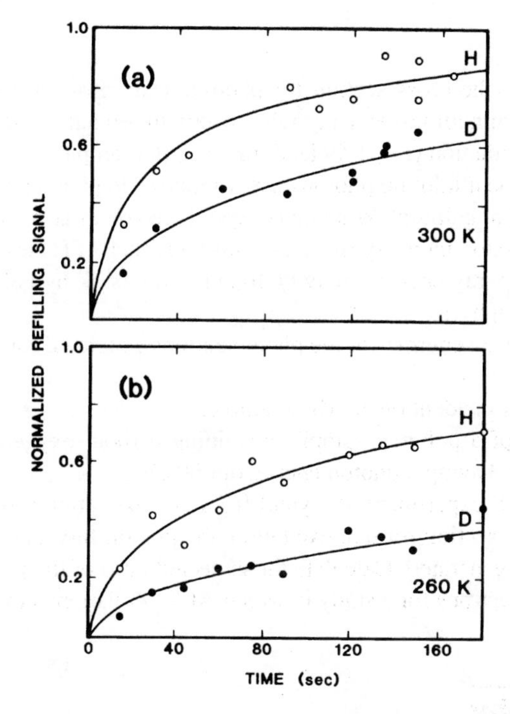

Figure 16.9 Surface diffusion of hydrogen and deuterium atoms on the Ru (001) surface. Curves indicate the rate of inmigration of adatoms into an area from which adatoms had been removed by laser-induced thermal desorption. (a) T = 300 K; (b) T = 260 K; (Source: C.K. Mak, J.L. Brand, B.G. Koehler and S.M. George, *Surface Sci.* 1987;188:312. Reprinted with permission.)

LID signal with time mirrors the decrease in unreacted ethylene on the surface. The results show clearly that the decomposition process follows first-order kinetics and that substitution of deuterium for hydrogen severely retards the reaction rate.

The experimental arrangement for surface diffusion studies is similar, except that in this case the laser irradiates the same spot on the surface with each successive pulse. After the surface is populated with the desired adsorbate, an initial pulse desorbs all of the material from the area of interest. Subsequent pulses, following increasing waiting periods, provide a measure of the amount of material that has diffused back into the depopulated area during the waiting period. (This experiment is essentially the inverse of the technique used to study carbon diffusion on platinum, discussed in Chapter 6.) Typical curves for the refilling process are shown in Figure 16.9, for hydrogen and deuterium diffusion on Ru (001). The curves represent the fit to the Fick's law calculation for the refilling process, showing the approach to complete refilling at long times.

These laser-induced desorption techniques offer an important addition to previous techniques for the study of surface processes, as they yield information in systems that do not produce spontaneous desorption of a product species in an isothermal experiment.

Problems

16.1. Calculate the cross section for photoionization of the carbon K shell by photons from an aluminum target x-ray tube, where the radiation has been filtered to permit only the K_α radiation ($E = 1.49$ keV) to reach the sample.

16.2. What is the minimum photon energy required to obtain photoelectron emission from the L shell of gallium? What metal would one use as an x-ray target to obtain maximum photoemission intensity from the gallium L shell? (Tables of electron binding energies and K_α x-ray energies may be found in the book by Feldman and Mayer listed in the Bibliography.)

16.3. Calculate the energies of the photoelectrons generated for the following situations:

a. Al K_α x-rays incident on an MgO sample.

b. Irradiation of a polymer sample containing carbon, oxygen and fluorine by synchrotron radiation having a photon energy of 600 eV.

16.4. In an XPS experiment, the yield from any excitation process will be proportional to the cross section for that excitation, the photon flux, and the escape depth of the photoelectron generated. Calculate the expected ratio of the peak heights for Ni 2p and Si 2p photoelectrons for a study in which Al K_α radiation is incident on an NiSi sample.

Bibliography

Anpo, M, and Matsuura, T, eds. *Photochemistry on Solid Surfaces*. Amsterdam: Elsevier, 1989.

Briggs, D, and Seah, MP. *Practical Surface Analysis by Auger and X-ray Photoelectron Spectroscopy*. New York: John Wiley & Sons, 1983.

Chuang, TJ. Laser-induced gas-surface interactions. *Surface Sci. Reports* 1983;3:1.

Chuang, TJ, Seki, H, and Hussla, I. Infrared photodesorption: vibrational excitation and energy transfer processes on surfaces. *Surface Sci.* 1985;158:525–552.

Cullity, BD. Elements of X-ray Diffraction, 2nd ed. Reading, Mass.: Addison-Wesley, 1978.

Ertl, G, and Kuippers, J. *Low Energy Electrons and Surface Chemistry*. Weinheim: Verlag Chemie International, 1974.

Feldman, LC, and Mayer, JW. *Fundamentals of Surface and Thin Film Analysis*. New York: North-Holland, 1986.

Friedhans'l, R. Surface structure determination by x-ray diffraction. *Surface Sci. Reports* 1989;10:3.

Ibach, H, ed. *Electron Spectroscopy for Surface Analysis: Topics in Current Physics*, Vol. 4. New York: Springer-Verlag. 1977.

Richtmyer, FK, Kennard, EH, and Cooper, JN. *Introduction to Modern Physics*, 6th ed. New York: McGraw Hill, 1969.

Siegbahn, K, Nordling, CN, Fahlman, A, Nordberg, R, Hamrin, K, Hedman, J, Johanson, G, Bergmark, T, Karlsson, SE, Lindgren, I, and Lindberg, B. *ESCA, Atomic, Molecular and Solid State Structure Studied by Means of Electron Spectroscopy*. Uppsala: Almquist & Wiksells, 1967.

Wandelt, K. Photoemission studies of adsorbed oxygen and oxide layers. *Surface Sci. Reports* 1982;2:1.

PART IV

Crystal Growth

CHAPTER 17

Crystal Nucleation and Growth

The nucleation and growth of crystalline solids from the vapor is a process that can be understood in terms of the kinetics of formation of the initial small particles of a solid phase by the clustering of atoms in a gas phase, followed by the interaction of further vapor atoms with the growing surface. The discussion of these processes will use the concept of the stability of small clusters relative to a bulk condensed phase, the concept of adsorptive equilibrium, the concept of surface diffusion of adatoms over the surface, and the atomistic model of the crystalline surface developed in Chapter 1.

Homogeneous Nucleation

In order to form a bulk condensed phase from a monatomic vapor phase, one must first consider the process of forming a small but stable cluster having the approximate density and atomic arrangement of the desired condensed phase. One can conceive of this process as taking place by a series of bimolecular collisions, each of which increases the size of the growing cluster by one molecule, thus:

$$\begin{gathered} B_1 + B_1 \rightarrow B_2 \\ B_2 + B_1 \rightarrow B_3 \\ \bullet \\ \bullet \\ \bullet \\ B_{i-1} + B_1 \rightarrow B_i. \end{gathered} \qquad 17.1$$

The net rate of such a process, at steady state, will be set by the steady state concentration of clusters of each size. This, in turn, is set, as we will see later in this section, by the energy of formation of the cluster relative to the monomolecular vapor phase.

The free energy of formation of a cluster of size i from individual molecules could, in principle, be calculated by summing the pairwise interactions that arise in the course of bringing the i molecules required to form the cluster close together. In practice, this calculation is not practical, and approximate methods must be used. A commonly used approximation is to assume that the clusters can be described in terms of macroscopic thermodynamic properties and that the total free energy of formation can be split into contributions from surface and volume free energy changes. For the case at hand, the net free energy change for the process

$$i\,B_1 \rightarrow B_i \tag{17.2}$$

is

$$\Delta G_i = \int_o^i \mu_i di - \int_o^i \mu_1 di. \tag{17.3}$$

For single molecules

$$\mu_1 = \mu_0(T) + kT \ln p. \tag{17.4}$$

For clusters, using LaPlace's equation, and recalling the arguments of Chapter 4,

$$\mu_i = \mu_\infty + \left(\frac{\mathcal{V}}{N_{Av}}\right)\left(\frac{2\gamma}{r}\right) \tag{17.5}$$

or

$$\mu_i = \mu_o(T) + kT \ln p_o + \left(\frac{\mathcal{V}}{N_{Av}}\right)\left(\frac{2\gamma}{r}\right). \tag{17.6}$$

Furthermore, since the cluster radius, r, is related to the cluster size, i, by

$$i = \frac{4\pi r^3}{3\Omega}, \tag{17.7}$$

where $\Omega = \mathcal{V}/N_{Av}$, we may write

$$\Delta G_i = \int_o^r \mu_i \left(\frac{4\pi}{\Omega}\right) r^2 dr - \int_o^r \mu_1 \left(\frac{4\pi}{\Omega}\right) r^2 dr \tag{17.8}$$

$$= \left(\frac{4\pi r^3}{3\Omega}\right)[kT \ln p_o - kT \ln p] + 4\pi r^2 \gamma, \tag{17.9}$$

or, defining ΔG_v, the change in bulk free energy accompanying condensation, as

$$\Delta G_v = \frac{-kT}{\Omega} \ln\left(\frac{p}{p_o}\right), \tag{17.10}$$

where p_0 is the equilibrium vapor pressure of the condensed phase,

$$\Delta G_i = 4\pi r^2 \gamma + \left(\frac{4\pi r^3}{3}\right) \Delta G_v. \tag{17.11}$$

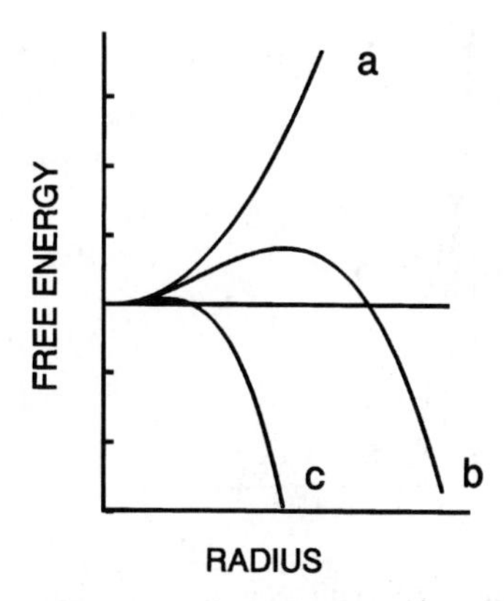

Figure 17.1 Free energy of formation of small clusters as a function of cluster size, for various values of the volume free energy change accompanying condensation, namely $\Delta G_v = 0$ (a), $\Delta G_v < 0$ (b), and $\Delta G_v << 0$ (c)

This final expression has in it a term associated with the surface energy of the cluster and a term associated with the volume free energy change, as mentioned earlier.

One can look at the form of this relation for ΔG_i as a function of i (or r) for various possible situations, as shown in Figure 17.1. For the case $\Delta G_v \geq 0$ ($p \leq p_0$), ΔG_i increases monotonically with increasing i, indicating that larger clusters are increasingly unstable. For $\Delta G_v < 0$ ($p > p_0$) ΔG_i will go through a maximum, and large clusters will be stable relative to the vapor phase. As shown in the figure, as the ratio p/p_0 (the supersaturation) increases, both the maximum ΔG_i and the value of i associated with this maximum decrease. The significance of this is that the cluster having the maximum ΔG_i represents a critical point in the condensation process: addition of one more molecule will lead to a cluster whose free energy decreases monotonically with further growth. This size, i^*, thus represents the critical nucleus for the condensation process.

The size of this critical nucleus may be found by differentiation of Equation 17.11 to be

$$i^* = -\left(\frac{32\pi}{3\Omega}\right)\left(\frac{\gamma}{\Delta G_v}\right)^3, \tag{17.12}$$

and its free energy of formation

$$\Delta G_i^* = \frac{16\pi\gamma^3}{3\Delta G_v^2}. \tag{17.13}$$

Having determined ΔG_i, one must now proceed to relate this to n_i, the metastable equilibrium concentration of clusters of size i. Here the argument is similar to that used in Chapter 1 to determine the equilibrium concentrations of surface adatoms or vacancies. Although the free energy of formation of the clusters is positive, for small sizes,

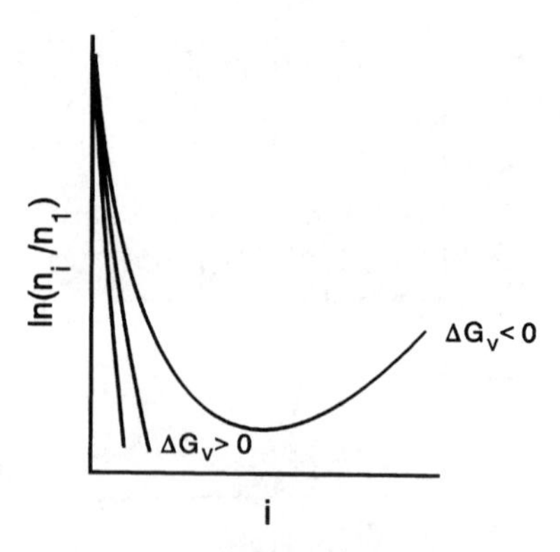

Figure 17.2 Equilibrium cluster concentrations as a function of cluster size, for various values of the free energy charge accompanying condensation.

there is a positive entropy of mixing associated with mixing the various sized clusters with the single molecules in the vapor. Thus, for the cluster formation process

$$\Delta G_{sys} = n_i \Delta G_i + kT\left[n_i \ln\left(\frac{n_i}{n_i + n_1}\right) + n_1 \ln\left(\frac{n_1}{n_i + n_1}\right)\right]. \qquad 17.14$$

At equilibrium, the derivative $\partial \Delta G_{sys}/\partial n_i$ must equal zero for all values of n_i. Assuming that $\sum_i i\, n_i << n_1$ leads to

$$n_i = n_1 \exp\left(\frac{-\Delta G_i}{kT}\right). \qquad 17.15$$

The form of the relation between n_i and i is shown in Figure 17.2. For $\Delta G_v \geq 0$, the curves are monotonically decreasing. For $\Delta G_v < 0$ there is a minimum in the concentration at $i = i^*$. Note that as ΔG_v becomes more negative, i^* decreases and n_{i^*} increases.

The next question is how to relate these cluster concentrations to the rate of formation of stable, growing droplets or crystallites, that is, to the nucleation frequency. This process has been treated by several approaches. The simplest, known as the Vollmer–Weber theory, simply assumes that the equilibrium concentration of critical sized clusters is maintained at all times and that once a cluster grows by one incremental molecule to become a cluster of $(i^* + 1)$, it will continue to grow without limit and become a stable particle. In this case, the nucleation rate will be simply the rate at which single molecules collide with clusters of size i^*, or

$$J = A^* \omega\, n_{i^*} \qquad 17.16$$

where A^* is the surface area of a cluster of size i^*, $4\pi r^{*2}$, ω is the impingement rate for condensation and is given by

$$\omega = \alpha_c \frac{p}{(2\pi mkT)^{1/2}}, \qquad 17.17$$

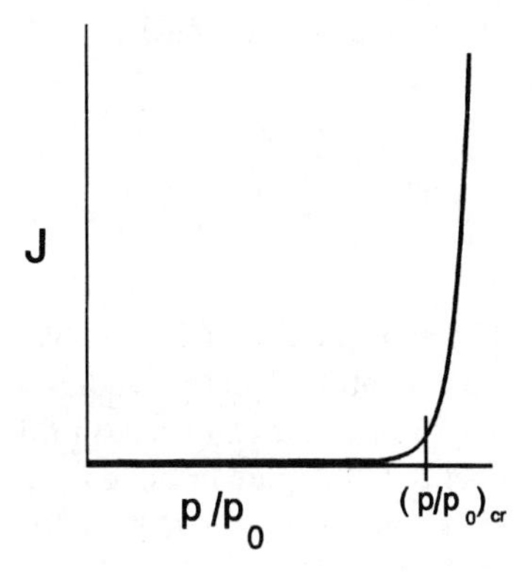

Figure 17.3 Nucleation rate as a function of supersaturation ratio, p/p_0, according to the Vollmer–Weber treatment.

in which α_c is the fraction of collisions that lead to sucessful incorporation into the cluster. The final expression for the nucleation rate is thus

$$J = \alpha_c(4\pi r^{*2})\left[\frac{p}{(2\pi mkT)^{1/2}}\right]n_1 \exp\left(\frac{-\Delta G_{i^*}}{kT}\right). \qquad 17.18$$

Note that this leads to a positive value of J for all negative values of ΔG_v but one that approaches zero as ΔG_v approaches zero.

The Vollmer–Weber theory is quite straightforward but has a number of shortcomings that limit its usefulness. The chief among these is the assumption that clusters of a few molecules can be described in terms of macroscopic values of surface and volume free energies. This problem is exacerbated in the case of crystal nucleation by the known dependence of surface energy on crystallographic orientation. Some attempts have been made to circumvent this problem, either by assuming a dependence of γ on r or by carrying out quantum mechanical calculations of cluster energy as a function of size. In some cases, these latter calculations have indicated oscillatory behavior in the energy vs. size relation, which have been observed in some experimental studies to be discussed in this section.

A second major shortcoming of the treatment presented is that it does not allow for depletion of the population of critical sized clusters due to their conversion to stable growing nuclei. Calculations that account for this effect, and for the possibility of dissolution of supercritical nuclei, lead to a correction factor to the Vollmer–Weber theory of a factor of about 10^{-2} at steady state.

Finally, consider the predictions of the theory developed as far as what one can see in a practical experiment. If one plots the expected value of J as a function of p/p_0, as shown in Figure 17.3, one observes a very low nucleation rate at small supersaturations, followed by a rapid rise over a very narrow range of p/p_0. As a practical matter, this rise is so abrupt that it appears that a "critical" value of the supersaturation is required to cause nucleation. If one assumes that the value $J = 1$ per cm^3 per sec lies within this critical region, then one may determine a critical supersaturation for a measurable rate of nucleation as

$$\ln\left(\frac{p}{p_o}\right)_{cr} = \left(\frac{16\pi\gamma^3\Omega^2}{3k^3 \ln B}\right)^{1/2} \left(\frac{1}{T}\right)^{3/2}, \qquad 17.19$$

where B is a constant of order 10^{65} cm^{-3}.

Experimental studies of homogeneous nucleation have been aimed primarily at verification of this, or similar, relations. Early studies of this process generally involved use of the so-called *Wilson cloud chamber*. In this device, a volume of gas containing the condensible species, either pure or mixed with a carrier gas, is suddenly cooled by adiabatic expansion. As discussed in Chapter 8, the temperature drop in such a process will be

$$T_f = T_i \left(\frac{V_i}{V_f}\right)^{(\gamma-1)} \qquad 17.20$$

In principle, then, one can carry out this process at increasing expansion ratios to determine the supersaturation necessary to cause rapid nucleation. More recent studies have generally used a free jet expansion to produce the adiabatic cooling and resultant supersaturation, again as described in Chapter 8. The net result of these studies is that the prediction of an apparent critical supersaturation for condensation has invariably been observed but that quantitative agreement with any of the theoretical treatments is lacking. Studies with free jet expansions can also be used, if the expansion "freezes" before the onset of bulk condensation, to investigate the relation between cluster size and cluster concentration. Results of these studies indicate that certain cluster sizes are particularly stable, as they appear in high concentrations relative to adjacent sized clusters.

Heterogeneous Nucleation

Hetrogeneous nucleation is the process of the formation of one condensed phase on the surface of a second, preexisting condensed phase. This process is commonly seen in the growth of thin films on substrates of a different material, or the condensation of steam on a boiler wall.

This process may be treated similarly to the homogeneous nucleation case treated in the previous section. The principal difference in this case is that one must treat the formation of a cluster in contact with a preexisting surface. In this case, making the same assumptions as before concerning the applicability of macroscopic thermodynamic properties to very small clusters, one arrives at a model for the nucleus as shown in Figure 17.4. The cluster is a segment of a sphere, with the angle θ determined by the Young equation, as described in Chapter 3. Going through the same argument as used previously, assuming clusters to be formed on the surface by a series of bimolecular collisions, one arrives at an expression for ΔG_i for this case of

$$\Delta G_i = \left(\frac{4\pi r^3}{3}\right) f(\theta)\Delta G_v + 4\pi r^2 f'(\theta)\gamma_{nv}, \qquad 17.21$$

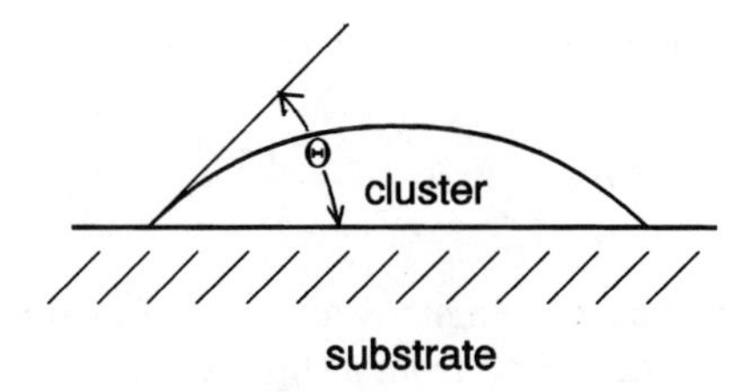

Figure 17.4 Schematic view of the formation of a small cluster on an existing solid surface. θ is the contact angle between the cluster and the surface.

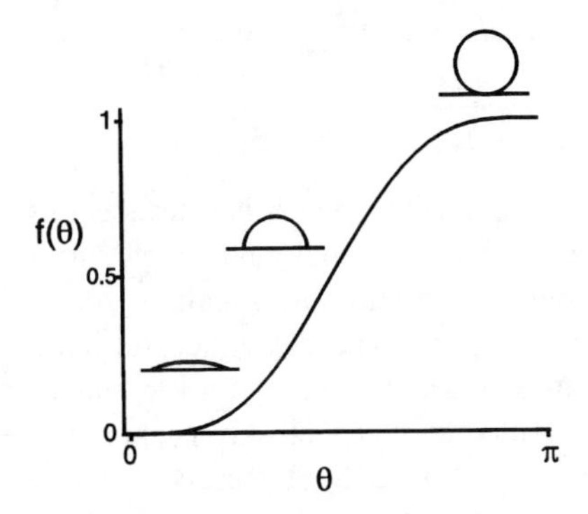

Figure 17.5 Variation of the efficiency factor for nucleation catalysis by a solid surface, $f(\theta)$, with the contact angle, θ.

where $f(\theta) = (2 + \cos\theta)(1 - \cos\theta)^2/4$ and γ_{nv} is the surface energy of the cluster. From this point, one can proceed as before to determine

$$\Delta G_{i*} = \frac{16\pi\gamma^3 f(\theta)}{3\Delta G_v^2}. \tag{17.22}$$

Note that this expression differs from that calculated for the case of homogeneous nucleation only by the factor $f(\theta)$. This factor is shown as a function of θ in Figure 17.5. For small values of θ, that is, good wetting, $f(\theta)$ is small. This greatly reduces ΔG_{i*}, and consequently $(p/p_0)_{cr}$.

The predominant mechanism for cluster growth in this case can be shown to be the addition of molecules to the cluster by surface diffusion from the adlayer. Thus, the final expression for the nucleation rate in this case is

$$J = L^* \, \omega \, n_{i*}, \tag{17.23}$$

where L^* is the perimeter length of the cluster and ω is the surface diffusion related impingement rate

$$\omega = n_a a \nu_0 \exp\left(\frac{-\Delta G_{sd}}{kT}\right). \tag{17.24}$$

The final expression for the nucleation rate is thus

$$J = L^* a v_o n_a^2 \exp -\left(\frac{\Delta G_{sd} + \Delta G_{i^*}}{kT}\right). \qquad 17.25$$

As in the previous case, there have been many attempts to verify the predictions of this treatment. For the case of condensation from a slightly supersaturated vapor, as would be encountered in the condenser of a boiler, the results are in reasonable agreement with theory. For the case of thin film growth by deposition from a vapor phase, agreement is much worse. This is most probably due to the very high imposed supersaturation in this case, which leads to a calculated value of $i^* \leq 1$.

Crystal Growth—Surface Model

In discussing the growth of an existing crystal surface, this section will use, from Chapter 1, the definitions of singular, vicinal, and rough surfaces based on the TLK model, and the concept of defect formation on singular surfaces, in particular the concept of equilibrium surface adatom and vacancy concentrations.

Recall that singular surfaces are found only for low-index orientations of the surface; that is, for surfaces oriented so that a closely packed, or nearly closely packed, plane of atoms is exposed at the surface. Such surfaces are essentially atomically flat and, at 0 K would look roughly as shown in Figure 1.1. At finite temperatures, the free energy of the system, $F = E - TS$, will be lowered by forming surface vacancy–adatom pairs as shown in Figure 1.20. The entropy increase associated with this process outweighs the reduction in total crystal binding energy for small concentrations of such defects. Thus, they are present on the surface at equilibrium in a concentration of

$$c_v = c_a = \exp\left(\frac{-\Delta G_{va}}{kT}\right), \qquad 17.26$$

where c_v and c_a are the fraction of surface sites associated with vacancies and adatoms, respectively, and ΔG_{va} is the free energy of formation of such a defect pair.

Recall, too, that vicinal surfaces are associated with crystallographic orientations finitely removed from singular orientations. Such surfaces can be represented, at 0 K, by the picture shown in Figure 1.2a. Here, the atomically flat terraces associated with the singular orientations are separated by ledges of monatomic height. The orientation of the surface relative to the closest singular orientation is given by the angle θ. The spacing between ledges is given by

$$S = \frac{h}{\tan\theta}, \qquad 17.27$$

where h is the height of a single ledge. If a surface is misoriented from the singular orientation in two directions, the ledges will have regularly spaced kinks to provide for the angle between the singular surface and the vicinal surface in this other direction. This is shown in Figure 1.2b. Again, as in the case of the singular surface, at finite temperatures the terraces will contain a finite concentration of adatom–vacancy pairs, and in addition there will be disorder along the ledges in the form of ledge vacancies and ledge adatoms, as shown in Figure 1.21. The concentration of these defects at equilibrium is given by

$$c_{la}^2 = c_{lv}^2 = c_k^4 = \exp\left(\frac{-\Delta G_{lva}}{kT}\right), \tag{17.28}$$

where c_{la}, c_{lv}, and c_k are the ledge site fractions of ledge adatoms, ledge vacancies, and kinks, respectively, and ΔG_{lva} is the free energy of formation of this defect pair.

The concentrations of these various defects can be estimated using the nearest neighbor model for the bond strength. For many crystals, near the melting point, typical concentrations are 5×10^{-5} site fraction for surface adatoms, and 0.2 ledge site fraction for kinks. Note that these defects represent intermediate configurations that an atom may pass through in the course of passing from the vapor to the bulk of the crystal, or vice versa, in the process of crystal growth or evaporation.

The Ideal Growth Rate

The impingement rate of molecules of a gas onto a surface exposed to this gas is given by

$$I = \frac{p}{(2\pi mkT)^{1/2}}. \tag{17.29}$$

This is a completely general relation and holds for all systems at all times, irrespective of whether the surface involved is growing, evaporating, or anything else.

Consider the case of a system composed of a crystal surface exposed to its own vapor, in thermodynamic equilibrium, and note that, since no net growth or evaporation is taking place at equilibrium,

$$E = I = \frac{p_o}{(2\pi mkT)^{1/2}}, \tag{17.30}$$

in which p_0 is the equilibrium vapor pressure of the solid. That is, at equilibrium, the evaporation flux equals the impingent flux.

If the material in the vapor phase is now pumped away, so that the impingement rate is reduced, unless the evaporation process changes the atomic configuration at the surface, the evaporation rate is unchanged. That is, the impingement of molecules from the vapor and the evaporation of molecules from the crystal surface are independent processes. In such a case, the net growth or evaporation rate, J_c or J_v, will be given by the difference between the actual impingement rate, which is controlled by the pressure in the vapor phase, and the equilibrium evaporation rate, which is proportional to the equilibrium vapor pressure of the crystal. That is

$$J_c \text{ (or } J_v) = \frac{p - p_o}{(2\pi mkT)^{1/2}}. \tag{17.31}$$

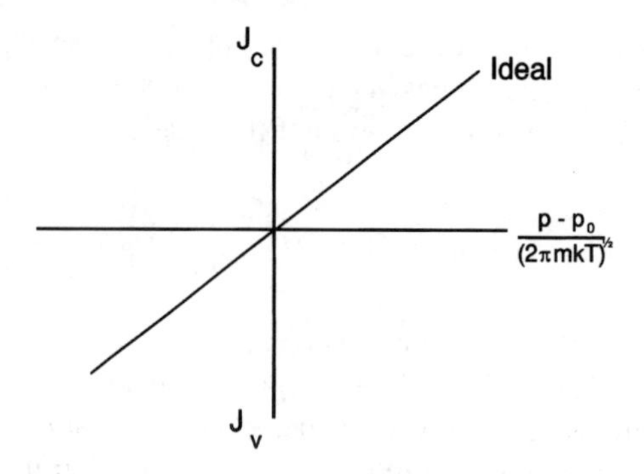

Figure 17.6 Ideal crystal growth or evaporation rate as a function of $(p - p_0)$.

The rate defined by this equation is known as the *ideal growth or evaporation rate* and represents the maximum rate at which a crystal can grow or evaporate at a given value of p and T. This rate is represented graphically in Figure 17.6. In this diagram net growth takes place in the first quadrant of the figure, and net evaporation in the third quadrant.

Consider whether this type of behavior will in fact be observed in real systems, beginning with the process of growth and evaporation at the clean surface of a monatomic liquid. In this case, any one position on the surface is equivalent to any other, and the configuration of the surface is not changed when an atom condenses or evaporates. Moreover, if the liquid is monatomic, there is no configurational problem associated with the evaporation or condensation processes. That is, the atom doesn't have to be oriented in any particular direction in order for these processes to occur. One would thus expect that the ideal relation developed above would be obeyed. This is, in fact, what is observed in practice.

Before leaving this case of the liquid surface, consider the atomic mechanism of the evaporation process, so that it can be compared later with the process on a solid surface. Using absolute reaction rate theory one may write

$$E = \nu^* n^* , \qquad 17.32$$

in which ν^* is the decomposition frequency of the activated complex formed in the evaporation process, and n^* is the concentration of this complex, which can be written as

$$n^* = n_L\left(\frac{f^*}{f}\right)\exp\left(\frac{-\Delta H_{vap}}{kT}\right), \qquad 17.33$$

in which n_L is the atom number density in the surface of the liquid, f^* and f are the molecular partition functions for the activated and equilibrium states, respectively, and ΔH_{vap} is the enthalpy change for the vaporization process. Inserting this expression into the expression for E,

$$E = \nu^* n_L\left(\frac{f^*}{f}\right)\exp\left(\frac{-\Delta H_{vap}}{kT}\right) \qquad 17.34$$

or

$$E = n_L \nu \exp\left(\frac{-\Delta H_{vap}}{kT}\right), \tag{17.35}$$

in which $\nu = \nu^* (f^*/f)$ is essentially the vibrational frequency of surface atoms perpendicular to the surface. This expression must of course be equal to the value of E developed from consideration of the vaporization–condensation equilibrium. Thus,

$$n_L \nu \exp\left(\frac{-\Delta H_{vap}}{kT}\right) = \frac{p_o}{(2\pi mkT)^{1/2}} \tag{17.36}$$

or

$$p_o = K_v T^{1/2} \exp\left(\frac{-\Delta H_{vap}}{kT}\right), \tag{17.37}$$

in which

$$K_v = (2\pi mk)^{1/2} n_L \nu. \tag{17.38}$$

This is the familiar phenomenological expression for the equilibrium vapor pressure.

Growth of "Rough" Solid Surfaces

Now return to consideration of the growth of crystalline surfaces. In the TLK model for crystal surfaces, rough surfaces were mentioned but not described. A *rough surface* would be one that was sufficiently disordered, or sufficiently rough on a molecular scale due to the structure of the crystal involved, that no delineation of surface areas into terraces separated by ledges was possible. On such a surface, all sites would effectively be kink sites and thus present effective sites for the incorporation of impinging atoms directly into the crystal lattice. For such a surface, one would expect ideal growth and evaporation behavior, as in the case of the liquid. However, there is some question as to whether such rough surfaces do exist on any real crystal, and discussion of this case has been included for completeness only.

Growth of Perfect Vicinal and Singular Surfaces

The TLK model can be used to describe the growth and evaporation process on perfect singular and vicinal surfaces. Consider first a vicinal surface on a crystal of finite size, as shown in Figure 17.7. Growth on such a surface proceeds by a sequence of steps involving adsorption from the vapor to form a surface adatom, diffusion of this adatom on the surface to a kink site, and incorporation into the crystal at the kink site. The reverse sequence is observed for evaporation.

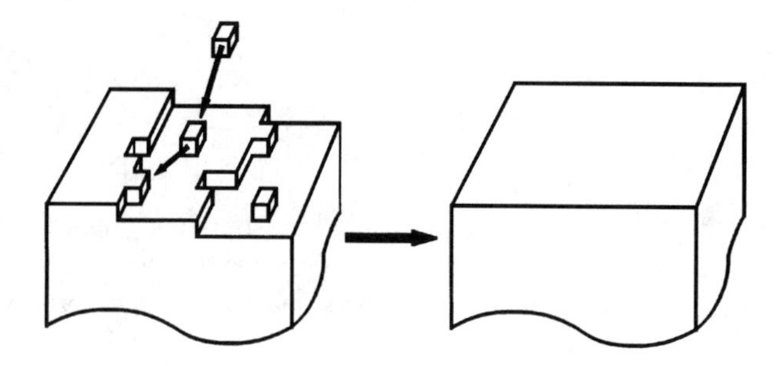

Figure 17.7 Schematic picture of the crystal growth process on a perfect vicinal surface.

The critical consideration in the operation of this growth mechanism is that there must be a high density of kink sites on the crystal surface in order to accommodate the surface adatoms. The necessary and sufficient condition for having a high kink density is the presence of ledges on the crystal surface, as the concentration of kinks due to thermal disordering of the ledges will be large enough at moderate temperatures to accommodate the adatom flux. The problem that arises is that on a vicinal surface, as shown in Figure 17.7, the ledges initially present on the surface will rapidly grow out to the edge of the crystal, no more ledges will be present on the resulting perfect singular surface, and consequently, no more potential growth sites exist. A mechanism is thus required for generating new ledges on this perfect singular surface.

To see what the required mechanism is, consider the situation on a perfect singular surface exposed to a vapor in which $p > p_0$. On such a surface there will be a population of surface adatoms, in equilibrium with the vapor but not with the crystal surface (due to the lack of growth sites). When this adlayer has formed, there will be no net growth rate, as $E = I$. The surface adlayer concentration, n_a, will be greater than that in equilibrium with the crystal surface, $n_{a_{eq}}$. The critical point in this situation is that although the adlayer cannot come into equilibrium with the crystal surface, it can come to equilibrium with small clusters of adatoms. This equilibrium is characterized in the same way as the equilibrium between a vapor and small three-dimensional particles, as discussed in the treatment of the nucleation process. For the present case, we may write the relation among μ, γ, and cluster size as

$$\mu_c - \mu_0 = -\Omega\gamma K \qquad 17.39$$

for the chemical potential in the adlayer relative to the chemical potential in the bulk crystal, where $\Omega = \mathcal{V}/N_{Av}$ is the volume per atom, and, for the case of disc-shaped clusters of monatomic height, h, $\gamma = \epsilon/h$, and K, the principal curvature of the disc edge is $1/r^*$. Here ϵ is the edge energy of the disc (essentially a surface energy per centimeter of perimeter) and r^* is the radius of the cluster which is in equilibrium with the existing pressure in the vapor phase. Making these substitutions,

$$\mu_c - \mu_o = \frac{\Omega\varepsilon}{hr^*}. \qquad 17.40$$

One can also express μ and μ_0 in terms of the vapor pressure of the crystal and the actual pressure in the vapor phase as

$$\mu = \mu_o + kT \ln\left(\frac{p}{p_o}\right) \tag{17.41}$$

or

$$\mu_c - \mu_o = kT \ln\left(\frac{p}{p_o}\right) = -\Delta G_v \Omega, \tag{17.42}$$

where $\Delta G_v = -kT/\Omega \ln (p/po)$ is the difference in Gibbs free energy between vapor and crystal, per unit volume of crystal. Equating these two expressions for $\mu - \mu_0$,

$$\frac{\Omega\varepsilon}{hr*} = -\Delta G_v \Omega \tag{17.43}$$

or

$$r* = \frac{-\varepsilon}{h\Delta G_v}. \tag{17.44}$$

These clusters of radius $r*$ are in unstable equilibrium with the adatom population. If the disc grows slightly, μ_c decreases and the disc becomes stable relative to the adlayer and grows further. If the disc shrinks slightly, μ_c increases and the disc becomes unstable relative to the adlayer and shrinks further, just as in the case of small droplets in a bulk vapor. The discs of size $r*$ are thus *critical nuclei* for the formation of a new layer on the crystal surface.

In order to determine the growth rate that will occur by this mechanism, which is called *two-dimensional nucleation and growth,* we must determine the rate of formation of stable clusters, that is, of clusters one atom larger than the $r*$ cluster. This will be the rate at which clusters of radius $r*$ grow by the addition of one incremental atom from the adlayer, which is given by

$$J = n_{r*} \omega Z, \tag{17.45}$$

in which J is the rate of stable cluster formation, $nr*$ the equilibrium surface concentration of clusters of $r*$, ω the frequency of addition of single adatoms to these clusters and Z is a nonequilibrium factor that accounts for the fact that if J is finite, the actual cluster population will be less then the equilibrium due to continued cluster growth. The equilibrium cluster concentration is given by

$$n_{r*} = n_a \exp\left(\frac{-\Delta G*}{kT}\right) \tag{17.46}$$

in which $\Delta G*$ is the Gibbs free energy of formation of a cluster from individual adatoms. If one assumes that the material in the cluster has the same volume free energy as the bulk of the crystal, then one may write that ΔG_i, the free energy of formation of a cluster of arbitrary size, r, is

$$\Delta G_i = 2\pi r \epsilon + \pi r^2 h\, \Delta G_v, \tag{17.47}$$

or, for clusters of size r^*, using $r^* = -\epsilon / h\, \Delta G_v$,

$$\Delta G^* = \frac{-\pi \varepsilon^2}{h \Delta G_v}. \tag{17.48}$$

The factor ω is simply the product of the probability that an adatom is adjacent to a cluster, which is

$$n_a\,(2\pi r^*)\,a, \tag{17.49}$$

and the jump frequency for the surface diffusion, which is

$$\nu_d \exp\left(\frac{-\Delta G_{sd}}{kT}\right). \tag{17.50}$$

In these equations, a is the jump distance for surface diffusion, ν_d is a surface vibrational frequency and ΔG_{sd} the activation free energy for the surface diffusive jump.

Making these substitutions,

$$J = Z(2\pi r^*) a\, n_a^2\, \nu_d \exp\left(\frac{-\Delta G_{sd}}{kT}\right) \exp\left(\frac{-\Delta G^*}{kT}\right). \tag{17.51}$$

It is found in practice that the preexponential part of this expression does not differ greatly from system to system and is on the order of e^{65} for most metals. The factor that exerts the predominant influence on J with changing p and T is ΔG^*, and more particularly the term $\Delta G_v = -kT/\Omega \ln (p/p_0)$, which is contained in ΔG^*. As was the case for three-dimensional nucleation discussed earlier in this chapter, the value of J defined above is small at values of (p/p_0) close to unity and rises abruptly at some critical value of this parameter, giving rise to what appears to be a *critical supersaturation* for the onset of two-dimensional nucleation. One may rearrange the expression for J, to put it in terms of $(p/p_0)_{crit}$, which is contained in ΔG_v, by assuming that the value $J = 1$ per cm^2 per sec lies in the range of (p/p_0) in which J is changing very rapidly, to obtain

$$\begin{aligned}(p / p_o)_{crit} &= \exp\left(\frac{\pi \varepsilon^2 \Omega}{65 h k^2 T^2}\right) \\ &= \exp\left(\frac{\pi h \Omega \gamma^2}{65 k^2 T^2}\right).\end{aligned} \tag{17.52}$$

The value of $(p/p_0)_{crit}$ is thus strongly dependent on both γ and T, with large values of γ and low temperatures being associated with large values of $(p/p_0)_{crit}$.

It now remains to relate this two-dimensional nucleation rate to the growth rate

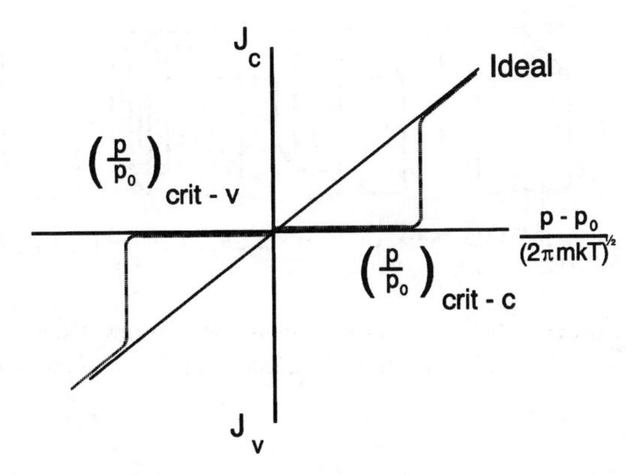

Figure 17.8 Growth or evaporation rates for two-dimensional nucleation and growth compared to the corresponding ideal rates.

of the crystal. The ratio of the observed growth rate in any case to the ideal growth rate for the same temperature and impingement rate is simply equal to the probability that an atom which strikes the surface will be adsorbed and will migrate over the surface far enough while it is adsorbed that it will encounter a growth site. For most cases, when an atom strikes the surface of a crystal of the same material, it will be adsorbed. Thus, the question of growth rate reduces to the question of the likelihood of finding a growth site. This probability is equal to the ratio of the mean distance an adatom travels over the surface by surface diffusion, $\bar{x}$, to the mean distance between ledges on the surface, λ. If λ is large relative to x, most adatoms will desorb before reaching ledges, and the growth rate will be low. In the alternative case, $x >> \lambda$, almost all adatoms will reach kink sites, and the growth rate will approach the ideal rate. It will not be necessary here to develop a quantitative relationship between J and λ but merely to use the fact that J is a strong function of p/p_0 to state that the growth rate will be very small (approximately zero) below $(p/p_0)_{crit}$ but will rise sharply toward the ideal rate as $(p/p_0)_{crit}$ is exceeded. The growth rate vs. p/p_0 may thus be plotted on the same type of curve as used for plotting the ideal growth rate, as shown in Figure 17.8. Note the sharp increase in growth rate at $(p/p_0)_{crit}$.

This behavior has been observed in practice but only in a few systems. It turns out that very few crystal surfaces are perfect, and the presence of certain types of imperfection allows growth at supersaturations much lower than $(p/p_0)_{crit}$. In some cases, though, crystallites can be formed that are perfect in one or two dimensions and imperfect in the remaining dimensions. In such cases, these crystals are observed to grow only in the imperfect directions at supersaturations below $(p/p_0)_{crit}$ and to thicken more or less uniformly at higher pressures.

The evaporation of perfect singular crystal surfaces may be treated exactly the same as growth, as the same problems are involved. In this case, the problem is one of creating ledges that can serve as sources for evaporating adatoms, and again these ledges can be formed by the two-dimensional nucleation of monolayer "holes" in the surface. The mathematical description in this case is essentially identical to that for the

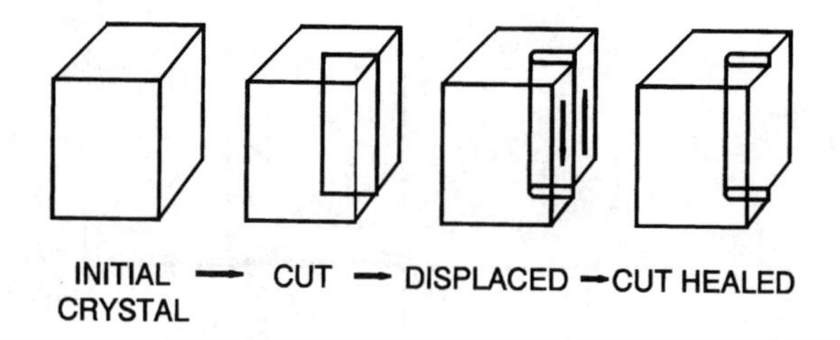

Figure 17.9 Conceptual process of the formation of a screw dislocation in a crystal, showing the development of a step on the crystal surface at the point of emergence of the dislocation line.

case of growth. The result of such a calculation is that there is a critical undersaturation for the nucleation of holes, given by

$$\left(\frac{p}{p_0}\right)_{crit} = \exp\left(\frac{-\pi\varepsilon^2\Omega}{B'hk^2T^2}\right) \qquad 17.53$$

in which B' is a constant on the order of 60–80. The overall evaporation rate is related to the rate of hole nucleation in the same way that the overall growth rate is related to disc nucleation. Thus the observed relation for growth rate vs. (p/p_0) will have the same form, as is shown in the third quadrant of the graph in Figure 17.8.

Growth of Imperfect Singular and Vicinal Surfaces

The treatment of crystal growth by two-dimensional nucleation predicts observable growth rates only at finite values of (p/p_0). In practice, however, most crystals are observed to grow at much smaller values of (p/p_0). This observation has been explained by Burton, Cabrera, and Frank [1951] in terms of a model involving lattice defects in the crystal as a self-perpetuating source of the ledges required for growth.

The defect involved is called a screw dislocation. It may be formed, conceptually, by taking a perfect crystal, cutting halfway through it, shearing one half of the cut surface relative to the other to get a displacement of about one interatomic distance, and rejoining the cut halves. This process is shown schematically in Figure 17.9.

Consider now how the configuration associated with the emergence of this defect at the crystal surface can provide a self-perpetuating growth site. Figure 17.10a shows a top view of a crystal of finite size containing a screw dislocation. When this surface is exposed to the vapor, atoms may adsorb, migrate over the surface, and attach themselves at the ledge associated with the site of emergence of the dislocation, as shown in Figure 17.10b. This still leaves a ledge available for further growth. The addition of succeeding increments of new material is shown in Figures 17.10c, d, e, and f. The important points to note are that the ledge always renews itself, and that, as growth proceeds, the initially straight ledge winds up into a spiral.

To determine the overall growth rate associated with this mechanism, one may

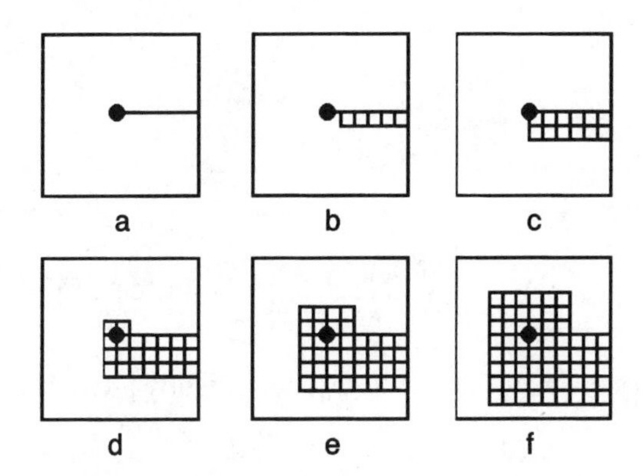

Figure 17.10 Schematic view of the development of a growth spiral by propagation of the step associated with the emergence of a screw dislocation on a crystal surface.

begin by developing a general relation for the propagation of a ledge. The ledge velocity, u, is given, in the general case, by

$$u = \Omega^{2/3} a\, \upsilon_d \exp\left(\frac{-\Delta G_{sd}}{kT}\right)(n_a - n_{a_{eq}}), \tag{17.54}$$

in which n_a and $n_{a_{eq}}$ are the existing and equilibrium adatom concentrations adjacent to the ledge, and all other terms are as previously defined. For the case of a straight ledge, assuming that the adlayer is in equilibrium with the vapor phase and that the ledge is in equilibrium with the crystal, one may write

$$u_\infty = \Omega^{2/3} a \upsilon_d \exp\left(\frac{-\Delta G_{sd}}{kT}\right)\left[\frac{p}{(2\pi mkT)^{1/2}}\tau_a - \frac{p_0}{(2\pi mkT)^{1/2}}\tau_a\right] \tag{17.55}$$

or

$$u_\infty = \Omega^{2/3} a \upsilon_d \exp\left(\frac{-\Delta G_{sd}}{kT}\right)\left[\frac{p}{(2\pi mkT)^{1/2}}\tau_a\left(1 - \frac{p_0}{p}\right)\right]. \tag{17.56}$$

Recall, however, that

$$\Delta G_v = -\left(\frac{kT}{\Omega}\right)\ln\left(\frac{p}{p_0}\right) = \left(\frac{kT}{\Omega}\right)\ln\left(\frac{p_0}{p}\right). \tag{17.57}$$

Thus,

$$\left(\frac{p_0}{p}\right) = \exp\left(\frac{\Omega \Delta G_v}{kT}\right). \tag{17.58}$$

Remembering, too, that

$$\frac{p}{(2\pi mkT)^{1/2}}\tau_a = n_a, \qquad 17.59$$

one may write

$$u_\infty = \Omega^{2/3} a v_d \exp\left(\frac{-\Delta G_{sd}}{kT}\right) n_a \left[1 - \exp\left(\frac{\Omega \Delta G_v}{kT}\right)\right]. \qquad 17.60$$

Finally, for the cases of small ΔG_v, i.e. (p/p_0) only slightly greater than unity,

$$1 - \exp\left(\frac{\Omega \Delta G_v}{kT}\right) \approx \frac{-\Omega \Delta G_v}{kT}. \qquad 17.61$$

Thus,

$$u_\infty = -\Omega^{2/3} n_a a v_d \exp\left(\frac{-\Delta G_{sd}}{kT}\right)\left(\frac{\Omega \Delta G_v}{kT}\right), \qquad 17.62$$

or

$$u_\infty = -u_o \left(\frac{\Omega \Delta G_v}{kT}\right), \qquad 17.63$$

in which

$$u_o = \Omega^{2/3} n_a a v_d \exp\left(\frac{-\Delta G_{sd}}{kT}\right). \qquad 17.64$$

One may alternatively write ΔG_v in terms of the chemical potentials at the ledge, μ_L, and in the vapor, μ_v, as

$$\Omega \Delta G_v = (\mu_L - \mu_v). \qquad 17.65$$

For the straight ledge considered thus far,

$$\mu_L = \mu_0 + kT \ln p_0, \qquad 17.66$$

$$\mu_v = \mu_0 + kT \ln p, \qquad 17.67$$

and thus

$$u_\infty = -\left(\frac{u_o}{kT}\right)\left[-kT \ln\left(\frac{p}{p_o}\right)\right]. \qquad 17.68$$

For the case of a curved ledge, however, μ_L will be different, as was seen in the case of two-dimensional nucleation, due to the effect of interface curvature on μ. For this case,

$$\mu_L = \mu_o + kT \ln p_o + \left(\frac{\gamma \Omega}{r}\right). \qquad 17.69$$

Thus, for a curved ledge of radius of curvature r

$$u_r = -\left(\frac{u_o}{kT}\right)\left[-kT\ln\left(\frac{p}{p_o}\right)+\left(\frac{\gamma\Omega}{r}\right)\right]. \tag{17.70}$$

Thus, as r decreases, u_r also decreases. Consequently, the smallest possible value of r will be that for which $u = 0$, which will be the case at the center of the spiral associated with a site of dislocation emergence. The growth rate is constrained to be zero here, as the spiral is pinned at the dislocation site. For this case

$$u_{r_\perp} = \left(\frac{-u_o}{kT}\right)\left[-kT\ln\left(\frac{p}{p_o}\right)+\left(\frac{\gamma\Omega}{r_\perp}\right)\right] = 0, \tag{17.71}$$

or

$$\frac{\gamma\Omega}{r_\perp} = kT\ln\left(\frac{p}{p_o}\right) = -\Omega\Delta G_v, \tag{17.72}$$

or

$$r_\perp = \frac{-\gamma}{\Delta G_v} = r^*, \tag{17.73}$$

where r^* is the same value that was found for the radius of the critical sized disc in the treatment of two-dimensional nucleation. In both cases r^* represents the radius of curvature at which the incremental decrease in volume free energy associated with growth is just balanced by the incremental increase in the surface energy associated with the ledge.

The relation obtained above may be rearranged to give

$$-\left(\frac{\gamma\Omega}{r^*}\right) = -kT\ln\left(\frac{p}{p_o}\right) \tag{17.74}$$

and substituted into the expression for u_r to give

$$\begin{aligned} u_r &= -\left(\frac{u_0}{kT}\right)\left[-\left(\frac{\gamma\Omega}{r^*}\right)+\left(\frac{\gamma\Omega}{r}\right)\right] \\ &= -\left(\frac{u_0}{kT}\right)\left(\frac{-\gamma\Omega}{r^*}\right)\left(1-\frac{r^*}{r}\right) \qquad (17.75) \\ &= -\left(\frac{u_0}{kT}\right)-\left[-kT\ln\left(\frac{p}{p_0}\right)\right]\left(1-\frac{r^*}{r}\right), \qquad (17.76) \end{aligned}$$

or

$$u_r = u_\infty\left(1-\frac{r^*}{r}\right). \tag{17.77}$$

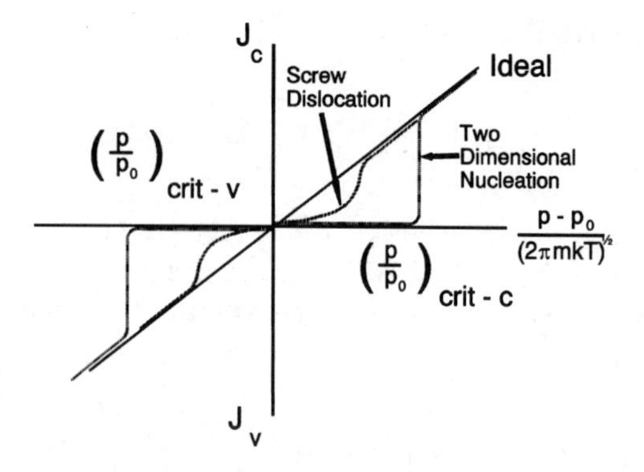

Figure 17.11 Growth and evaporation rates for growth dominated by ledges associated with points of emergence of screw dislocations at the crystal surface.

Thus, the greater the radius of curvature, for a given supersaturation, the more closely the ledge growth rate approaches that of a straight ledge. If we apply this relation to the growing spiral around an emergent screw dislocation, the closer the ledge is to the pinning point at the center of the spiral, the smaller is r for that ledge. Consequently, the ledge growth rate decreases as the center of the spiral is approached. Near the center of the spiral r is close to r^*, and u_r is small. However, since the turns of the spiral are short in this region, the amount of growth required to sweep out another turn of the spiral is also small. On the other hand, far from the center of the spiral the growth rate, u_r, is large, but the amount of growth required to sweep out one turn of the spiral is also large. The net result of these competing factors is that the steps of the spiral maintain constant spacing as growth proceeds. This conclusion may be reached by a rigorous mathematical argument, the result of which, in terms of the steady state spacing between ledges, is

$$\lambda = 4\pi r^* = -\left(\frac{4\pi\gamma}{\Delta G_v}\right) = \frac{4\pi\gamma}{(kT/\Omega)\ln(p/p_0)}. \qquad 17.78$$

Finally, consider the relation between the spiral growth rate previously developed, and the overall growth rate of the crystal relative to the ideal growth rate. Again, as in the case of growth by two-dimensional nucleation, the critical factor is the ratio of the ledge spacing, λ, to the mean free path for adatom diffusion on the surface, $\bar{x}$. At values of (p/p_0) only slightly larger than unity, the ledges will be very widely spaced ($\lambda >> \bar{x}$) and the growth rate will thus be small. As (p/p_0) is increased, λ decreases in such a way that one passes over to the condition $\lambda < \bar{x}$ quite abruptly, at a value of (p/p_0) considerably smaller than that required for appreciable growth by two-dimensional nucleation. The form of the growth rate vs. supersaturation plot for this case is shown in Figure 17.11, with the previously obtained results for ideal growth and growth by two-dimensional nucleation also shown for comparison. Again, as in the previous cases, evaporation can be treated similarly to growth, leading in this case to negative evaporation spirals propagating into the crystal This behavior is also shown in Figure 17.11.

Problems

17.1. Consider a system consisting of carbon vapor at p, T. Assume that the basal plane of graphite has the lowest surface energy of any face of the crystal, γ_c, and that all faces perpendicular to the basal plane have the same, higher surface energy, γ_e. Derive an expression for the free energy of formation of graphite clusters of critical size in terms of the appropriate surface and volume free energies.

17.2. Consider a system composed of a liquid droplet in contact with a bulk vapor phase of the same material for the two following cases:

a. The droplet is homogeneous throughout.

b. The droplet contains a small inert particle, *M*.

Assume that the net work of forming the droplet–particle interface is zero, that the pressure in the gas phase is constant, p^{α}, that $(p_0{}^{\beta})_{\infty}$ is the vapor pressure in equilibrium with the bulk liquid, and that the radii of the two droplets are equal. Develop an expression for the free energy of formation of the droplet containing the particle and compare it to that for the homogeneous droplet.

17.3. Develop an expression for the rate of nucleation of stable islands of a two-dimensional condensed phase from a two-dimensional gas adsorbed on the surface of another material. Assume that the substrate is energetically uniform, that the two-dimensional condensed phase is one monolayer thick, and that the adsorbed phase can come to equilibrium with a bulk vapor phase.

17.4. Consider the growth of a whisker crystal of a metal. The whisker grows with a square cross section 10 nm on a side. There is a single screw dislocation that passes through the crystal and emerges on two opposite faces. The other, orthogonal faces are perfect singular surfaces. You may assume that the supersaturation in the vapor phase is above that required for efficient incorporation of material on the ledges associated with the sites of dislocation emergence but below that required for growth by a two-dimensional nucleation mechanism. Develop an expression for the growth rate of the whisker, dl/dt, for the two following cases:

a. Growth occurs only by direct impingement of vapor atoms on the faces where the dislocations emerge.

b. Growth occurs by mechanism *a* plus the surface diffusion of atoms along the perfect faces to the faces where the dislocations emerge.

17.5. Prove that the spacing between the ledges associated with a screw dislocation growth source is constant and is given by

$$\lambda = 4\pi r^*,$$

where r^* is the critical radius of a two-dimensional nucleus.

Bibliography

Argide, C, and Rhead, GE. Adsorbed layer and thin film growth modes monitored by Auger electron spectroscopy. *Surface Sci. Reports* 1989;10:6–7

Bassett, GA, Menter, JW, and Pashley, DW. *Structure and Properties of Thin Films*. New York: John Wiley and Sons, 1959.

Burton, WK, Cabrera, N, and Frank, FC. The growth of crystals and the equilibrium structure of their surfaces. *Trans. Roy. Soc. (London)* 1951;A243:299–358.

Doremus, RH, Roberts, BW, and Turnbull, D, eds. *Growth and Perfection of Crystals*. New York: John Wiley and Sons, 1958.

Hirth, JP, and Pound, GM. *Condensation and Evaporation, Nucleation and Growth Kinetics*. London: Pergamon Press, 1963.

Hudson, JB, and Sandejas, JS. Observation of adsorption and crystal nucleation by mass-spectrometric techniques. *J. Vac. Sci. Technol.* 1967;4:230-238.

Rhodin, T, and Walton, D. Nucleation and growth processes on solid surfaces. In: W.D. Robertson, and N.A. Gjostein, eds. *Metal surfaces*. Metals Park, Ohio: American Society for Metals, 1963:259–286.

Ruttner, E, Goldfinger, P, and Hirth, JP, eds. *Condensation and Evaporation of Solids*. New York: Gordon and Breach, 1964.

Turnbull, D. Phase changes. In: F. Seitz, ed. *Solid State Physics* New York: Academic Press, 1956;3:226–309.

Wegener, P. Gasdynamics and homogeneous nucleation advances. *Colloid and Interface Sci.* 1977;7:325–417.

Index